HYDROTREATING TECHNOLOGY FOR POLLUTION CONTROL

CHEMICAL INDUSTRIES

A Series of Reference Books and Textbooks

Consulting Editor

HEINZ HEINEMANN

1. *Fluid Catalytic Cracking with Zeolite Catalysts,* Paul B. Venuto and E. Thomas Habib, Jr.
2. *Ethylene: Keystone to the Petrochemical Industry,* Ludwig Kniel, Olaf Winter, and Karl Stork
3. *The Chemistry and Technology of Petroleum,* James G. Speight
4. *The Desulfurization of Heavy Oils and Residua,* James G. Speight
5. *Catalysis of Organic Reactions,* edited by William R. Moser
6. *Acetylene-Based Chemicals from Coal and Other Natural Re-sources,* Robert J. Tedeschi
7. *Chemically Resistant Masonry,* Walter Lee Sheppard, Jr.
8. *Compressors and Expanders: Selection and Application for the Process Industry,* Heinz P. Bloch, Joseph A. Cameron, Frank M. Danowski, Jr., Ralph James, Jr., Judson S. Swearingen, and Marilyn E. Weightman
9. *Metering Pumps: Selection and Application,* James P. Poynton
10. *Hydrocarbons from Methanol,* Clarence D. Chang
11. *Form Flotation: Theory and Applications,* Ann N. Clarke and David J. Wilson
12. *The Chemistry and Technology of Coal,* James G. Speight
13. *Pneumatic and Hydraulic Conveying of Solids,* O. A. Williams
14. *Catalyst Manufacture: Laboratory and Commercial Preparations,* Alvin B. Stiles
15. *Characterization of Heterogeneous Catalysts,* edited by Francis Delannay
16. *BASIC Programs for Chemical Engineering Design,* James H. Weber
17. *Catalyst Poisoning,* L. Louis Hegedus and Robert W. McCabe
18. *Catalysis of Organic Reactions,* edited by John R. Kosak
19. *Adsorption Technology: A Step by Step Approach to Process Evaluation and Application,* edited by Frank L. Slejko

20. *Deactivation and Poisoning of Catalysts,* edited by Jacques Oudar and Henry Wise
21. *Catalysis and Surface Science: Developments in Chemicals from Methanol, Hydrotreating of Hydrocarbons, Catalyst Preparation, Monomers and Polymers, Photocatalysis and Photovoltaics,* edited by Heinz Heinemann and Gabor A. Somorjai
22. *Catalysis of Organic Reactions,* edited by Robert L. Augustine
23. *Modern Control Techniques for the Processing Industries,* T. H. Tsai, J. W. Lane, and C. S. Lin
24. *Temperature-Programmed Reduction for Solid Materials Characterization,* Alan Jones and Brian McNichol
25. *Catalytic Cracking: Catalysts, Chemistry, and Kinetics,* Bohdan W. Wojciechowski and Avelino Corma
26. *Chemical Reaction and Reactor Engineering,* edited by J. J. Carberry and A. Varma
27. *Filtration: Principles and Practices, Second Edition,* edited by Michael J. Matteson and Clyde Orr
28. *Corrosion Mechanisms,* edited by Florian Mansfeld
29. *Catalysis and Surface Properties of Liquid Metals and Alloys,* Yoshisada Ogino
30. *Catalyst Deactivation,* edited by Eugene E. Petersen and Alexis T. Bell
31. *Hydrogen Effects in Catalysis: Fundamentals and Practical Applications,* edited by Zoltán Paál and P. G. Menon
32. *Flow Management for Engineers and Scientists,* Nicholas P. Cheremisinoff and Paul N. Cheremisinoff
33. *Catalysis of Organic Reactions,* edited by Paul N. Rylander, Harold Greenfield, and Robert L. Augustine
34. *Powder and Bulk Solids Handling Processes: Instrumentation and Control,* Koichi Iinoya, Hiroaki Masuda, and Kinnosuke Watanabe
35. *Reverse Osmosis Technology: Applications for High-Purity-Water Production,* edited by Bipin S. Parekh
36. *Shape Selective Catalysis in Industrial Applications,* N. Y. Chen, William E. Garwood, and Frank G. Dwyer
37. *Alpha Olefins Applications Handbook,* edited by George R. Lappin and Joseph L. Sauer
38. *Process Modeling and Control in Chemical Industries,* edited by Kaddour Najim
39. *Clathrate Hydrates of Natural Gases,* E. Dendy Sloan, Jr.
40. *Catalysis of Organic Reactions,* edited by Dale W. Blackburn
41. *Fuel Science and Technology Handbook,* edited by James G. Speight
42. *Octane-Enhancing Zeolitic FCC Catalysts,* Julius Scherzer
43. *Oxygen in Catalysis,* Adam Bielański and Jerzy Haber
44. *The Chemistry and Technology of Petroleum: Second Edition, Revised and Expanded,* James G. Speight
45. *Industrial Drying Equipment: Selection and Application,* C. M. van't Land

46. *Novel Production Methods for Ethylene, Light Hydrocarbons, and Aromatics*, edited by Lyle F. Albright, Billy L. Crynes, and Siegfried Nowak
47. *Catalysis of Organic Reactions*, edited by William E. Pascoe
48. *Synthetic Lubricants and High-Performance Functional Fluids*, edited by Ronald L. Shubkin
49. *Acetic Acid and Its Derivatives*, edited by Victor H. Agreda and Joseph R. Zoeller
50. *Properties and Applications of Perovskite-Type Oxides*, edited by L. G. Tejuca and J. L. G. Fierro
51. *Computer-Aided Design of Catalysts*, edited by E. Robert Becker and Carmo J. Pereira
52. *Models for Thermodynamic and Phase Equilibria Calculations*, edited by Stanley I. Sandler
53. *Catalysis of Organic Reactions*, edited by John R. Kosak and Thomas A. Johnson
54. *Composition and Analysis of Heavy Petroleum Fractions*, Klaus H. Altgelt and Mieczyslaw M. Boduszynski
55. *NMR Techniques in Catalysis,* edited by Alexis T. Bell and Alexander Pines
56. *Upgrading Petroleum Residues and Heavy Oils*, Murray R. Gray
57. *Methanol Production and Use*, edited by Wu-Hsun Cheng and Harold H. Kung
58. *Catalytic Hydroprocessing of Petroleum and Distillates*, edited by Michael C. Oballah and Stuart S. Shih
59. *The Chemistry and Technology of Coal: Second Edition, Revised and Expanded,* James G. Speight
60. *Lubricant Base Oil and Wax Processing*, Avilino Sequeira, Jr.
61. *Catalytic Naphtha Reforming: Science and Technology*, edited by George J. Antos, Abdullah M. Aitani, and José M. Parera
62. *Catalysis of Organic Reactions*, edited by Mike G. Scaros and Michael L. Prunier
63. *Catalyst Manufacture,* Alvin B. Stiles and Theodore A. Koch
64. *Handbook of Grignard Reagents,* edited by Gary S. Silverman and Philip E. Rakita
65. *Shape Selective Catalysis in Industrial Applications: Second Edition, Revised and Expanded,* N. Y. Chen, William E. Garwood, and Francis G. Dwyer
66. *Hydrocracking Science and Technology*, Julius Scherzer and A. J. Gruia
67. *Hydrotreating Technology for Pollution Control: Catalysts, Catalysis, and Processes,* edited by Mario L. Occelli and Russell Chianelli
68. *Catalysis of Organic Reactions*, edited by Russell Malz

ADDITIONAL VOLUMES IN PREPARATION

Synthesis of Porous Materials: Zeolites, Clays, and Nanostructures, edited by Mario L. Occelli and Henri Kessler

Methane and Its Derivatives, Sunggyu Lee

HYDROTREATING TECHNOLOGY FOR POLLUTION CONTROL

Catalysts, Catalysis, and Processes

edited by
Mario L. Occelli
Georgia Tech Research Institute
Georgia Institute of Technology, Atlanta
Atlanta, Georgia

Russell Chianelli
University of Texas at El Paso
El Paso, Texas

Marcel Dekker, Inc.   New York•Basel•Hong Kong

Library of Congress Cataloging-in-Publication Data

Hydrotreating technology for pollution control : catalysts, catalysis, and processes / edited by Mario Occelli, Russell Chianelli.
   p. cm. — (Chemical industries ; v. 67)
  Includes index.
  ISBN 0-8247-9756-6 (hardcover : alk. paper)
  1. Hydrotreating catalysts—Congresses. 2. Petroleum—Refining—Congresses. 3. Cracking process—Congresses. I. Occelli, Mario. II. Chianelli, Russell. III. Series.
TP690.4.H93 1996
665.5'33—dc20
                                                  96-23962
                                                    CIP

The publisher offers discounts on this book when ordered in bulk quantities. For more information, write to Special Sales/Professional Marketing at the address below.

This book is printed on acid-free paper.

Copyright © 1996 by MARCEL DEKKER, INC. All Rights Reserved.

Neither this book nor any part may be reproduced or transmitted in any form or by any means, electronic or mechanical, including photocopying, microfilming, and recording, or by any information storage and retrieval system, without permission in writing from the publisher.

MARCEL DEKKER, INC.
270 Madison Avenue, New York, New York 10016

Current printing (last digit):
10 9 8 7 6 5 4 3 2 1

PRINTED IN THE UNITED STATES OF AMERICA

Preface

Interest in environmentally related catalysis continues to grow, especially in the United States, where environmental science and technology have advanced at a rapid pace. Cities and urban areas in the United States have benefitted from continuing improvement in air quality because the fuels now burned are cleaner and the cars now driven are less polluting than those of the past. Much of the improved fuel quality can be attributed to better technology for removal of sulfur and other pollutants from petroleum-based fuel stocks. The heart of this technology has been—and continues to be—catalysts based on transition metal sulfides. All projections indicate that evermore stringent requirements for removal of sulfur and nitrogen will be demanded by refiners and regulators throughout the world.

This volume reflects the editors' belief that greater improvements in sulfide catalysts will be required to meet the demands of higher fuel quality; and that these improvements can be only met through an intedisciplinary effort based on high-quality science and engineering. The papers were chosen for inclusion as representative of the best in the various disciplines of catalysis by transition metal sulfides, especially concerning applications for making improved catalysts and improving processes. The bulk of the papers were taken from the Second Symposium on Advances in Hydrotreating Catalysts, 208th ACS National Meeting, Washington, D.C., August 21-25, 1994. The remainder were solicited from individuals covering areas not included in the symposium.

We have included papers on the theoretical basis for activity and selectivity of hydrotreating catalysts, experimental measurement of catalyst activity and selectivity of hydrotreating reactions in model and real feeds, new catalytic materials, structural models of hydrotreating catalysts and their characterization, reaction mechanism and pathways, and biological processing of hydrocarbon feeds. The material is presented in the hope that it will stimulate interdisciplinary research efforts to further understand and tailor hydrotreating catalysts for the growing demands for more efficient hydrotreating processes and catalysts, leading to cleaner transportation fuels.

Mario L. Occelli
Russell Chianelli

Contents

Preface	*iii*
Contributors	*ix*
1. A Unified Theory of Periodic and Promotion Effects in Transition Metal Sulfide Hydrodesulfurization Catalysts *Timo S. Smit and Keith H. Johnson*	1
2. A Review of Some Specific Phases as Possible Active Hydrotreating Constituents *H. D. Simpson*	17
3. Catalytic Sites and Heterolytic Reaction Mechanisms in Hydrotreating Catalysis *S. Kasztelan*	29
4. Microstructural Characterization of Molybdenum Disulfide and Tungsten Disulfide Catalysts *M. Del Valle, M. J. Yáñez, M. Avalos-Borja, and S. Fuentes*	47
5. STM and AFM Studies of Transition Metal Chalcogenides *S. P. Kelty, A.F. Ruppert, R. R. Chianelli, J. Ren, and M.-H. Whangbo*	71
6. HREM Characterization of MoS_2, MoS_2:Co, and MoS_2:Fe *A. Vázquez-Zavala and M. José Yacamán*	85

7. On the Identification of "Co-Sulfide" Species in Sulfided Supported Co and CoMo Catalysts 95
   M. W. J. Crajé, V. H. J. de Beer, J. A. R. van Veen, and A. M. van der Kraan

8. Properties of MoS_2-Based Hydrotreating Catalysts Modified by Electron Donor or Electron Acceptor Additives (P, Na, Cs, F) 115
   R. Hubaut, O. Poulet, S. Kasztelan, E. Payen, and J. Grimblot

9. Modeling HDS Catalysis with Sulfido Bimetallic Clusters 129
   M. David Curtis

10. Evidence for H_2S as Active Species in the Mechanism of Thiopene Hydrodesulfurization 147
    J. Leglise, J. van Gestel, and J.-C. Duchet

11. Role of the Sulfiding Procedure on Activity and Dispersion of Nickel Catalysts 159
    J. P. Janssens, A. D. van Langeveld, R. L. C. Bonné, C. M. Lok, and J. A. Moulijn

12. Ruthenium Sulfide Based Catalysts 169
    Michèle Breysse, Christophe Geantet, Michel Lacroix, Jean Louis Portefaix, and Michel Vrinat

13. Comparison of Sulfided CoMo Al_2O_3 and $NiMo/Al_2O_3$ Catalysts in Hydrodesulfurization of Gas Oil Fractions and Model Compounds 183
    Xiaoliang Ma, Kinya Sakanishi, Takaaki Isoda, and Isao Mochida

14. Competitive Coversion of Nitrogen and Sulfur Compounds in Naphtha with Transition Metal Sulfide Catalysts 197
    Shuh-Jeng Liaw, Ajoy Raje, Rongguang Lin, and Burtron H. Davis

15. The Effect of Additives and Impregnation Stabilizers on Hydrodesulfurization Activity 211
    E. P. Dai and L. D. Neff

16. Molten Salt Preparation of Mixed Transition Metals on Zirconia: Application to Hydrotreating Reactions 235
    P. Afanasiev, C. Geantet, M. Breysse, and T. des Courieres

Contents

17. The Role of Hydrogen in the Hydrogenation and Hydrogenolysis of Aniline on the Nickel Single Crystal Surfaces: Its Implication on the Mechanisms of HDN Reactions ... 253
    Sean X. Huang, Daniel A. Fischer, and John L. Gland

18. Hydrodenitrogenation of Carbazole on an Alumina-Supported Molybdenum Nitride Catalyst ... 263
    Masatoshi Nagai, Toshihiro Miyao, and Shinzo Omi

19. Aromatics Hydrogenation over Alumina Supported Mo and NiMo Hydrotreating Catalysts ... 277
    N. Marchal, D. Guillaume, S. Mignard, and S. Kasztelan

20. Applications of Pillared and Delaminated Interlayered Clays as Supports for Hydrotreating Catalysts ... 291
    M. F. Wilson, J.-P. Charland, E. Yamaguchi, and T. Suzuki

21. Catalytic Hydrotreating with Pillared Synthetic Clays ... 313
    Ramesh K. Sharma and Edwin S. Olson

22. Residual Oil Hydrotreating Catalyst Rejuvenation by Leaching of Foulkant Metals: Effect of Metal Leaching on Catalyst Characteristics and Performance ... 327
    A. Stanislaus, M. Marafi, and M. Absi-Halabi

23. Characterization of Coke on Spent Resid Catalyst from Ebullating Bed Service and its Effect on Activity ... 337
    Per Zeuthen, Barry H. Cooper, Fred Clark, and David Arters

24. Microbial Removal of Organic Sulfur from Fuels: A Review of Past and Present Approaches ... 345
    Matthew J. Grossman

Index ... *361*

Contributors

M. Absi-Halabi Kuwait Institute for Scientific Research, Safat, Kuwait

P. Afanasiev Institut de Recherches sur la Catalyse, Villeurbanne, France

David Arters Amoco R&D, Naperville, Illinois

R. L. C. Bonné Unilever Research Port Sunlight Laboratory, Wirral, England

M. Avalos-Borja Instituto de Física, UNAM, Laboratorio de Ensenada, Ensenada, B.C., Mexico

Michèle Breysse Institut de Recherches sur la Catalyse, Villeurbanne, France

J.-P. Charland CANMET Energy Research Laboratories, Ottawa, Ontario, Canada

R. R. Chianelli* Exxon Research and Engineering Co., Annandale, New Jersey

Fred Clark Amoco R&D, Naperville, Illinois

Barry H. Cooper Haldor Topsøe Research Laboratories, Lyngby, Denmark

M. W. J. Crajé Delft University of Technology, Delft, The Netherlands

M. David Curtis The University of Michigan, Ann Arbor, Michigan

**Current affiliation:* University of Texas at El Paso, El Paso, Texas

E. P. Dai TEXACO R&D-Port Arthur, Port Arthur, Texas

Burtron H. Davis Center for Applied Energy Research, University of Kentucky, Lexington, Kentucky

V. H. J. de Beer Eindhoven University of Technology, Eindhoven, The Netherlands

M. Del Valle Centro de Investigación Científica y de Educación Superior de Ensenada, Ensenada, B. C., Mexico

T. des Courieres ELF, Antar, France

J.-C. Duchet Université de Caen, Caen, France

Daniel A. Fischer National Institute of Standards and Technology, Gaithersburg, Maryland

S. Fuentes Instituto de Física, UNAM, Laboratorio de Ensenada, Ensenada, B.C., Mexico

Christophe Geantet Institut de Recherches sur la Catalyse, Villeurbanne, France

John L. Gland University of Michigan, Ann Arbor, Michigan

J. Grimblot Université des Sciences et Technologies de Lille, Villeneuve d'Ascq, France

Matthew J. Grossman Exxon Research and Engineering Co., Clinton, New Jersey

D. Guillaume Institut Français du Pétrole, Rueil-Malmaison, France

Sean X. Huang University of Michigan, Ann Arbor, Michigan

R. Hubaut Université des Sciences et Technologies de Lille, Villeneuve d'Ascq, France

Takaaki Isoda Institute of Advanced Material Study, Kyushu University, Fukuoka, Japan

J. P. Janssens Delft University of Technology, Delft, The Netherlands

Contributors xi

Keith H. Johnson   Massachusetts Institute of Technology, Cambridge, Massachusetts

S. Kasztelan Institut Français du Pétrole, Rueil-Malmaison, France

S. P. Kelty Exxon Research and Engineering Co., Annandale, New Jersey

J. Leglise Université de Caen, Caen, France

Michel Lacroix Institut de Recherches sur la Catalyse, Villeurbanne, France

Shuh-Jeng Liaw Center for Applied Energy Research, University of Kentucky, Lexington, Kentucky

Rongguang Lin Center for Applied Energy Research, University of Kentucky, Lexington, Kentucky

C. M. Lok Unilever Research Port Sunlight Laboratory, Wirral, England

Xiaoliang Ma   Institute of Advanced Material Study, Kyushu University, Fukuoka, Japan

M. Marafi Kuwait Institute for Scientific Research, Safat, Kuwait

N. Marchal Institut Français du Pétrole, Rueil-Malmaison, France

S. Mignard Institut Français du Pétrole, Rueil-Malmaison, France

Toshihiro Miyao Tokyo University of Agriculture and Technology, Koganei, Tokyo, Japan

Isao Mochida   Institute of Advanced Material Study, Kyushu University, Fukuoka, Japan

J. A. Moulijn Delft University of Technology, Delft, The Netherlands

Masatoshi Nagai Tokyo University of Agriculture and Technology, Koganei, Tokyo, Japan

L. D. Neff TEXACO R&D-Port Arthur, Port Arthur, Texas

Edwin S. Olson   University of North Dakota, Energy & Environmental Research Center, Grand Forks, North Dakota

Shinzo Omi Tokyo University of Agriculture and Technology, Koganei, Tokyo, Japan

E. Payen Université des Sciences et Technologies de Lille, Villeneuve d'Ascq, France

Jean Louis Portefaix Institut de Recherches sur la Catalyse, Villeurbanne, France

O. Poulet* Université des Sciences et Technologies de Lille, Villeneuve d'Ascq, France

Ajoy Raje Center for Applied Energy Research, University of Kentucky, Lexington, Kentucky

J. Ren North Carolina State University, Raleigh, North Carolina

A.F. Ruppert Exxon Research and Engineering Co., Annandale, New Jersey

Kinya Sakanishi Institute of Advanced Material Study, Kyushu University, Fukuoka, Japan

Ramesh K. Sharma University of North Dakota, Energy & Environmental Research Center, Grand Forks, North Dakota

H. D. Simpson Unocal Corporation, Brea, California

Timo S. Smit† Massachusetts Institute of Technology, Cambridge, Massachusetts

A. Stanislaus Kuwait Institute for Scientific Research, Safat, Kuwait

T. Suzuki Sumitomo Metal Mining Company Ltd., Chiba, Japan

A. M. van der Kraan Delft University of Technology, Delft, The Netherlands

J. van Gestel Université de Caen, Caen, France

A. D. van Langeveld Delft University of Technology, Delft, The Netherlands

**Current affiliation:* Centre de recherches Total, Harfleur, France
†*Current affiliation:* Akzo Nobel Central Research Institute, Arnhem, The Netherlands

Contributors

J. A. R. van Veen Koninklijke/Shell-Laboratorium (Shell Research B. V.), Amsterdam, The Netherlands

A. Vázquez-Zavala Instituto de Física, UNAM, México D. F., México

Michel Vrinat Institut de Recherches sur la Catalyse, Villeurbanne, France

M.-H. Whangbo North Carolina State University, Raleigh, North Carolina

M. F. Wilson CANMET Energy Research Laboratories, Ottawa, Ontario, Canada

M. José Yacamán Instituto de Fisica, UNAM, México D. F., México

E. Yamaguchi Sumitomo Metal Mining Company Ltd., Chiba, Japan

M. J. Yáñez Centro Regional de Investigaciones Básicas y Aplicadas de Bahía Blanca, Argentina

Per Zeuthen Haldor Topsøe Research Laboratories, Lyngby, Denmark

HYDROTREATING TECHNOLOGY FOR POLLUTION CONTROL

A Unified Theory of Periodic and Promotion Effects in Transition Metal Sulfide Hydrodesulfurization Catalysts

Timo S. Smit[1] and Keith H. Johnson
Department of Materials Science and Engineering
Massachusetts Institute of Technology
77 Massachusetts Avenue
Cambridge, MA 02139

Abstract

Electronic structure calculations on model catalyst clusters and catalyst-thiophene complexes have made it possible to develop a unified theory of periodic and promotion effects in TMS HDS catalysis, with direct implications for catalyst design. Both effects are above all electronic: the quality of the active site directly affects the activity of the catalyst.
Periodic effects can be explained through differences in the rate of the rate-limiting desulfurization step in the HDS process. A theoretical activity parameter I, based on the strength of the interaction between metal d and sulfur $3p$ electrons in monometallic TMS, correlates well with HDS activity. I reflects the ability of the catalyst to "capture" a sulfur atom from the heterocycle. I depends strongly on the oxidation state of the transition metal and on the metal-sulfur bond length.
The promotion effect in mixed TMS, such as Co(Ni)-Mo-S arises from electron transfer from Co(Ni) to Mo. This corresponds to the removal of σ^* metal d - sulfur $3p$ electrons from Co(Ni), resulting in a high intrinsic activity of these elements. Molybdenum sulfide acts as a support and promotor for Co(Ni).

1. Introduction

Transition metal sulfides (TMS) are widely used as hydrodesulfurization (HDS) catalysts for thiophenic species occurring in oil. Although the removal of sulfur is of great industrial importance and has been studied

---

[1]Author to whom correspondence should be addressed. Present address: Akzo Nobel Central Research Institute, P.O. Box 9300, 6800 SB Arnhem, The Netherlands. E-mail: timo.t.s.smit@akzo.nl

extensively, it is still not known how an HDS catalyst functions at a fundamental level. Open questions in HDS catalysis include (1) the nature of the dependence of the activity of monometallic TMS on the position of the metal atom in the periodic table (periodic effect) and (2) the origin of the unusually high activity of certain mixed TMS (promotion effect). The rational design of new (classes of) catalysts requires a thorough understanding of these effects, the topic of this paper.

A systematic investigation of the activity of unsupported monometallic TMS for the HDS of dibenzothiophene was first carried out by Pecoraro and Chianelli [1], see figure 1. A similar curve was later obtained by Vissers et al. [2] for carbon supported catalysts. An explanation of this activity curve has proven to be a challenging task. This is due in part to the uncertainty in its shape, since the normalization of activity data is not a trivial matter. Pecoraro and Chianelli nonetheless conclude that periodic effects are above all electronic.

In a recent paper [3] we treated two competing explanations of the periodic effect, both based on electronic structure calculations on model catalysts. Nørskov and co-workers [4,5] calculated the metal-sulfur bond strength (normalised per mole of sulfur atoms) for monometallic TMS having the CsCl structure. It was found that the lowest values for the sulfur bond strength are obtained for the most active sulfides. Since it is easy to form sulfur surface vacancies, thought to be responsible for the catalytic activity, when the sulfur bond strength is low, Nørskov and co-workers claim that most of the variation in catalytic activity across the periodic table can be explained through different numbers of such vacancies. Harris and Chianelli [6], on the other hand, had previously explained the periodic effect emphasizing the character of the active site, rather than the number of such sites. An activity parameter, based on the covalency of metal-sulfur bonding orbitals and the occupancy of the highest occupied molecular orbital of octahedral MS_6^{-n} clusters, correlates well with catalytic activities.

Like Haris and Chianelli, we concluded that periodic effects are above all electronic, rather than structural. However, we offered a different theoretical activity parameter. The differences manifest themselves most clearly when it comes to explaining the promotion effect. In this paper, we treat this alternative activity parameter, emphasizing the information it provides about the HDS process. We begin with a brief review of the electronic structure of MS_6^{-n} clusters.

2. The electronic structure of MS_6^{-n} clusters

Obviously, an octahedral MS_6^{-n} cluster (point group O_h) is by no means representative of the environment of the metal cations at the surface of the catalyst. It is also well-known that not all monometallic TMS possess octahedral symmetry, so even as a model for the bulk catalyst, an MS_6^{-n} cluster is the simplest option imaginable. Yet this simple model provides valuable information about the interaction between the metal and sulfur atoms in bulk sulfides, and this, in turn, provides information on the binding of the sulfur atom in thiophenic species to the exposed metal atom at the surface of the catalyst. Only by performing calculations on the same geometry for all TMS are we in a position to construct an activity parameter which reflects the activity of one sulfide relative to another. This does not affect the basic chemistry.

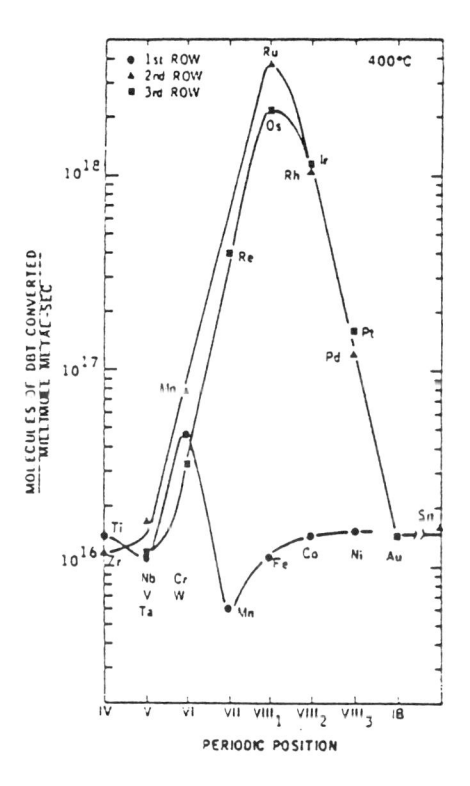

Figure 1. The activity of monometallic TMS for HDS of dibenzothiophene. From ref. [1].

A schematic representation of a spin-restricted molecular orbital diagram of an octahedral TMS cluster, obtained from calculations employing the scattered-wave method, modified for overlapping spheres [7], is given in figure 2. At the bottom of this diagram one finds metal d - sulfur $3p$ bonding orbitals. Higher in energy there is a manifold of orbitals which are dominated by sulfur $3p$ contributions. At the top of this sulfur band, there is a non-bonding sulfur orbital which may be used as a reference state since it is the same for all TMS clusters. Still further up in energy there are metal d - sulfur $3p$ antibonding orbitals, first π (irreducible representation t_{2g}, sixfold degenerate for spin-restricted calculations), then σ (e_g, fourfold degenerate). When going from left to right in the periodic table, the first obvious change in the electronic structure is the number of electrons in the cluster. The MS_6^{-n} clusters all carry a net charge: sulfur is formally in the -2 oxidation state and the metal atom in the oxidation state representative of the TMS. The charge surplus is compensated in the calculations by surrounding the cluster with a Watson sphere [8] in order to mimick the stabilizing Madelung potential of a crystalline environment. The oxidation state of the metal atom and other information are given in table I. Consider, for example, the second row transition metals. The stable catalytic phase for Mo is MoS_2, so Mo has a formal charge of +4. The entire cluster, then, has a net charge of -8. This means that the HOMO for MoS_2 is the t_{2g}-π^* orbital, occupied by 2 electrons.

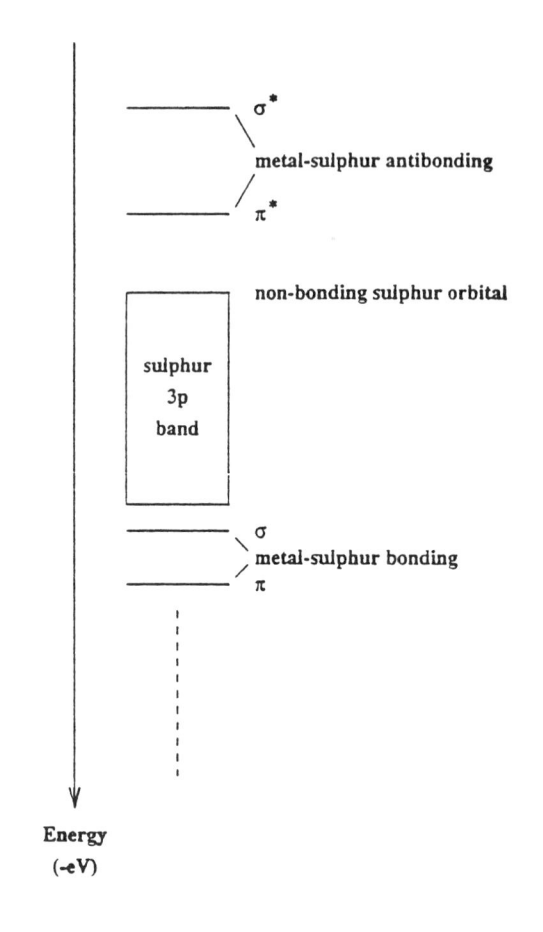

Figure 2. A schematic representation of a spin-restricted molecular orbital diagram for MS_6^{-n} clusters. See text for details.

Going one element to the right in the periodic table (Tc), one adds one electron to the cluster, so the t_{2g}-π^* orbital is now occupied by three electrons. More electrons are added for Ru and Rh, at which point the t_{2g} orbital is fully occupied. Consequently, at Pd the e_g-σ^* orbital is occupied for the first time. The t_{2g} orbital is of course fully occupied, e_g's occupancy is 2. Generally speaking, in going from left to right in the periodic table, antibonding electrons are added to the cluster. Hence it is no surprise that the heat of formation of bulk sulfides also decreases in this

Unified Theory of Effects and Catalysts

direction. Since maximum catalytic activity is observed for Ru, in the middle of the series, this suggests that the metal-sulfur bond strength should be neither too strong nor too weak for good HDS catalysts. This is the well-known principle of Sabatier [9] and is consistent with the intuitive notion that a good HDS catalyst should be able to capture the sulfur atom in the heterocycle, yet easily regenerate sulfur surface vacancies. However, intermediate values for the heat of formation do not imply good catalytic activity. As an example, consider MnS. Although the heat of formation is similar to that of RuS_2, MnS is one of the least active catalysts. Hence information on the metal-sulfur bond strength alone, partially given by the number of antibonding metal-sulfur electrons, is not sufficient for predicting catalytic activity.

Not only the number of electrons in the cluster, but many other features of the electronic structure vary across the periodic table. For example, when going from left to right across the second and third transition metal series, the metal content of the metal-sulfur bonding orbitals increases, the energy of these orbitals decreases with respect to the non-bonding reference state at the top of the sulfur band, the width of the sulfur band increases, the metal-sulfur antibonding orbitals come down in energy and the sulfur content in these orbitals increases. Since the formal d electron count is insufficient, these other elements of the electronic structure have to be included in the definition of a theoretical activity parameter.

3. The activity parameter of Harris and Chianelli

Harris and Chianelli [6] observed that while the formal d electron count in the clusters increases from left to right in the periodic table, the number of electrons in the HOMO, n, first increases up to and including Rh, but then decreases when going to Pd. In fact, a plot of n against the position of the metal atom in the periodic table has the same general shape as the experimental activity curve. n is multiplied by a metal-sulfur *covalent* bond strength B to arrive at a theoretical activity parameter A_2. B is defined as the product of the metal-sulfur covalency (in turn defined as the metal content of the metal-sulfur bonding orbitals) and the net number of bonding electrons (e.g. for Rh the net number of π bonding electrons is 0 but the net number of σ bonding electrons is 4). A_2 correlates well with the experimental activities.

We believe that n should not be included in the definition of a theoretical activity parameter. n has no physical meaning, since the HOMO is a t_{2g}

Table I. The metal-sulfur bond length, metal oxidation state, and number of σ^* and π^* electrons in MS_6^{-n} clusters.

TMS	metal ion oxidation state	metal-sulfur bond length (a.u.)	number of σ^* electrons	number of π^* electrons
TiS_2	+4	4.57	-	0
V_2S_3	+3	4.48	-	1
Cr_2S_3	+3	4.57	-	3
MnS	+2	4.89	2	3
FeS	+2	4.27	-	6
Co_9S_8	+2	4.38	1	6
Ni_3S_2	+2	4.54	2	6
ZrS_2	+4	4.84	-	0
NbS_2	+4	4.67	-	1
MoS_2	+4	4.57	-	2
TcS_2	+4	4.50	-	3
RuS_2	+4	4.45	-	4
Rh_2S_3	+3	4.48	-	6
PdS	+2	4.76	2	6
HfS_2	+4	4.84	-	0
TaS_2	+4	4.61	-	1
WS_2	+4	4.55	-	2
ReS_2	+4	4.38	-	3
OsS_2	+4	4.42	-	4
IrS_2	+3	4.48	-	6
PtS	+2	4.53	2	6

orbital for certain clusters, but an e_g orbital for others. Consider also the first row TMS. It is well-known that the effects of spin are important and have to be included in the scattered-wave calculations. At the top of the molecular orbital diagram one now finds two t_{2g}-π^* orbitals (one for spin up and one for spin down), each threefold degenerate. Similarly there are now two e_g orbitals, each twofold degenerate. Hence as a result of performing spin-unrestricted calculations, the maximum occupancy of the HOMO has been reduced by a factor two. If n were to be included in the definition of a theoretical activity parameter, one would automatically find that the first row TMS are much less active than the second and third row TMS. Although this is in fact the case, the observed correlation between this activity parameter and the experimental activities would then be completely coincidental, and lack physical meaning. Prins et al. [10] have also pointed out that as a consequence of the definition of the activity parameter, Zr^{4+} and Hf^{4+} ions would have no activity whatsoever,

4. The definition of an alternative activity parameter, I

Given our objections against the use of n in the definition of a theoretical activity parameter, we have also carried scattered-wave calculations on first, scond and third row TMS. While n has no physical meaning, the total number of d electrons no doubt has an effect on the overall electronic structure. We assume that it is the interaction between metal d and sulfur $3p$ electrons which is important in the HDS process. Our aim is to define a parameter which takes into account the character and occupancy of all the orbitals which have some metal d and sulfur $3p$ character, i.e. all the orbitals which are depicted in figure 2. We propose the following definition of a theoretical activity parameter I:

$$I = - \sum_{I \geq M\text{-}S \text{ bonding}} n_i \times (E_i - E_{non-bonding})$$

Why does a parameter which is defined solely in terms of the energies and occupancies of the orbitals provide information about the character of these orbitals? To answer this question, consider the π^* orbital. From left to right in the periodic table, the energy of this orbital decreases and the sulfur content increases (for second and third row TMS). The orbital topology, figure 3a, indicates why this should be so. The orbital is antibonding between the metal and sulfur atoms, but bonding between neighbouring sulfur atoms. The higher the sulfur content of the orbital, the more the metal-sulfur repulsion will be compensated by sulfur-sulfur attraction. When a thiophene molecule is desulfurized, the sulfur atom probably binds to the catalyst at the site of an exposed metal atom. As binding occurs, this atom will interact not only with the exposed metal atom, but also with the neighbouring sulfur atoms in the catalyst. The energy of the π^* orbital thus contains information on the overall effect of the metal-sulfur and sulfur-sulfur interactions which are occur during the HDS process. The same is true for all the other orbitals included in the definition of I. The individual terms in the summation represent the effect of individual orbitals on the binding of a sulfur atom to an MS_5^{-m} cluster.

Suppose, then, that the metal-sulfur π bonding orbital lies 2.8 eV below the top of the sulfur band. This orbital contributes 6x2.8=16.8 units to the activity parameter, since it is fully occupied. A partially occupied σ^* metal-sulfur orbital which lies 2 eV above the sulfur band decreases the

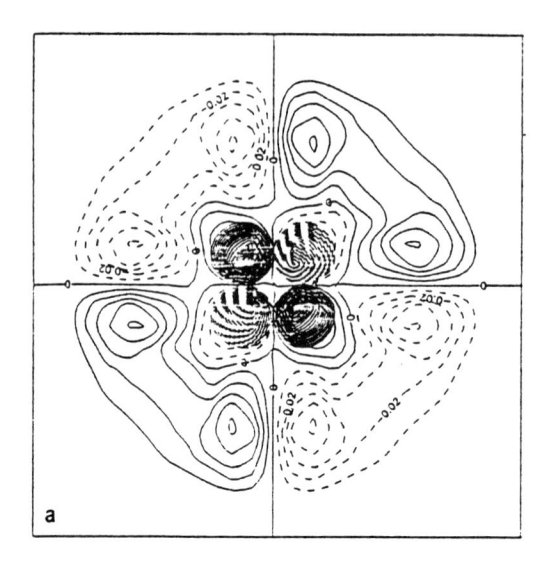

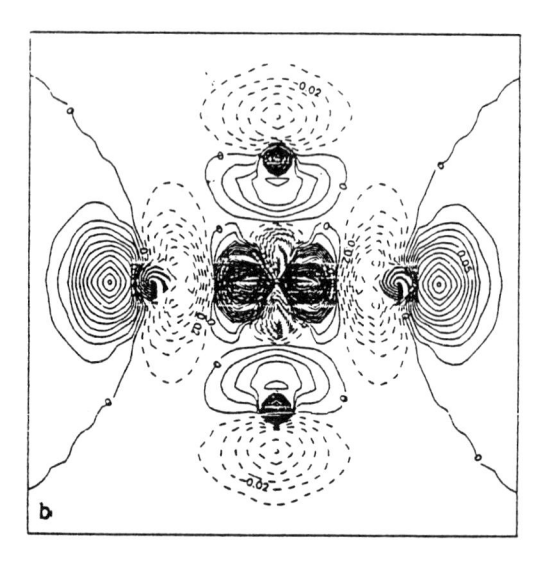

Figure 3. The topology of the π^* orbital for RhS_6^{-9} (a), antibonding between the metal and sulfur atoms, but bonding between neighbouring sulfur atoms. Solid lines indicate positive phase for the wave-function, dotted lines correspond to negative phase. In (b) a contour plot for the σ^* orbital is given, clearly antibonding between metal and sulfur.

value of I by 4 units if its occupancy is 2. Hence there is some similarity between I and the parameter B defined by Harris and Chianelli [6], but I includes *all* the orbitals of figure 2 and eliminates the need for arbitrary definitions. Hence we feel that I accurately represents the interaction strength between metal d and sulfur $3p$ orbitals. All the electrons contribute to I, not just those in the HOMO. Hence there is no inconsistency in performing spin-unrestricted calculations for first row TMS and spin-restricted calculations for second and third row TMS.

5. The correlation between I and the experimental activities

Density functional scattered-wave calculations have been carried out on first, second and third row TMS clusters. The exchange-correlation functional of Hedin and Lundqvist [11] has been used for second and third row TMS. For first row TMS the Ceperley-Alder [12] functional was used. Also, spin-unrestricted calculations were carried for the first row elements, spin-restricted calculations for the second and third row elements. Relativistic effects have been included for third row transition metals. I is plotted in figure 4. It is immediately apparent that the general features of the experimental activity curve (figure 1) are reasonably reproduced. The second and third row TMS are seen to be more active than the first row sulfides. A maximum occurs for Ru and Os. Also, a distinct minimum occurs for Mn in the first row TMS. This sharp decrease in the value of I coincides with the occupation of a σ^* orbital. Sharp decreases in the value of I are also observed when going from Rh to Pd and Ir to Pt in the second and third row respectively. Pd and Pt are also the first elements in their series for which the e_g-σ^* orbital is occupied (the fact that it happens for Mn in the first row is due to the lifting of the spin degeneracy). It appears, then, that the occupation of a σ^* orbital results in a sharp decrease in the value of I and, since I is taken to represent the catalytic activity of a TMS, it results in a sharp decrease in catalytic activity. Again, the topology of the σ^* orbital, shown in figure 3b, shows why this is physically reasonable. The character of a σ^* orbital is unambiguously antibonding between the metal and sulfur atoms. Metal-sulfur attraction is not compensated by sulfur-sulfur attraction. This is, of course, the reason why the π^* orbitals lie lower in energy than the σ^* orbitals. Suppose then that the thiophene molecule binds to the catalyst through the ring sulfur atom. The net effect of a π^* orbital will be fairly small, but a σ^* orbital will strongly repel the sulfur atom of the heterocyclic ring.

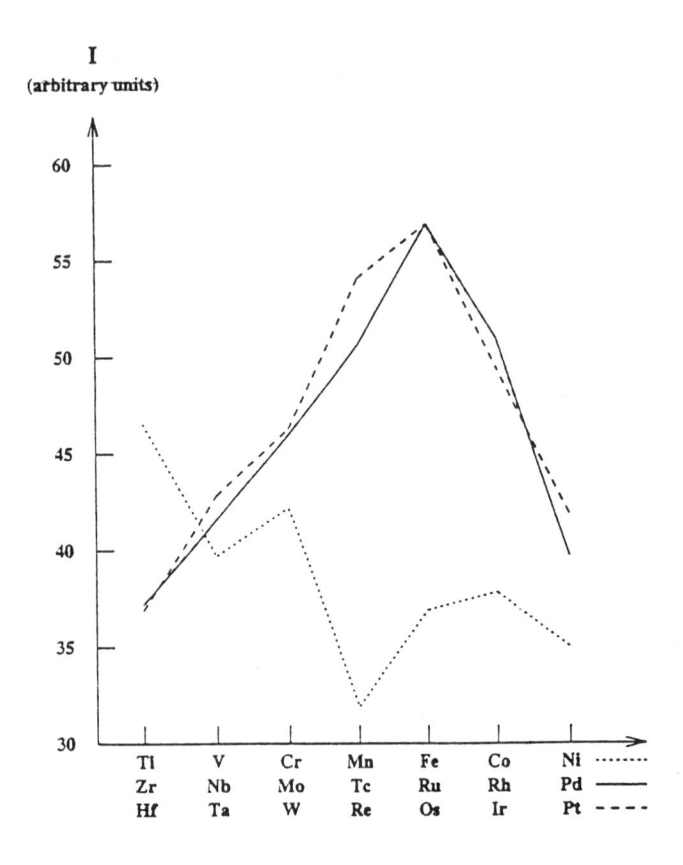

Figure 4. The theoretical activity parameter *I* vs. the position of the metal atom in the periodic table.

6. Information about the HDS mechanism provided by the activity parameter

The reasonably good correlation between *I* and the experimental activities does not imply that the adsorption mechanism of a thiophene molecule into surface sulfur vacancies (i.e. exposed metal atoms) involves the formation of metal-sulfur bonds. Rather, it indicates that the rate-limiting step in the HDS process must involve the interaction between the sulfur atom in the heterocycle on the one hand and the catalyst on the other. More information is needed to determine which step in the overall process is rate-limiting.

Scattered-wave calculations on MS_5^{-m}-thiophene complexes show that the thiophene molecule and the MS_5^{-m} cluster act much like individual subunits. Mixing between the thiophene orbitals and the catalyst orbitals is limited. Rather than strong metal-S_T interactions (where S_T denotes the sulfur atom in the heterocycle), one observes strong sulfur-sulfur interactions [13]. *I* therefore provides little information on the thiophene *adsorption* mechanism. Even if the η_1-adsorption mechanism is operative, the formation of strong metal-sulfur bonds is not observed. Therefore, the correlation between *I* and the experimental activity curve in combination with calculations on adsorption complexes indicates that adsorption is not rate-limiting and may occur in a perpendicular or parallel fashion or anything in between. In fact, in a recent paper [14] we have shown that there is not even a need for an exposed metal atom for adsorption to occur: a sulfur-sulfur adsorption mechanism is possible.

If, however, *dihydrothiophene*-MS_5^{-m} complexes are investigated, one finds that the dihydrothiophene molecule and the catalyst cluster have lost their individual character and that strong metal-sulfur interactions occur [15]. The observed correlation between *I* and the experimental catalytic activities tells us that the rate-limiting step in the HDS process involves the interaction between transition metal *d* electrons and sulfur 3*p* electrons. Hence hydrogenation of the thiophene molecule must occur before or in conjunction with the rate-limiting step in the HDS process. Also, differences in the degree of orbital mixing for different metal atoms in dihydrothiophene-MS_5^{-m} clusters indicates that desulfurization is the rate-limiting step in the HDS process. However, the emphasis should not be on the cleavage of the C-S_T bonds in thiophene, but instead on the ability of the catalysts to capture the sulfur atom of the heterocyclic ring when cleavage of the C-S_T bonds occurs. Desulfurization must occur at exposed metal atoms, but *I* shows that sulfur-sulfur interactions are also important.

How is this interpretation different from conclusions reached by previous theoretical studies of the HDS mechanism? Thiophene is a planar molecule, point group symmetry C_{2v}. The lowest unoccupied molecular orbital (LUMO) of thiophene is an antibonding combination of *p* contributions from the various atoms in the heterocyclic ring, perpendicular to the plane of the molecule. If this orbital were occupied at any point during the HDS process, it would clearly activate the carbon-sulfur bonds. Some studies have therefore used the occupation of the LUMO of thiophene as a criterion for selecting the thiophene adsorption mechanism [16].

Obviously, activation of the C-S$_T$ bonds is necessary at some point during the HDS process for cleavage to occur. However, we do not concern ourselves with the details of the activation mechanism. It need not occur during adsorption, but may very well occur as a result of hydrogenation of the carbon atoms adjacent to sulfur, or through some other mechanism. What is important for the activity of a catalytic site is the ability of the catalyst to capture the sulfur atom, when it is "released" by the heterocycle. Suppose that the C-S$_T$ bonds are activated, but that there is no site on the surface of the catalyst which is available for binding this sulfur atom. Desulfurization would still not occur. Consequently, desulfurization must be the rate-limiting step in the overall HDS process.

7. Factors that influence the value of the activity parameter

While the correlation between the theoretical activity parameter I and the observed catalytic activities is quite reasonable, it is interesting to examine the factors that influence the value of I. The two factors which are most important are the oxidation state of the transition metal and the metal-sulfur bond length.

The oxidation state directly affects the number of electrons in the cluster, and therefore the occupancy of the π^* and σ^* orbitals. Electrons which occupy these orbitals have the effect of lowering the value of I. Also, the energy of the other orbitals are influenced by the total number of electrons.

The metal-sulfur bond length affects the energies of the orbitals. A larger bond length pushes the bonding orbitals up in energy and the antibonding orbitals down.

In all the clusters we have assigned a formal oxidation state of -2 to the sulfur atom. The oxidation state of the metal atom is representative of the TMS, as given by the stoichiometry. In other words, Ru has a formal oxidation state of +4, since RuS$_2$ is the corresponding sulfide. RuS$_2$ has the pyrite structure and one may argue that it consists of Ru^{+2} and S$_2^{2-}$ units. If one were to choose an oxidation state of +2 for Ru, one should also adjust the oxidation state of sulfur. Such difficulties are avoided by relying on the stoichiometry of the sulfide for the determination of the net charge of the cluster. It should nonetheless be emphasized that the effects of a change in the oxidation state on the value of I is significant, a weakness in this theory.

With the exception of Pd, the metal-sulfur bond length is the experimental value. Pd is in a four-fold environment and the metal-sulfur bond length in an octahedral cluster has to be adjusted appropriately. This has been done on the basis of the Shannon atomic radii [17].

8. The activity parameter and promotion effects

It is well known that the addition of small quantities of cobalt and nickel to MoS_2 has a strong promoting effect. The activity of the combined system is much higher than the simple addition of the activities of the individual components. These mixed sulfides consist of three phases: MoS_2, Co_9S_8 (Ni_3S_2) and a third, mixed Co(Ni)-Mo-S phase, thought to be responsible for the high catalytic activity. In this phase the Co(Ni) atoms are believed to decorate the edges of MoS_2 slabs [18]. A synergistic promotion effect results from the close contact between Mo and Co(Ni) atoms. Harris and Chianelli [19] explain this effect on the basis of electron transfer from Co(Ni) to Mo. This increases the d-electron density on Mo and therefore changes the occupancy of the HOMO in an MoS_6^{-n} cluster from 2 to 3(4). The value of the activity parameter goes up accordingly. The theoretical activity parameter of Harris and Chianelli therefore implies that Mo is the active site, Co(Ni) is the promotor element.

If Mo indeed withdraws electrons from Co(Ni), the alternative activity parameter I leads to exactly the opposite conclusions. The I value of Mo decreases due to the addition of antibonding electrons. The I value of Co(Ni), on the other hand, goes up significantly due to the removal of σ^* electrons from the cluster. In other words, the new activity parameter implies that Co(Ni) is the active site in sulfided Co(Ni)Mo catalysts, Mo acting as a promotor and support.

The question whether Co(Ni) or Mo is the active site/promotor is still disputed. Instead of one element being the promotor and the other one the active site, it has recently been suggested that Co(Ni) and Mo together constitute the HDS site.

The Harris and Chianelli activity parameter also predicts that copper acts as a poison for MoS_2, by withdrawing electrons from Mo. What does I predict about this effect? If Cu withdraws electrons from Mo, then this should increase the activity of the surface Mo ions. If anything, Cu should have an electronic promotion effect, although this may of course be hidden by structural poisoning effects (dispersion). We believe that a definitive answer has not yet been given.

9. Conclusions

Periodic effects in HDS catalysis by TMS are electronic. Desulfurization is the rate-limiting step in the overall HDS process. A theoretical activity parameter I, which is a measure of the ability of the catalyst to "capture" a sulfur atom in the heterocycle, correlates well with experimental catalytic activities. The promotion effect in Co(Ni)-Mo-S is also electronic. The intrinsic activity of Co(Ni) is greatly enhanced through the transfer of σ^* electrons to Mo. Co(Ni) is the active site in these mixed sulfides, Mo acts as a promotor and support.

10. Acknowledgements

One of the authors (TSS) wishes to acknowledge helpful discussions with S. Eijsbouts, J.N. Louwen and E.T.C. Vogt. This research has been made possible through generous financial support from Akzo Nobel NV.

Bibliography

[1] T.A. Pecoraro and R.R. Chianelli. *J. Cat.*, 67:430 (1981)
[2] J.P.R. Vissers, C.K. Groot, E.M. van Oers, V.H.J. de Beer, and R. Prins. *Bull. Soc. Chim. Belg.*, 93:813 (1984)
[3] T.S. Smit and K.H. Johnson. *Cat. Lett.*, 28:361 (1994)
[4] J.K. Nørskov, B.S. Clausen, and H. Topsøe. *Cat. Lett.*, 13:1 (1992)
[5] H. Topsøe, B.S. Clausen, N.Y. Topsøe, J. Hyldtoft, and J.K. Nørskov, in Symposium on the Mechanism of HDS/HDN Reactions, Amer. Chem. Soc., Div. of Petr. Chem., Chicago IL, August 1993, p. 683
[6] S. Harris and R.R. Chianelli. *J. Cat.*, 86:400 (1984)
[7] K.H. Johnson. *Adv. Quant. Chem.*, 7:143 (1973)
[8] R.E. Watson. *Phys. Rev.*, 111:1108 (1958)
[9] P. Sabatier. *Ber. Deutsche Chem. Ges.*, 44:2001 (1911)
[10] R. Prins, V.H.J. de Beer and G.A. Somorjai. *Cat. Rev. Sci. Eng.*, 31:1 (1989)
[11] L. Hedin and B.I. Lundqvist. *J. Phys. C.*, 4:2064 (1971)
[12] D.M. Ceperley and B.J. Alder. *Phys. Rev. Lett.*, 45:566 (1980)
[13] T.S. Smit and K.H. Johnson. *Chem. Phys. Lett.*, 212:525 (1993)

[14] T.S. Smit and K.H. Johnson. *J. Mol. Cat.*, 91:207 (1994)
[15] T. S. Smit, Ph.D Thesis, Mass. Inst. of Techn., May 1994
[16] M.C. Zonnevylle, R. Hoffmann, and S. Harris. *Surf. Sci.*, 199:320 (1988)
[17] R.D. Shannon. *Acta Cryst.*, A32:751 (1976)
[18] H. Topsøe, B.S. Clausen, R. Candia, C. Wivel, and B. Mørup, *Bull. Soc. Chim. Belg.*, 90:1189 (1981)
[19] S. Harris and R.R. Chianelli, *J. Cat.*, 98:17 (1986)

A Review of Some Specific Phases as Possible Active Hydrotreating Constituents

H. D. SIMPSON
Unocal Corporation, Fred L. Hartley Research Center, P. O. Box 76, Brea, California 92621 USA

ABSTRACT

This work reviews compounds such as $CoMo_2S_4$ and $NiMo_3S_4$, which contain the proper elements in proportions that closely match the active phases in commercial hydrotreating catalysts. The compounds are believed to be inactive by themselves, but the packing density of the Mo atoms in them is similar to that in MoS_2. This was useful years ago in making preliminary estimates of the upper limits of the amounts of catalytic metals that could be dispersed on alumina supports. The estimates were important in guiding experimental programs that eventually led to some of today's most active high metals hydrotreating catalysts.

In related work, Sanderson's principles were used, together with chemisorption models based on structural units of $CoMo_2S_4$, to estimate the thermochemistry of model catalytic reaction cycles for the denitrogenation of pyridine. The study suggested that some metal-metal bonding in the active phase is desirable, and that the chemisorbed pyridine must have an entropy content more representative of a liquid than a solid in order for the assumed reaction cycles to be feasible. Future studies using, e.g., modern density functional approaches together with EXAFS structural data from CoMoS and NiMoS phases might prove to be even more insightful.

INTRODUCTION

Currently accepted models of the active sites in hydrotreating catalysts center around the so-called CoMoS and NiMoS phases. Although considerable progress has been made towards defining these phases in recent years, their structures are still not known precisely. A definitive assessment can be found in Topsoe, et al.[1]

Before the advent of the CoMoS era, substantial progress in visualizing the active phases in hydrotreating catalysts was made in some laboratories by relying on well characterized mixed sulfides for which structural data were available. The compound $CoMo_2S_4$ was found to be particularly useful, because it contains the proper elements in proportions that match those in many commercial hydrodesulfurization (HDS) catalysts. There was considerable intuitive appeal to the idea that compounds which could be synthesized in the laboratory might be very much like the active phases formed on catalysts during the activation step.

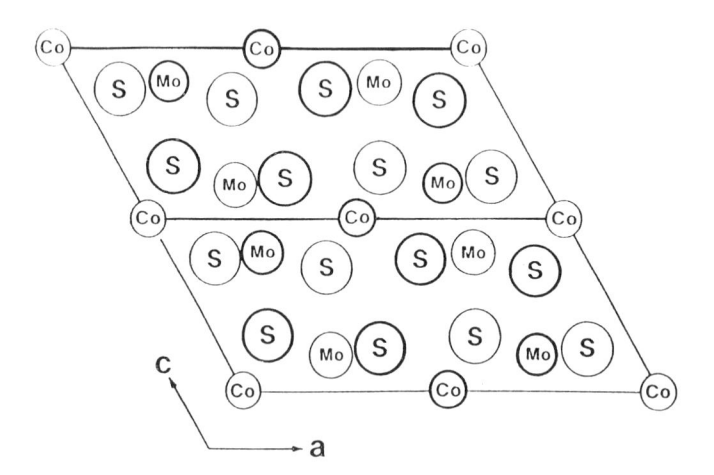

Fig. 1

Projection of $CoMo_2S_4$ along the b axis. Space group: C2/m;

a = 13.09Å; b = 3.23Å; c = 5.90Å; β = 118.9°.

(Adapted from Anzenhofer and DeBoer, Acta Cryst. B25: 1419 (1969)).

Specific Phases as Hydrotreating Constituents

This work reviews information presented previously on these structures[2], and attempts to make comparisons with CoMoS/NiMoS. Also reviewed is the thermochemistry of model catalytic cycles for the denitrogenation of pyridine based on the structural surroundings of Mo in $CoMo_2S_4$.[3]

$CoMo_2S_4$ AND MoS_2

A projection of the $CoMo_2S_4$ structure along the **b** axis of its monoclinic unit cell is shown in Fig. 1, and a projection of the hexagonal representation of MoS_2 along the **a** axis is shown in Fig. 2.

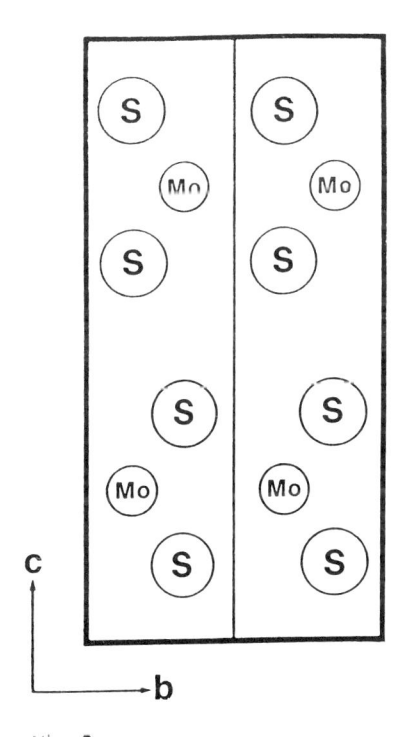

Fig. 2

Projection of MoS_2 along the a axis. Space group C6/mmc; a = 3.02Å; c = 12.30Å. Since the space group is hexagonal, b = a.

(Adapted from R. W. G. Wyckoff, Crystal Structures, Second ed., Interscience, New York, 1963, Vol. 1, pp. 280-282).

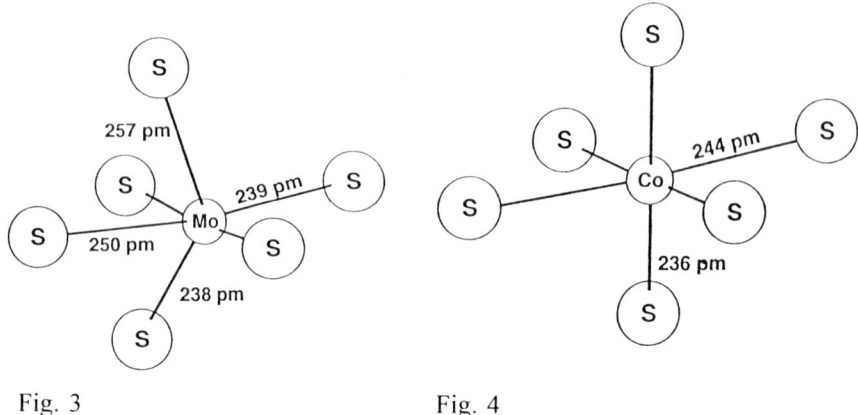

Fig. 3

Fig. 4

Mo environment in $CoMo_2S_4$.

Co environment in $CoMo_2S_4$.

Simple calculations based on these views show that four Mo atoms are supported over about 68 Å² in the former structure, and over about 78 Å² in the latter. Thus, the packing densities of the Mo atoms in what might be viewed as monolayers from the two structures are fairly similar. Relative to MoS_2, the packing of Mo in the $CoMo_2S_4$ structure is tightened somewhat by the effects of the Co atoms. Mo is coordinated to six sulfur atoms in both structures. In $CoMo_2S_4$, the Co atoms are also coordinated to six sulfurs. The distorted octahedral environments around the metal atoms are shown in Figs. 3 and 4.

Dispersed monolayers of MoS_2 and $CoMo_2S_4$ on alumina supports would provide the equivalent of about 25 wt.% MoO_3 on supports with about 300 m²/g of surface area, the amount found in today's best vacuum gas oil hydrotreating catalysts.

$NiMo_3S_4$

A projection of $NiMo_3S_4$ is shown in Fig. 5. In this structure, octahedral Mo metal entities are connected through shared sulfur atoms to mobile Ni atoms along the edges of the unit cell.

From the information in Fig. 5, it is estimated that a monolayer of $NiMo_3S_4$ would be equivalent to even higher levels of MoO_3 and NiO than are used in today's best catalysts. Thus, improved activities might be possible if $NiMo_3S_4$ should indeed be an active phase, and if it can be loaded on supports in high coverage.

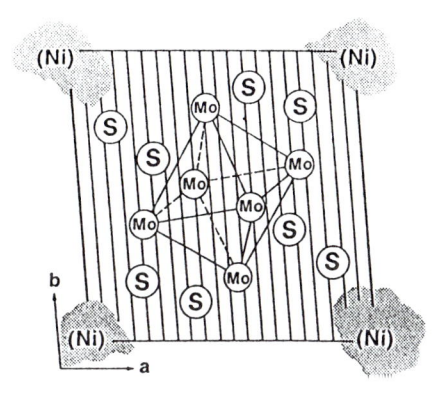

Fig. 5

Projection of NiMo$_3$S$_4$ along the c axis. Space group $R\bar{3}$; a = 6.46Å; α = 94.7°. Since the space group is rhombohedral, a = b = c.

(From J. Guillevic et al., J. Solid State Chem. 1:158 (1973)).

MODEL HDN CYCLES WITH CoMo$_2$S$_4$

In reference 3, the hydrodenitrogenation (HDN) of pyridine was modeled assuming no participation of SH groups, and assuming that the active sites were Mo atoms. The principal chemisorption complexes that were used are illustrated in Figs. 6 and 7.

The indicated interatomic distances were used to estimate heats of formation (ΔH°$_f$), employing Sanderson's methods calibrated on the individual substances. Gibbs free energies (ΔG°$_f$) were estimated from ΔH°$_f$ by analogy from the known heats and free energies of the individual substances. The estimates for C$_5$H$_5$N - [Co$_{0.5}$MoS$_{1.67}$] were: ΔH°$_f$ = -122 kpm (kcal/mole), and ΔG°$_f$ = -100 → -119 kpm, depending on the entropy of the chemisorbed pyridine. The estimates for H - [Co$_{0.5}$MoS$_2$] were: ΔH°$_f$ = -92 → -112 kpm, depending on whether a Mo-Mo bond exists. ΔG°$_f$ was taken to be 97% of ΔH°$_f$.

The steps that were used in the model are listed below:

Hydrogen Dissociation

1/2 CoMo$_2$S$_4$ + 1/2 H$_2$ → H - [Co$_{0.5}$MoS$_2$]   (1)

Pyridine Chemisorption

$$1/2\ CoMo_2S_4 + C_5H_5N + 0.33\ H_2 \rightarrow C_5H_5N - [Co_{0.5}MoS_{1.67}]$$
$$+ 0.33\ H_2S \quad (2)$$

Reaction and Regeneration

$$C_5H_5N - [Co_{0.5}MoS_{1.67}] + 9.34\ H - [Co_{0.5}MoS_2] + 0.33\ H_2S$$
$$\rightarrow 10.34\ Co_{0.5}MoS_2 + NH_3 + C_5H_{12} \quad (3)$$

Free energies of reaction, ΔG_r, were computed as a function of temperature for the reactions, using the standard heats, free energies and heat

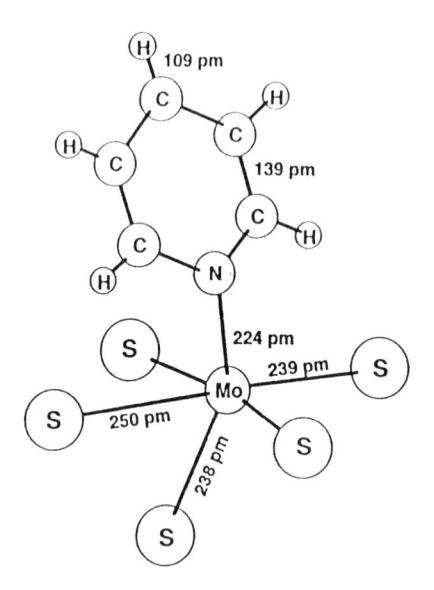

Fig. 6

Portion of the chemisorption complex $C_6H_5N - [Co_{0.5}MoS_{1.67}]$. It has been assumed that nitrogen coordinates to Mo in place of the most weakly bonded sulfur.

Specific Phases as Hydrotreating Constituents

capacities (some estimated with Kopp's rule) of the substances. Inferences from the study were:

1. Eq. 1 is in near-thermochemical balance, and some Mo-Mo bonding that appears to exist in $CoMo_2S_4$ is alternatively broken and remade to drive the catalytic cycle. A feasible process is obtained if a Mo-Mo bond is broken during hydrogen chemisorption/dissociation, and re-formed during the reaction and regeneration step.

2. The feasibility of eq. 3 is low at higher temperatures (where HDN is known to proceed rapidly) if the chemisorbed pyridine is assumed to have an ordered, solid-state arrangement. A more realistic model is obtained by assuming that the pyridine has the higher entropy content characteristic of a liquid phase.

3. The use of gas-phase hydrogen instead of the hydrogen chemisorption complex in Eq. 1 is not feasible. This might be one of the reasons why high pressures are needed in commercial processes.

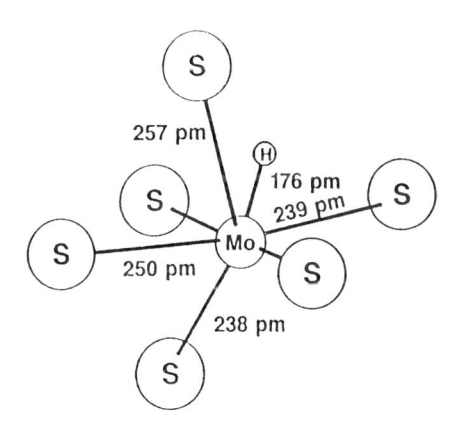

Fig. 7

Portion of the chemisorption complex H - $[Co_{0.5}MoS_2]$. It has been assumed that hydrogen coordinates to Mo with extra electron density transferred from Co, and that no Mo - S bonds are broken.

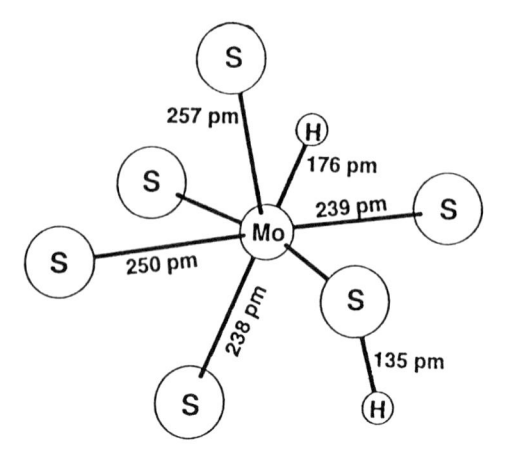

Fig. 8

Portion of the chemisorption complex H - [Co$_{0.5}$MoS(SH)].

As stated, the above model did not consider the use of SH groups, i.e., the dissociation of hydrogen by the sulfur atoms in the sulfide structure instead of by the metal atoms. Additional work to address this was done with a related model, by substituting the complex H - [Co$_{0.5}$MoS(SH)] for the complex H - [Co$_{0.5}$MoS$_2$] in the first model. A pictorial representation of part of the revised complex is shown in Fig. 8.

The estimated value of ΔH°$_f$ for H - [Co$_{0.5}$MoS(SH)] is -87 kpm. As before, ΔG°$_f$ is assumed to be 97% of ΔH°$_f$. The revised equations representing the model are shown below.

Hydrogen Dissociation

$$1/2 \ CoMo_2S_4 + H_2 \rightarrow H - [Co_{0.5}MoS(SH)] \tag{1a}$$

Pyridine Chemisorption

$$1/2 \ CoMo_2S_4 + C_5H_5N + 0.33 \ H_2 \rightarrow C_5H_5N - [Co_{0.5}MoS_{1.67}]$$
$$+ \ 0.33 \ H_2S \tag{2a}$$

Specific Phases as Hydrotreating Constituents

Reaction and Regeneration

$$C_5H_5N - [Co_{0.5}MoS_{1.67}] + 4.67\ H - [Co_{0.5}MoS(SH)] + 0.33\ H_2S$$
$$\rightarrow 5.67\ Co_{0.5}MoS_2 + NH_3 + C_5H_{12} \quad\quad (3a)$$

The estimated free energies of these reactions are shown in the following tabulation:

Reaction	ΔG_r (kcal) 400°F	800°F
1a	+5	+2
2a	-75	-84
3a	-52	-36

These results are very similar to the ones in the original model. The hydrogen dissociation step is in very close thermodynamic balance when only the metal atoms are involved, and when the sulfur atoms are involved also. Thus, hydrogen dissociation is probably the limiting step. Again, this may explain why such high hydrogen pressures are needed in commercial hydrotreating operations.

COMPARISON WITH CoMoS/NiMoS

In reference 1, Topsoe, et al. compare the spectroscopic properties and EXAFS results of NO chemisorption complexes of supported catalysts with those of inorganic clusters synthesized in the laboratory. $CoMo_2S_4$ was also considered in the study. On the basis of that work, it appears that the Co atoms in CoMoS catalysts are bonded to four-sulfur faces at the edges of MoS_2 crystallites, in pseudo-tetrahedral configurations. The promotional effect of Co is explained in terms of increased electron density transferred to the neighboring S and Mo atoms. The average overall coordination of the Co atoms by sulfur is about five, at a distance of about 223 pm. In $CoMo_2S_4$ the EXAFS fit (which was difficult because of the variation among the different distances), showed the Co to be surrounded by about five sulfurs also, but at an average distance of 240 pm. This distance agrees very favorably with the average of the distances from the x-ray structure.

From other EXAFS studies, Bouwens, et al. concluded that, in CoMoS supported on carbon, the Co coordination is octahedral[4], or distorted 5- to 6-fold.[5] The Co atoms appear to also be coordinated with some Mo atoms at a distance of 280 pm.[5]

In general, the sulfur coordination of both Co and Mo by sulfur atoms in CoMoS catalysts is less than six.[6]

EXAFS studies have also been carried out on NiMoS supported on carbon.[7,8] In these instances, the Ni atoms appear to be situated in square and tetragonal pyramidal sites with sulfurs at a distance of 221 pm. Some Mo atoms are located 282 pm from the Ni.

Comparison between the structural aspects of $CoMo_2S_4$ and $NiMo_3S_4$, and the foregoing information about CoMoS/NiMoS, indicates that the structure and bond distances in the former are sufficiently different from those in CoMoS and NiMoS to justify ruling them out as the active phases in actual hydrotreating catalysts. However, $CoMo_2S_4$ does appear to be closely related to the active phases. The major differences may well derive from the fact that we are trying to compare a bulk substance with highly dispersed surface counterparts. The longer Co-S bond distances in $CoMo_2S_4$ (236-244 pm, compared to ca. 221 pm in CoMoS) may largely be a consequence of the full coordination in the bulk substance. If it is imagined that one of the sulfurs in Fig. 4 is pulled away, the remaining Co-S distances would contract and the nature of the distortion of the reduced motif would change. Also, the Mo and Co atoms in $CoMo_2S_4$ are within about 300 pm of each other, similar to the Co-Mo distances reported in some of the referenced work.

The usefulness of the $CoMo_2S_4$ structure in the original work done in our laboratories largely stemmed from an engineering effort to make estimates of maximum supportable loadings of catalytic metals. As it turns out, this substance was similar enough to CoMoS and NiMoS to give estimates that were quite good. The data on the $NiMo_3S_4$ structure was discovered later in our program. If we had used that substance in our original work, our estimates about maximum loadings would have been too high relative to present knowledge, because of its more metallic nature.

In short, well characterized specific phases will probably rarely, if ever, actually be the same as the active substances in working catalysts, due to the distortional effects imposed by the support and perhaps the reaction environment. However, using them as models of the active substances can lead to good progress in the absence of more sophisticated information.

PROGNOSIS

Much of the work described in this report was done before our present knowledge about CoMoS/NiMoS became available. Because of the use of bulk compounds as a basis, and because of the inherent limitations associated with adapting Sanderson's techniques to what is actually a complex problem, the thermochemical results reported here and in reference 3 must necessarily be regarded as approximate and tentative. However, now that we do have some structural information on what appear to be the actual working catalytic phases, it might be worthwhile using this information to conduct thermochemical studies on more refined models of catalytic cycles. Such techniques as the modern density functional approaches can probably be employed to make improved estimates of the fundamental thermochemical properties of the chemisorption complexes.

ACKNOWLEDGEMENT

The author wishes to thank Elsevier Science Publishers B. V., for permission to present the material from references 2 and 3. Figures 1, 2 and 5 were reproduced from reference 2.

REFERENCES

1. H. Topsoe, B. S. Clausen, N. Topsoe, E. Pedersen, W. Niemann, A. Muller, H. Bogge, and B. J. Lengeler, Chem. Soc., Faraday Trans. 1, 83:2157 (1987).

2. H. D. Simpson, in Studies in Surface Science and Catalysis, 50, (M. L. Occelli and R. G. Anthony eds.), Elsevier, Amsterdam, 1989, pp. 133-146.

3. H. D. Simpson in Studies in Surface Science and Catalysis, 38, (J. W. Ward ed.), Elsevier, Amsterdam, 1987, pp. 399-413.

4. S. Bouwens, D. C. Koningsberger, V. H. J. de Beer, and R. Prins, Catalysis Letters, 1:55 (1988).

5. S. Bouwens, J. A. R. van Veen, D. C. Koningsberger, V.H.J. de Beer, and R. Prins, J. Phys. Chem. 95:123 (1991).

6. R. Prins, V. H. J. de Beer, and G. A. Somorjai, Catal. Rev. - Sci. Eng., 31:1 (1989).

7. W. Niemann, B. S. Clausen, and H. Topsoe, Catalysis Letters, 4:355 (1990).

8. S. Bouwens, D. C. Koningsberger, V. H. J. de Beer, S. P. A. Louwers, and R. Prins, Catalysis Letters, 5:273 (1990).

Catalytic Sites and Heterolytic Reaction Mechanisms in Hydrotreating Catalysis

S. Kasztelan

Kinetics and Catalysis Division, Institut Français du Pétrole
B.P. 311, 92506 Rueil-Malmaison Cedex, France

Abstract

The nature of the surface sites and species of MoS_2 or WS_2 based hydrotreating catalysts are examined in the light of the heterolytic reaction mechanism hypothesis. The hydrotreating reactions are proposed to occur in the form of elimination, addition and substitution reactions on a surface composed of coordinatively unsaturated Mo^{4+} ions, sulfide, H and SH species. The active sites are proposed to be pairs of acid-base and acid-nucleophile sites involved with the rate determining step of the reaction.

I INTRODUCTION

The active phase of sulfided molybdenum or tungsten based hydrotreating catalysts is able to perform a variety of catalytic reactions and in particular the carbon-heteroatom bond cleavage in hydrodesulfurization (HDS), hydrodenitrogenation (HDN) and hydrodeoxygenation (HDO) as well as the hydrogenation (HYD) of double bonds and aromatic rings [1-10].

All of these reactions are performed by unpromoted and Co or Ni promoted Mo(W) disulfide catalysts in bulk or supported forms [8-12]. In addition, $MoS_2(WS_2)$ is an anisotropic compound and all of these reactions are considered to occur on the edge planes of the bidimensional layers of the sulfide particles of which the most probable ones proposed in the literature are the $(\bar{1}010)$ and $(10\bar{1}0)$ edge planes [11-17].

On these edge planes there are coordinatively unsaturated transition metal ions [11-15,18-21], sulfide ions, sulfhydryl groups [22-26] and possibly $(S-S)^{2-}$ species [27]. Hydrogen species are also present [28-32] in the form of SH groups and possibly in the form of atomic or hydride species. Presence of the later species is suggested by theoretical considerations [33] and by ^{1}H NMR [34] and inelastic neutron scattering studies of RuS_2 [35]. The oxidation state of the Mo ions is proposed to be either 5+ [36], 4+ [30], 3+ [12], 2+ [37], depending on the characterization techniques employed.

Many different reaction mechanisms and rate laws have been proposed for HDS/HDN/HYD reactions to account for the presence of these sites and species. Remarquably, there is no consistency between the reaction mechanisms proposed in the literature [1-7,38]. For example, redox reaction mechanisms have been often prefered for HDS [1,12,14] whereas elimination or substitution reactions are invoked for the C-N bond cleavage in HDN [4-6,39-43]. HYD is often written as an addition of hydrogen species generated by homolytic dissociation [1,3,6,38].

Kinetic models for HYD/HDS/HDN reactions over sulfides catalysts are most often based on the molecular adsorption of hydrogen and hydrogen sulfide and sometimes on their homolytic dissociation. Kinetic models for one type and two types of actives sites are usually compared and in general two sites models are preferred [1-6]. These models fail in general to account for the various effects of H_2S on the HDS/HDN/HYD reactions which stem from an inadequate description of reaction mechanisms and rate laws [6].

In this work an attempt is made to provide a link between the nature of the surface species, the mechanisms of HDS, HDN and HYD reactions and their kinetics by considering the possibility that heterolytic reaction mechanisms occur on a surface composed of coordinatively unsaturated Mo ions in the 4+ state, sulfide ions, H and SH species.

II RESULTS AND DISCUSSION

II.1 Surface Species on MoS$_2$ Edge Planes

The active phase of Mo or W based hydrotreating catalysts can be described by small particles of MoS$_2$ or WS$_2$ containing one or several layers or slabs. Of particular relevance to this work is the fact that a MoS$_2$ slab, whatever its shape and size, must remain electroneutral. A slab can be obtained by cutting a pieces into an infinite MoS$_2$ layer in which there is only Mo^{4+} and S^{2-} ions. The geometrical description of an ideal hexagonal MoS$_2$ slab allows to compute the number of each component of the slab vs the slab size with the formula reported in Table 1 [15].

Table 1 : Number of Mo and S ions in an ideal hexagonal MoS$_2$ slab vs the slab size.

Nb of Mo ions along one side: n	4	7	10
Length=3.2(2n-1) in Å	22.5	42	61
Mo$_{total}$=3n^2-3n+1	37	127	271
S$_{total}$=6n^2	96	294	600
S$_{basal}$=6n^2-12n+6	54	216	486
Mo$_{basal}$=3n^2-9n+7	19	91	217
Mo$_{edge}$=6n-12	12	30	48
S$_{edge}$=12n 6	42	78	114
Mo$_{corner}$	6	6	6

If all of the molybdenum ions of the slab are kept saturated by sulfur ions, the S/Mo will be larger than 2 and the slab will be negatively charged whereas the infinite layer, with its hexagonal hole, will be sulfur deficient and positively charged. To keep the electroneutrality of the slab, the oxidation number of Mo need to be changed. This situation is illustrated in Table 2 by the chemical formula 2.1 of a model hexagonal slab containing 127 Mo's and 294 S^{2-} ions (see Table 1) showing that the S/Mo ratio is 2.3 and that the excess negative charge is compensated by Mo^{5+} ions.

A stoechiometric slab will therefore possess a number of coordinative unsaturations on the edges (symbol V). As shown by formula 2.2, for the 127 Mo's stoichiometric slab, there are 38 S^{2-} and 40 V on the edges, the number of V being equal to the number of S^{2-} plus two coordinative unsaturations due to the MoS$_2$ structure.

When exposed to hydrogen sulfide, the electroneutrality of this slab will be respected if there is heterolytic dissociation of H$_2$S according to reaction 1. Saturation of the coordinative unsaturations (except two of them) will lead to edges saturated by SH$^-$ groups as indicated by formula 2.3. The S/Mo ratio is still 2.3 and the H's associated to the SH groups give an H/Mo ratio of 0.6, a value close to experimental values [28-30].

Table 2 : Chemical formula of an hexagonal MoS$_2$ slab containing 127 Mo ions (the core of the slab is represented in brackets).

	Formula	S/Mo	H/Mo	Mo^{x+}
2.1	$[Mo_{127}^{4/5+} S_{216}^{2-}] S_{78}^{2-}$	2.3	0	4.6
2.2	$[Mo_{127}^{4+} S_{216}^{2-}] S_{38}^{2-} V_{40}$	2	0	4
2.3	$[Mo_{127}^{4+} S_{216}^{2-}] SH_{76}^{-} V_{2}$	2.3	0.6	4
2.4	$[Mo_{127}^{4+} S_{216}^{2-}] (S_{2}^{2-})_{76} V_{2}$	2.9	0	4
2.5	$[Mo_{127}^{4+} S_{216}^{2-}] SH_{38}^{-} H_{38}^{-} V_{2}$	2	0.6	4
2.6	$[Mo_{127}^{4+} S_{216}^{2-}] SH_{1}^{-} H_{75}^{-} V_{2}$	1.7	0.6	4
2.7	$[Mo_{127}^{3/4+} S_{216}^{2-}] SH_{38}^{-} H_{38}^{-} V_{2}$	2	0.6	3.7
2.8	$[Mo_{127}^{3/4+} S_{216}^{2-}] S_{1}^{2-} V_{77}$	1.7	0	3.4

The saturation of the edges of an electroneutral slab can also be reached by (S-S)$^{2-}$ groups [27] giving a large S/Mo ratio but no change of Mo oxidation number and no hydrogens as shown by formula 2.4. When exposed to hydrogen, heterolytic adsorption according to reaction 2 is required to keep all of the Mo's in a 4+ state

Catalytic Sites and Heterolytic Reaction Mechanisms

with formation of a singly charged negative species H$^-$. The H/Mo ratio is still 0.6 (formula 2.5) but half of it is in the form of SH$^-$ (S^{2-} H$^+$) groups and half in the form of H$^-$ species.

$$H_2S + Mo^{4+}\text{-}V + Mo^{4+}\text{-}S^{2-} \rightarrow Mo^{4+}\text{-}SH^- + Mo^{4+}\text{-}S^{2-}H^+ \quad (1)$$

$$H_2 + Mo^{4+}\text{-}V + Mo^{4+}\text{-}S^{2-} \rightarrow Mo^{4+}\text{-}H^- + Mo^{4+}\text{-}S^{2-}H^+ \quad (2)$$

All of the edge sulfur species, except the last one, can be removed by substituting each SH$^-$ by an H$^-$ (formula 2.6). This allows to keep all of the Mo's in the 4+ state and the electroneutrality of the slab. The S/Mo ratio is now 1.7 although there is no reduction of the Mo's. The H/Mo ratio is still 0.6 but all of the H's are now H$^-$.

Reduction or oxidation of Mo^{4+} can be a way to compensate the excess of electrostatic charges. Formula 2.1 illustrates the oxidation of Mo^{4+} to Mo^{5+} required to accommodate for S^{2-} ions in excess of the stoichiometry. Mo^{3+} can be formed by electron transfer during hydrogen adsorption according to reaction 3 (formula 2.7). If such electron transfer occurs, a mean Mo oxidation number of 3.7 will be found. Eventually, if all of the hydrogens covering the edges in formula 2.6 are in the form of Mo^{3+}-H rather than Mo^{4+}-H$^-$ or if all of these hydrogens are removed according to reaction 4, then the mean Mo oxidation number of all of the Mo's of the slab will be 3.4 (formula 2.8) [30]. Note, however, that if only the edge and corner Mo's are reduced while the internal or basal Mo's remain Mo^{4+}, edge and corner Mo's with oxidation number of 2+ are needed [18].

$$H_2 + Mo^{4+}\text{-}V + Mo^{4+}\text{-}S^{2-} \rightarrow Mo^{3+}\text{-}H + Mo^{4+}\text{-}S^{2-}H^+ \quad (3)$$

$$2\, Mo^{4+}\text{-}H^- \rightarrow 2\, Mo^{3+}\text{-}V + H_2 \quad (4)$$

Formula 2.2, 2.3, 2.5 and 2.6 indicate that on a surface containing only Mo^{4+} ions a limited number of acidic sites (H$^+$ from S^{2-}H$^+$, coordinatively unsaturated metal centers such as Mo^{4+}-V), basic sites (SH$^-$, S^{2-}) and nucleophilic species (SH$^-$, H$^-$) can coexist.

These surface species are present on the same edge planes and therefore several pairs of sites can be sorted out such as :
- acid-base pair of sites :
    $(S^{2-}H^+/S^{2-})$, (V/S^{2-}), $(S^{2-}H^+/SH^-)$
- acid-nucleophile pair of sites :
    $(S^{2-}H^+/SH^-)$, $(S^{2-}H^+/H^-)$, (V/SH^-), (V/H^-)

II.2 Elimination, Addition and Substitution Reaction Mechanisms

HDS/HDN/HYD reactions can be analyzed in terms of elimination, addition and substitution (EAS) reaction mechanisms which involve generally heterolytic bond dissociation and pairs of acid-base sites or acid-nucleophile sites [44]. The species present on the edge surfaces identified previously, although they may not represent all of the species which can exist on the surface of MoS_2, can at least allow EAS reactions to occur.

To illustrate this point, simplified reactions are proposed in Table 3 for a reactant RX with R, a phenyl group and X, a SH, OH or NH_2 group [45]. Table 3 shows that hydrogenation can be either the successive addition of an hydride followed by a proton (HYD1, reaction 3.1) or the addition of a proton followed by an hydride (HYD2, reaction 3.2) [46,47].

The C_{sp}^2-X bond cleavage, can occur via an aromatic nucleophilic substitution S_NAr as shown by reaction 3.3. The C_{sp}^3-X bond cleavage can occur via E2 elimination with S^{2-} as a base (reaction 3.4) or via S_N2 nucleophilic substitution with SH^- as the nucleophile (reaction 3.5). In both cases the reactant must be at least partially hydrogenated and is symbolized by RXH_2. Incidently, as shown by reaction 3.4 , the C-X bond cleavage can also in some instance be apparently a C_{sp}^2-X bond cleavage by an E2 elimination where the product is an aromatic.

A catalytic cycle for HDN by E2 elimination previously proposed is shown in Figure 1 [42]. It involves the pair of sites M^{4+}-V/M^{4+}-S^{2-} in which the Lewis acid site is prefered to the Bronsted site, $S^{2-}H^+$, because of the well known weak acidity of SH groups. Figure 1 also shows that E2 on this pair of sites will be poisoned by

Catalytic Sites and Heterolytic Reaction Mechanisms

H$_2$S. The nucleophilic substitution cycle shown in Figure 2 involves the pair of acid-nucleophile sites M^{4+}-V/M^{4+}-SH$^-$ and will be promoted by H$_2$S [43]. The catalytic cycle for HYD1 represented in Figure 3 involves the pair of sites M^{4+}-V/M^{4+}-H$^-$ and will be poisoned by H$_2$S.

Table 3 : Hydrogenation, C-S and C-N bond cleavage via addition, elimination and substitution reactions (X=SH, NH$_2$, OH).

HYD1 :	H$^-$ + RX → XRH$^-$ + H$^+$ → XRH$_2$	[3.1]
HYD2 :	H$^+$ + RX → XRH$^+$ + H$^-$ → XRH$_2$	[3.2]
S$_N$Ar :		[3.3]

$$H^- + C_6H_5X \rightarrow C_6H_6 + X^-$$

E2 : [3.4]

$$S^{2-} + C_6H_5(H)(XH) \rightarrow C_6H_6 + X^- + SH^-$$

S$_N$2 : [3.5]

$$SH^- + C_6H_5(H)(XH) \rightarrow C_6H_5(H)(SH) + X^-$$

Other catalytic cycles can be designed for other pair of sites, but these three examples indicate that EAS reaction mechanisms may provide a consistant interpretation of the HDS/HDN/HYD reactions without necessarily involving Mo ions in a state lower or higher than 4+.

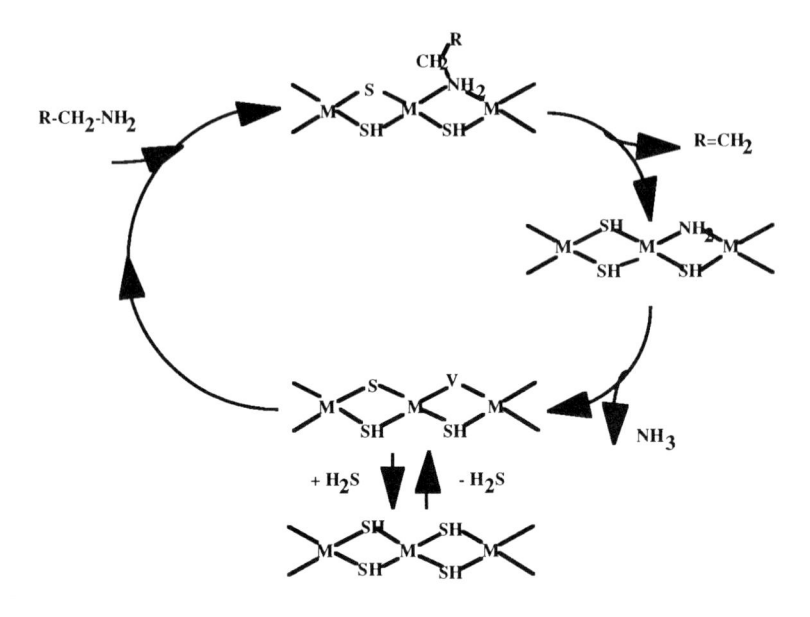

Figure 1 : Catalytic cycle for hydrodenitrogenation by elimination.

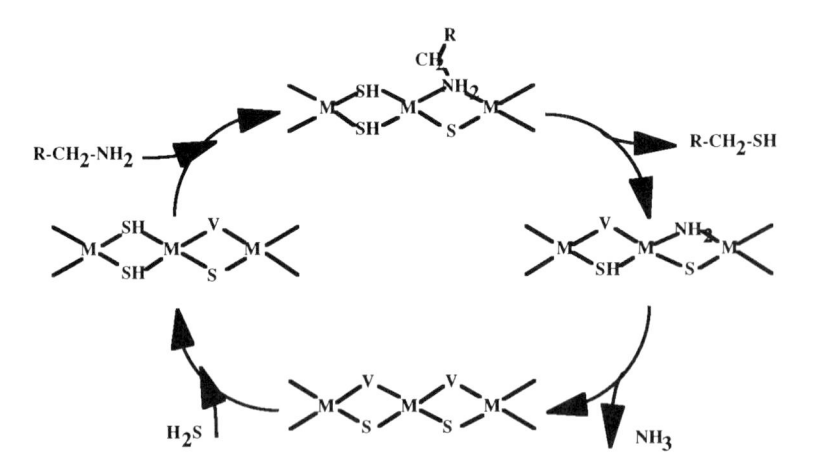

Figure 2 : Catalytic cycle for hydrodenitrogenation by nucleophilic substitution.

Catalytic Sites and Heterolytic Reaction Mechanisms 37

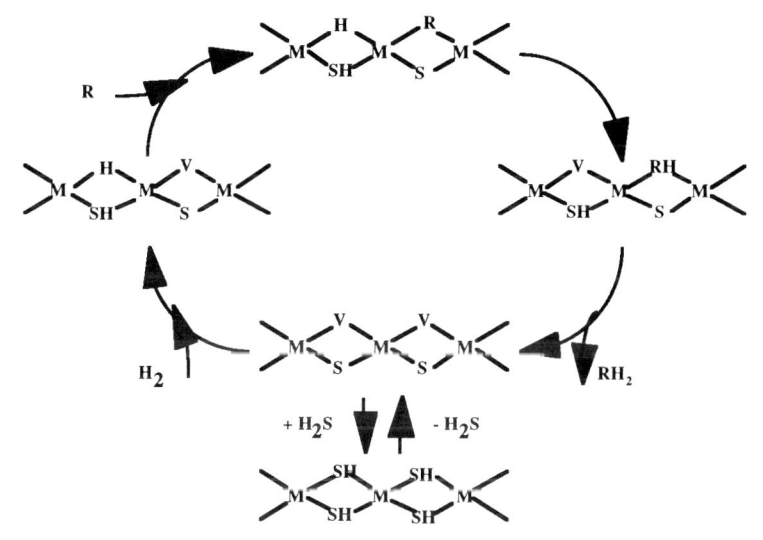

Figure 3 : Catalytic cycle for hydrogenation by the sequence H + SH.

II.3 Kinetics of Elimination, Addition and Substitution Reactions.

The kinetic rate laws must reflect the qualitative considerations indicated above. A set of elementary reactions for the EAS reactions considered in Table 3 is presented in Table 4. A representation without electrostatic charges is employed in this work. The virgin surface of the catalyst corresponding to formula 2.2 is composed of coordinatively unsaturated Mo^{4+} ions, symbolized by ∗-V, and a sulfur ion, symbolized by •-S, host of the protons generated by the hydrogen and hydrogen sulfide heterolytic dissociation according to reaction 1 and 2.

The rate laws for the different rate determining steps (rds) considered are computed using the classical Langmuir-Hinshelwood-Hougen-Watson (LHHW) method and are reported in Table 5. The computation is made by writing the equations for the equilibrium constants K_i and one conservation equation for each type of sites: $[\ast\text{-V}]_o$ and $[\bullet\text{-S}]_o$ with $[\ast\text{-V}]_o = [\bullet\text{-S}]_o$ because saturation of one of these two types of sites leads to no more heterolytic dissociation.

Table 4: Hydrogenation and carbon-heteroatom bond cleavage by elimination, addition and substitution reactions

Adsorption:
H_2 + *-V + •-S ↔ *-H + •-SH	K_{H_2}, r_1	[4.1]
H_2S + *-V + •-S ↔ *-SH + •-SH	K_{H_2S}	[4.2]
HX + *-V + •-S ↔ *-X + •-SH	K_{HX}	[4.3]
RX + *-V ↔ *-XR	K_{RX}, r_4	[4.4]
RH_2X + *-V ↔ *-XRH_2	K_{RH_2X}	[4.5]

Hydrogenation HYD1:
*-XR + *-H → *-V + *-XRH	K_H, r_6	[4.6]
*-XRH + •-SH → *-XRH_2 + •-S	r_7	[4.7]

Hydrogenation HYD2:
*-XR + •-SH → *-V + •-SXRH	K_H, r_8	[4.8]
•-SXRH + *-H → *-XRH_2 + •-S	r_9	[4.9]

Nucleophilic aromatic substitution (C_{sp}^2-X bond cleavage)
*-XR + *-H → *-V + *-X + RH↑ r_{10} [4.10]

Elimination E2 (C_{sp}^3-X bond cleavage)
*-XRH_2 + •-S → *-X + •-SH + RH↑ r_{11} [4.11]

Nucleophilic substitution S_N2 (C_{sp}^3-X bond cleavage)
*-XRH_2 + *-SH → *-X^- + *-V + RH_2SH↑ r_{12} [4.12]

Site conservation equations:
$[*-V]_0$ = [*-V] + [*-H] + [*-SH] + [*-X] + [*-XR] + [*-XRH_2] + [*-XRH]
    [4.13]
$[•-S]_0$ = [•-SH] + [•-S]   [4.14]

Cation/anion balance equation:
$[•-SH^-]$ = [*-H] + [*-SH] + [*-X] + [*-XRH]   [4.15]

Approximation
$[*-V]_0 = [•-S]_0$ [4.16]

Catalytic Sites and Heterolytic Reaction Mechanisms

Table 5: Rate Laws for hydrogenation and carbon-heteroatom bond cleavage by EAS reactions ($\alpha_i = K_i P_i$) and orders of reactions when $\alpha_i << 1$.

Rate Laws	n_{RX}, n_{H2}, n_{H2S}
Adsorption of RX : $$r_4 = k_4 \frac{P_{RX}}{(1+(\alpha_{H2}+\alpha_{HX}+\alpha_{H2S})^{1/2})}$$	1,0,0
Adsorption of H$_2$: $$r_1 = k_1 \frac{P_{H2}}{((\alpha_{H2}+\alpha_{HX}+\alpha_{H2S})^{1/2}+(1+\alpha_{RX})^{1/2})^2}$$	0,1,0
Hydrogenation HYD1: $$r_6 = k_6 \frac{\alpha_{RX}\alpha_{H2}(1+\alpha_{RX})^{-1/2}(\alpha_{H2}+\alpha_{HX}+\alpha_{H2S})^{-1/2}}{((\alpha_{H2}+\alpha_{HX}+\alpha_{H2S})^{1/2}+(1+\alpha_{RX})^{1/2})^2}$$ $$r_7 = k_7 \frac{K_H \alpha_{RX} \alpha_{H2}}{((\alpha_{H2}(1+K_H\alpha_{RX})+\alpha_{HX}+\alpha_{H2S})^{1/2}+(1+\alpha_{RX})^{1/2})^2}$$	1,1,-1/2 1,1,0
Hydrogenation HYD2 : $$r_8 = k_8 \frac{\alpha_{RX}(1+\alpha_{RX})^{-1/2}(\alpha_{H2}+\alpha_{HX}+\alpha_{H2S})^{1/2}}{((\alpha_{H2}+\alpha_{HX}+\alpha_{H2S})^{1/2}+(1+\alpha_{RX})^{1/2})^2}$$ $$r_9 = k_9 \frac{K_H \alpha_{RX} \alpha_{H2}}{((\alpha_{H2S}+\alpha_{HX}+\alpha_{H2})^{1/2}(1+K_H\alpha_{RX})^{1/2}+(1+\alpha_{RX})^{1/2})^2}$$	1,0,-1/2 1,1,0
Elimination E2 : $$r_{11} = k_{11} \frac{\alpha_{RXH2}}{((\alpha_{H2}+\alpha_{HX}+\alpha_{H2S})^{1/2}+(1+\alpha_{RXH2})^{1/2})^2}$$	1,0,0
Nucleophilic substitution S$_N$2 : $$r_{12} = k_{12} \frac{\alpha_{RXH2}\alpha_{H2S}(1+\alpha_{RXH2})^{-1/2}(\alpha_{H2}+\alpha_{HX}+\alpha_{H2S})^{-1/2}}{((\alpha_{H2}+\alpha_{HX}+\alpha_{H2S})^{1/2}+(1+\alpha_{RXH2})^{1/2})^2}$$	1,0,1/2
Ratio $r_6/E2$ or $r_6/SN2$ when $\alpha_i << 1$ $$\frac{r_6}{r_{11}} = \frac{k_6 \alpha_{RX} \alpha_{H2}}{k_{11}\alpha_{RXH2}(\alpha_{H2}+\alpha_{HX}+\alpha_{H2S})^{1/2}}$$ $$\frac{r_6}{r_{12}} = \frac{k_6 \alpha_{RX} \alpha_{H2}}{k_{12}\alpha_{RXH2}\alpha_{H2S}}$$	-,1, 1/2 -,1,-1

To solve the system of equations, one more equation is needed namely equation 4.15, the balance of the cations and anions generated by heterolytic dissociation of hydrogen and hydrogen sulfide because all of the positive species are evidently compensated by negative species. This equation is in fact equivalent to the electrostatic charge conservation equation proposed previously [45-47]. It appears that to compute the rate laws, it is indifferent whether electrostatic charges are explicitely written or not as species like Mo^{4+}-H^- can also be written Mo^{3+}-H or $Mo^{(4-\delta)+}$-$H^{-\delta}$.

The rate laws reported in Table 5 have a form slightly different from conventional LHHW rate laws. There are many cases of simplification of these rate laws that can be considered depending on the relative values of the $\alpha_i = K_i P_i$. At high reaction temperature such as 350°C, it can be considered that the α_i are lower than 1 which lead to the orders of reaction proposed in Table 5. Within this hypothesis, the hydrogenation by the rate determining addition of the H from the SH followed by the H from the Mo-H, leads to an order of one relatively to the hydrocarbon, an order of zero relatively to hydrogen and an order of +1/2 relatively to H_2S (r8). This is not found experimentally and this addition sequence must therefore be ruled out.

Conversely, the hydrogenation by the rate determining addition of the first H from Mo-H (r_6) followed by the H from the SH group, lead to an order of one relatively to hydrogen, an order of one relatively to the hydrocarbon and an order of -1/2 relatively to H_2S as usually observed [1,3,38]. If the addition of the second hydrogen is the rds (r_7) a zero order relatively to H_2S is found. Such orders of reaction have been measured for toluene hydrogenation at 6MPa and 350°C over MoS_2/alumina [47] and Ni-MoS_2/alumina catalysts [48]. The addition sequence H/SH is therefore more likely.

Recently, a key step in the HDS of thiophene by ruthenium-thiophene complex have been proposed to be a hydride attack [49]. It is possible that this sequence is favored by adsorption of the unsaturated molecule on a Lewis acid center with formation of an intermediate with a decreased electron density or even positively charged following electron transfer by an electrophilic attack from the coordinatively unsaturated transition metal. This is

Catalytic Sites and Heterolytic Reaction Mechanisms

schematically represented in Figure 4. In terms of rate laws it may be noted that it is equivalent to write Mo^{4+}-R or Mo^{3+}-R$^+$ for the adsorbed reactant.

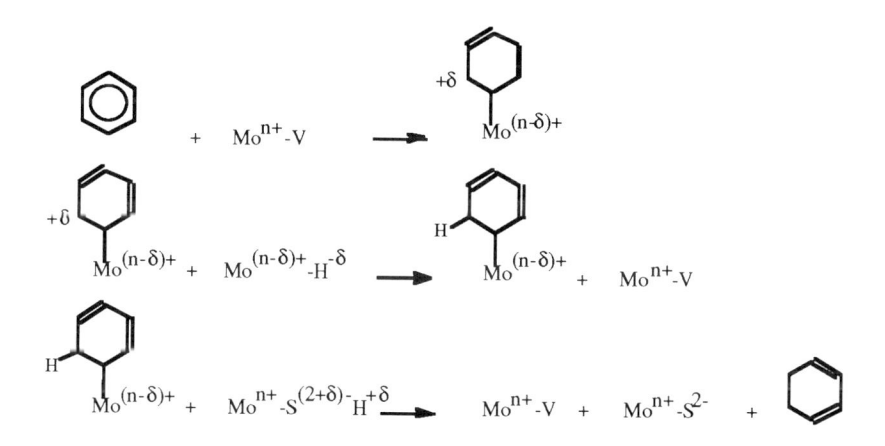

Figure 4 : Possible first hydrogenation step of an aromatic ring by the sequence H + SH.

HDN and HDS reactions can occur by E2 or S_N2 reaction mechanisms. Assuming that $\alpha_i \ll 1$, the rate law r_{11} for the E2 reaction mechanism is not dependant on P_{H_2S} at the numerator. Hence the promoting effect of H_2S in HDN can not be explained by an E2 mechanism on a pair of sites M^{4+}-V/M^{4+}-S^{2-}. On the contrary, the rate law r_{12} for the S_N2 mechanism where SH is the nucleophile, has a P_{H_2S} term in the numerator and, at low P_{H_2S}, a promoting effect of H_2S will be found. It is interesting to note that both E2 and S_N2 are not dependant on P_{H_2} at the numerator. There are indeed results where this is found such as HDS at low temperature or high hydrogen pressure [3,38].

HYD and HDS reactions are often claimed to occur on different sites. However, these reactions can occur according a different sequence of elementary steps and can have different rate determining step. If hydrogenation is an addition reaction and HDS an E2 elimination reaction, the adsorption site for the reactants may be the same (a coordinatively unsaturated Mo^{4+} ion) but the pairs of sites involved with the rds are different because the rds are different. It is of interest to note in Table 5 that the ratio rate of hydrogenation r_6 over rate of E2 (or SN2) is dependant on the partial pressure of H_2 and H_2S, an effect often claimed to prove the existence of different sites.

This analysis of the possible species present on the edge planes and of the heterolytic reaction mechanism suggest that a description of sites in terms of pair of acid-basic sites and acid-nucleophile sites may be a key to understand these reactions on sulfide catalysts. Such a description of catalytic sites for hydrotreating reaction is already apparent in reference 50 where reactants have been proposed to react with a pair of electron donating and electron withdrawing sites. Then the adsorption site of the hydrocarbon compound can be the same but the pair of sites on which the rate determining step occurs can be different.

It must be pointed out, however, that the description of sites and in particular adsorption sites made in this work may be too simplified. Single point adsorption on one coordinative unsaturation of Mo^{4+} has been considered in this work but multipoint adsorption as well as steric hindrance caused by the structure of the reactant or the stacking of the slabs for example [51] will require a more complex adsorption site description.

Interestingly, the EAS heterolytic reaction mechanisms analyzed in this work lead to rate laws with equations different from conventional LHHW rate laws and in particular they can account for promoting or inhibiting effect of hydrogen sulfide on the rate of reaction. In addition, all three types of reactions considered (addition, E2, SN2) for HDS/HDN/HYD reactions occur on the same surface and can coexist thus providing a coherent framework. More detailed kinetic studies of HYD/HDS/HDN reactions of model molecules are however needed to determine the validity of these rate laws.

IV CONCLUSION

In this work an attempt has been made to analyze the nature of the surface species, of the reaction mechanisms and of the kinetic rate laws for HDS/HDN/HYD reactions in the framework of the heterolytic reaction mechanism hypothesis. It is proposed that HYD/HDS/HDN over sulfides catalysts can be described by elimination, addition or substitution reaction mechanisms involving heterolytic dissociation of hydrogen, hydrogen sulfide and of the C-S and C-N bonds. These reactions can be different by their rate determining step which involve a pair of acid-base or acid-nucleophile sites on a surface containing coordinatively Mo^{4+} ions, sulfide ions, H and SH species. It is also shown that new form of rate laws are obtained which may account for the promoting or inhibiting effect of hydrogen sulfide on HDS/HDN/HYD reactions.

REFERENCES

1. B.C. Gates, J.R. Katzer and G.C.A. Schuit, Chemistry of Catalytic Process, McGraw Hill (1979).
2. P. Ratnasamy, S. Sivasanker, Catal. Rev. Sci. Eng., 22 (1980) 401.
3. M.L. Vrinat, Appl. Catal., 6 (1983) 137.
4. T.C. Ho, Catal. Rev. Sci. Eng., 30 (1988) 117.
5. G. Pérot, Catal. Today, 10 (1991) 447.
6. M.J. Girgis and B.C. Gates, Ind. Eng. Chem. Res., 30 (1991) 2021.
7. B. Delmon, Stud. Surf. Sci. Catal., 53 (1989) 1.
8. R. Prins, V.H.J. DeBeer and G.A. Somorjai, Catal. Rev. Sci. Eng., 31 (1989) 1.
9. H. Topsoe, B.S. Clausen, N.Y. Topsoe and P. Zeuthen, Stud. Surf. Sci. Catal., 53 (1989) 77.
10. H. Knozinger, in Proc. 9th Int. Cong. Catal., Calgary, 1 (1988) 20.
11. R.R. Chianelli and M. Daage, Stud. Surf. Sci. Catal., 50, (1989) 1.
12. R.J.H. Voorhoeve and J.C.M. Stuiver, J. Catal., 23 (1971) 228 ; ibid. 23 (1971) 236, ibid. 23 (1971) 243 .
13. A.L. Farragher, Adv. Coll. Inter. Sci., 11 (1979) 3.
14. G.C.A. Schuit, Int. J. Quantum Chem., 12 (1977) 43.
15. S. Kasztelan, H. Toulhoat, J. Grimblot and J.P. Bonnelle, Appl. Catal., 13 (1984) 127.
16. F.E. Massoth, G. Muralidhar, Proceed. 4th Int. Conf. Chemistry and uses of Molybdenum, H.F. Barry and P.C.H. Mitchell, Eds., Climax Molybdenum Co, Ann Arbor, MI, (1982) 343.
17. C.B. Roxlo, M. Daage, A.F. Ruppert and R.R. Chianelli, J. Catal., 100 (1986) 176.
18. A. Wambeke, L. Jalowiecki, S. Kasztelan, J. Grimblot and J.P. Bonnelle, J. Catal., 109, (1988) 320.

19. N.Y. Topsoe, H. Topsoe, J. Catal., 84 (1983) 386.
20. Y. Okamoto, H. Tomioka, Y. Katoh, T. Imanaka, S. Teranishi, J. Phys. Chem., 84 (1980) 1833.
21. S. Kasztelan, A. Wambeke, L. Jalowiecki, J. Grimblot, J.P. Bonnelle, Bull. Soc. Chim. Belg., 96, (1987) 1003.
22. J. Maternova, Appl. Catal., 3 (1982) 3.
23. N.Y. Topsoe, H. Topsoe, F.E. Massoth, J. Catal., 119 (1989) 252.
24. E. Payen, S. Kasztelan, J. Grimblot, J.P. Bonnelle, Catal. Today, 4, (1988) 57.
25. V. Stuchly, L. Beranek, Appl. Catal., 35 (1987) 35.
26. L. Portela, P. Grange, B. Delmon, Stud. Surf. Sci. Catal., 71 (1993) 559.
27. J. Polz, H. Zeilinger, B. Muller, H. Knozinger, J. Catal., 120 (1989) 22.
28. T. Komatsu, W.K. Hall, J. Phys. Chem., 95, (1991) 9966.
29. T. Komatsu, W.K. Hall, J. Phys. Chem., 96, (1992) 8131.
30. G.B. McGarvey, S. Kasztelan, J. Catal., 148 (1994) 149.
31. D.G. Kalthodt, S.W. Weller, J. Catal., 95, (1985) 455.
32. W.P. Dianis, Appl. Catal., 30, (1987) 99.
33. A.B. Anderson, Z.Y. Al-Saigh, W.K. Hall, J. Phys. Chem., 92, (1988) 803.
34. M. Lacroix, S. Yuan, M. Breysse, C. Dorémieux-Morin and J. Fraissard, J. Catal., 138 (1992) 409.
35. H. Jobic, G. Clugnet, M. Lacroix, S. Yuan, C. Mirodatos, M. Breysse, J. Am. Chem. Soc., 115, (1993) 3654.
36. B.G. Silbernagel, T.A. Pecoraro and R.R. Chianelli, J. Catal., 72 (1982) 380.
37. B. Muller, A.D. Van Langeveld, J.A. Moulijn, H. Knozinger, J. Phys. Chem., 97 (1993) 9028.
38. J.F. Le Page, Applied Heterogeneous Catalysis, Editions Technip, Paris, 1987.
39. N. Nelson, and R.B. Levy, J. Catal., 58, (1979) 485.
40. L. Vivier, V. Dominguez, S. Kasztelan, and G. Pérot, J. Mol. Catal., 67, (1991) 267.
41. S.H. Yang, C.N. Satterfield, Ind. Eng. Chem., Proc. Des. Dev., 23, (1984) 20.
42. J.L. Portefaix, M. Cattenot, M. Guerriche, J. Thivolle-Cazat and M. Breysse, Catal. Lett., 10, (1991) 473.
43. L. Vivier, P. D'Araujo, G. Pérot and S. Kasztelan, Preprints, Div. Petrol. Chem., ACS, 37(3), (1992) 710.
44. H. Noller, and W. Kladnig, Catal. Rev.-Sci. Eng., 13, (1976) 149.
45. S. Kasztelan, Preprints, Div. Petr. Chem., ACS, 38(3), (1993) 642.

46. S. Kasztelan, C. Rend. Acad. Sci., Paris, 311-II, (1990) 1309.
47. S. Kasztelan, D. Guillaume, Ind. Eng. Chem. Res., 31, (1992) 2497.
48. N. Marchal, D. Guillaume, S. Mignard, S. Kasztelan, This issue.
49. R.J. Angelici, Acc. Chem. Res., 21, (1988) 387.
50. C. Moreau, J. Joffre, C. Saenz, P. Geneste, J. Catal., 122 (1990) 448.
51. M. Daage, R.R. Chianelli, A.F. Ruppert, Stud. Surf. Sci. Catal., 71 (1993) 572.

Microstructural Characterization of Molybdenum Disulfide and Tungsten Disulfide Catalysts

M. Del Valle, M.J. Yáñez*, M. Avalos-Borja** and S. Fuentes**

Programa de Posgrado en Física de Materiales, Centro de Investigación Científica y de Educación Superior de Ensenada, Ensenada, B.C., México, and Facultad de Ciencias Químicas, UABC, Tijuana, B.C., México.
*On leave from Centro Regional de Investigaciones Básicas y Aplicadas de Bahía Blanca, Argentina.
**Instituto de Física, UNAM, Laboratorio de Ensenada, Apartado Postal 2681, Ensenada, B.C., México.

ABSTRACT

High resolution electron microscopy (HREM) and image digitization techniques have been used to study the microstructure of unsupported molybdenum disulfide and tungsten disulfide catalysts. The microstructural information involving the $\{100\}$, $\{101\}$, $\{102\}$, $\{103\}$ and $\{310\}$ crystalline planes of MoS_2 and WS_2 catalysts is compared with that of $\{002\}$ planes. As a result, defects such as kinks, steps and microporosity are found in $\{100\}$ and $\{10l\}$ planes. The effect of the preparation temperature on WS_2 catalysts is to vary the crystallinity of observed particles. Changes in the microstructure of unsupported MoS_2 catalysts have been determined by HREM following the catalytic reaction of thiophene hydrodesulfurization (HDS). The increase in the average length of MoS_2 crystallites is greater than the increase in average number of crystallite stacking layers as a result of the test reaction. Variations in catalytic activity are discussed in terms of a mechanism relating active sites with different planes.

I. INTRODUCTION

Better knowledge of the structure of molybdenum disulfide catalysts is important for the improvement of present industrial applications, given stricter environmental regulations and continuing economic demands for greater efficiency. In general, the structure of metal dichalcogenides is highly anisotropic and many physical and chemical properties are derived from this character (1). The anisotropy of molybdenum disulfide increases when prepared by low temperature methods, leading to the use of the term "poorly crystalline" to describe the resulting microstructure which is highly folded, 5-6 layers thick and 10-100 nm in length (2).

Liang et al. (3), calculating the X-ray scattering intensities of model structures of "poorly crystalline" MoS_2 and comparing them with experimental patterns, have found that experimental results cannot be explained using ideal crystallite models. Agreement with experimental diffraction patterns is only aproached when defects (rotation, shifting and folding of planes) are added to the model.

Transmission electron microscopy (TEM) has been succesfully used by several authors for the characterization of hydrotreating catalysts. In particular, the technique has allowed the discovery of some important features of the "poorly crystalline" structure of MoS_2-based catalysts (4-15). Since state of the art electron microscopes can give a point to point resolution of about 0.17 nm, HREM has now begun to reveal the characteristics of solid substances at the atomic level. Applied to the study of sulfide catalysts, HREM is yielding new and more detailed information about their microstructure (8, 9).

Chemical reactivity relates to the type of crystal surface offered by the catalyst. Surfaces composed of basal planes have been shown to be relatively inert unless defects are introduced, while edge planes are found to provide significant chemical reactivity (16, 17). In many cases, the catalytic activity of metal sulfides is promoted by the presence of a second transition metal (1). There is good evidence to conclude that cobalt or nickel are placed on edge planes of the layered structure of molybdenum disulfide or tungsten disulfide, creating new and catalytically more efficient sites (18-21).

Previous studies on the effect of the method of preparation, the promoter concentration (8) and the HDS reaction (22) on the number and length of stacking layers (002 planes) in MoS_2 crystallites have found that changing the preparation method (impregnated thiosalt decomposition, ITD, or homogeneous sulfide precipitation, HSP) influences the length of the stacking layers (L), whereas the number of stacking layers (N) remains about the same. The introduction of cobalt slightly decreases both N and L. In general, these results indicate that the reordering induced by those parameters in the microstructure affects the length more markedly than the stacking. This finding may be explained as owing to the higher chemical reactivity of edge planes compared to basal planes.

The present work applies image digitization techniques to further the HREM study of the microstructure of unsupported molybdenum disulfide and tungsten disulfide catalysts. The work also offers improved HREM evidence of the modifications induced in the microstructure of WS_2 crystallites by the preparation temperature, and in MoS_2 crystallites by the HDS reaction.

II. EXPERIMENTAL

Molybdenum disulfide and tungsten disulfide catalysts were prepared by decomposition of ammonium thiomolybdate, ATM (23) and ammonium thiotungstate, ATT (24), respectively. This is done by heating the thiosalt under a 15% hydrogen sulfide in hydrogen gas mixture at 673 K for 4 hours. Catalyst samples were ground with a mullite mortar and pestle and were ultrasonically dispersed in heptane. A drop of the suspension was air-dried in a carbon-coated electron microscope grid for examination in a JEOL 4000EX transmission electron microscope (0.18 nm point-to-point resolution). The test reaction was the HDS of thiophene, which was carried out in a conventional flow microreactor under atmospheric pressure, in the temperature range 603-673 K. The reaction products were analyzed by gas chromatography, as detailed in reference 23. Surface areas were determined by the BET method, using nitrogen as adsorbate at 77 K. Samples were first outgassed at room temperature under vacuum (1.33 X 10^{-3} Pa) and then heated at 673 K for two hours prior to analysis. Elemental analysis of catalysts was carried out in a JEOL JSM-5300 scanning electron microscope equipped with a Kevex Quantum Superdry Analyzer. The elemental analysis was measured on wafers obtained by pressing the catalyst powder and averaging the results over ten different fields.

III. RESULTS AND DISCUSSION

In order to compare the microstructure of unsupported MoS_2 and WS_2 catalysts with their crystalline parents, a series of computer-simulated HREM images of a MoS$_2$-2H crystal were first obtained with the aid of a commercial computer program (25). Similar images correspond to WS$_2$-2H, since it is isostructural with crystalline MoS$_2$ and has nearly the same cell dimensions. The catalysts were then analized by HREM, followed by the use of image digitization techniques on the high resolution micrographs to improve the visibility of lattice planes.

The computer-simulation of HREM images is based on the so-called *multislice method*. Briefly, said method follows the evolution of a plane wave (incident electron beam) as it is traversing the crystal until it reaches the bottom part of the sample, while considering dynamical effects such as multiple diffraction, and diffraction back into the transmitted beam. At this point the program simulates the effect of the objective lens with its proper set of aberration coefficients (spherical, chromatic, etc.) for the specified microscope configuration, as well as experimental parameters like the amount of error in focusing (defocus). After all the crystallographic information (space group, atomic positions, etc.), instrumental parameters (accelerating voltage, aberration coefficients, etc.) and experimental parameters (crystal orientation, specimen thickness and amount of defocus) are input, the computer program then renders a "theoretical" image, or a series of images if a given parameter is varied.

A sample of the resulting simulated HREM images is shown in Fig. 1. The series of simulated images in Fig. 1a correspond to a MoS$_2$ crystal in [001] orientation as seen with an "electron microscope" equivalent to the JEOL 4000 EX. In the figure, same row images have an equal sample thickness (1.2, 2.4, 3.6 and 4.8 nm), roughly equivalent to a thickness of 1, 2, 3 and 4 unit cells, while same column images have equal defocus conditions (-5, -10, -15, -20 and -25 nm). In like fashion, a large assortment of images can in fact be produced, depending on the experimental conditions. In some cases the resulting "dots" are black while in others they are white, due to the interference of waves (electrons) across the microscope column. Along the [001] orientation, the molybdenum and sulfur atoms "overlap", forming six-"motif" rings with a large hole in the center. In the first micrograph (thickness 1.2 nm, defocus -5 nm), the black dots correspond to the holes in the rings and not to any particular molybdenum or sulfur atom, whereas in the last row micrographs (defocus of -25 nm) it is the white dots which correspond to the above mentioned "holes".

Fig. 1b shows a series of simulated images for a side view ([010] orientation) of the MoS$_2$-2H crystal, using the same instrumental specifications and for the same series of thicknesses and defocus conditions set out in Fig. 1a. Among the images, a nearly "structural" one is found for the conditions of 1.2 nm thickness

Microstructural Characterization of Catalysts

and -15 nm defocus. Here, the large "white" dots correspond to sulfur atoms forming two adjacent columns, while the small, barely visible, white dots correspond to the molybdenum atoms. When the crystal is misoriented from a low index like [010] or [100], the fine detail is lost and only the broad dark and white fringes spaced by 0.61 nm are apparent. These fringes are visible in most of the micrographs published on this material. In general, *perfect crystals* of MoS_2 are expected to give high resolution electron micrographs which closely resemble the computer simulated images under specific conditions. The same discussion holds for crystalline WS_2.

Three catalysts were obtained and studied by HREM. The MoS_2 catalyst was prepared at 673 K, while WS_2 catalysts were obtained at 673 and 1073 K. Under these conditions, the X-ray diffraction patterns of all three materials are similar to that of the "poorly crystalline" structure reported by Liang et al. (3). A high resolution micrograph of the MoS_2 catalyst prepared at 673 K is shown in Fig. 2. Typically, the material shows no long range order, either in the basal or the stacking planes (confirmed by formation of broad rings in the electron diffraction pattern). Overall, the observed structure is different from any of the uniform images simulated in Fig. 1. However, lattice distances corresponding to (002) planes are easily identified. Other planes, such as (101), (102), and (103) are also found in this micrograph, suggesting that crystalline microdomains exist in the catalyst.

The (002) planes in MoS_2 crystallites are easily identified by TEM or HREM for two main reasons: 1) the lattice distance is large enough (0.615 nm) to be resolved by most modern transmission electron microscopes, and 2) there are a great number of possibilities for a given crystal (360° rotation about the *c* axis) to give lattice resolution, the only requirement being that the direction of the electron beam must be parallel to these planes. In contrast, (100) planes are very hard to resolve by TEM due to the small lattice spacing (0.27nm), and also because their detection is determined by only a few possible orientations for a given crystal with respect to the electron beam. Molybdenum as well as tungsten disulfide catalysts differ markedly from the crystalline material in that they are composed of flake-like crystallites, where the number of planes along the *a* or *b* direction is higher than in the *c* direction. This is due to the layered structure of MoS_2, where (002) intralayer Mo-S covalent bonding is strong, while (002) interlayer S-S Van der Waals bonding is weak, allowing neighboring sets of planes to be rotated, shifted, or separated with respect to each other.

Electron microscopy studies of metal sulfide catalysts have mostly discussed the observed structures with respect to their (002) planes. Yet the theoretical work of Liang et al. (3) implies that the characterization of planes such as the (100) and the (10*l*) family can offer additional information about the

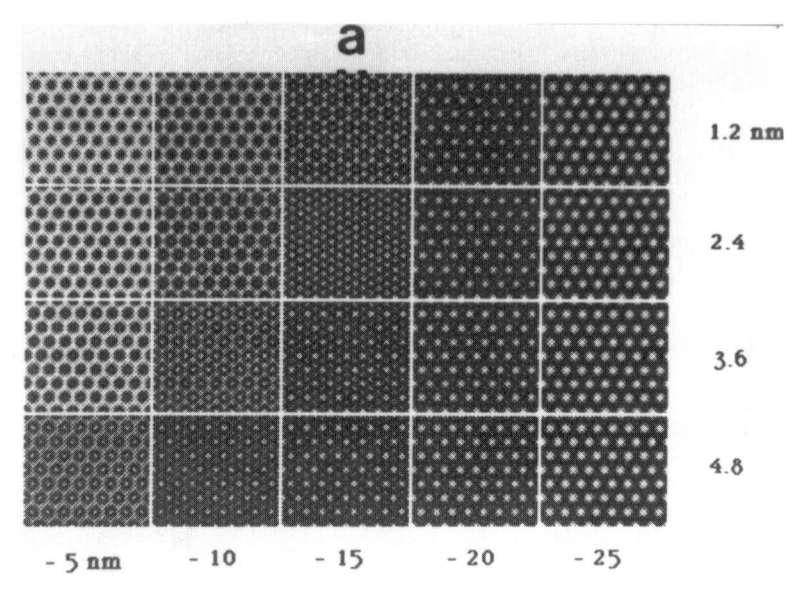

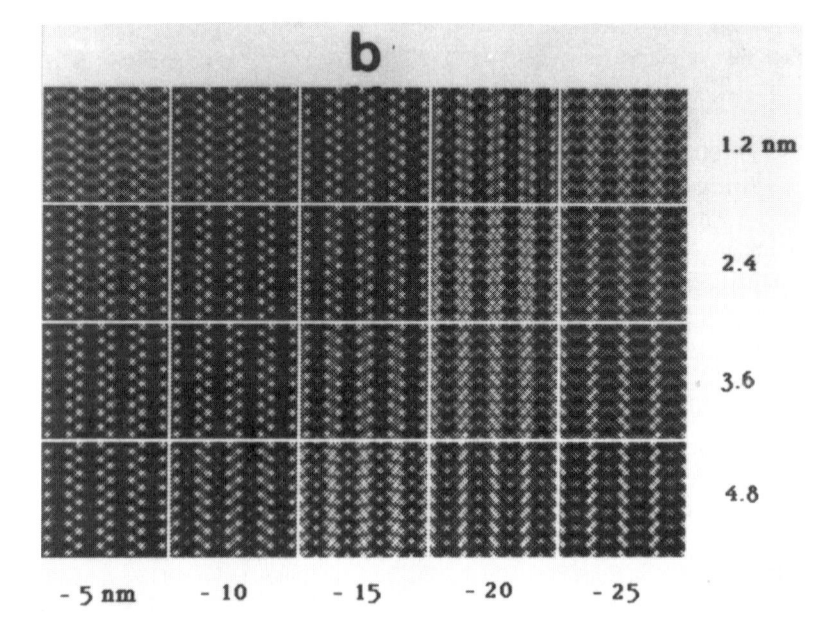

Fig. 1 Simulated HREM images corresponding to different defocus conditions (rows) and sample thicknesses (columns) for a MoS$_2$ crystal in a) [001] orientation and b) [010] orientation.

Microstructural Characterization of Catalysts 53

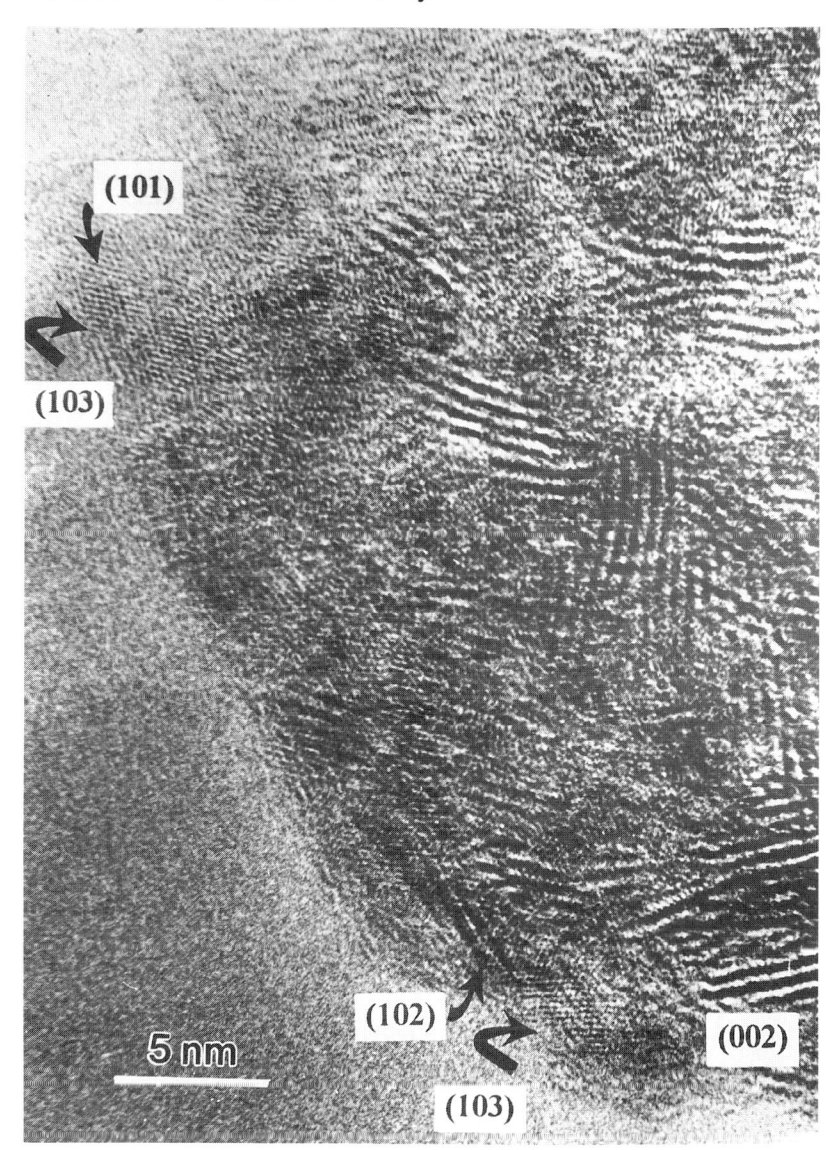

Fig. 2 High resolution micrograph of MoS_2 catalyst prepared at 673 K.

intracrystalline microstructure of particles. For example, the intensity of X-ray diffraction peaks due to these planes is very sensitive to interlayer rotation about the c-axis, which hardly affects the (002) peak. Overall, the (10l) family of planes is found to mix pure intralayer (100) with interlayer defects in imperfect crystallite models.

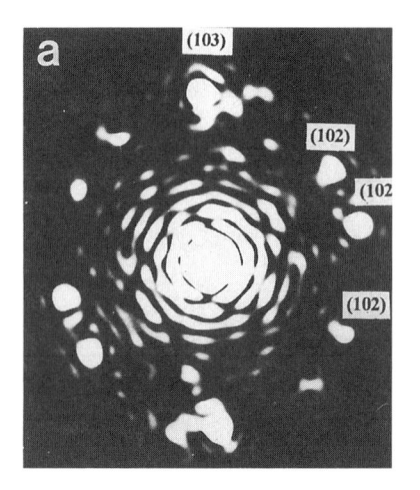

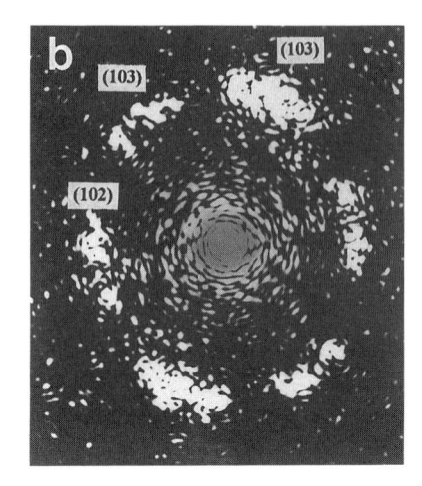

Fig. 3 Optical diffractograms of WS_2 catalyst prepared at 673 K,

showing a) three different orientations of (102) planes, and

b) arcs indicating slightly misoriented domains with the same

crystallinity.

The patterns in Fig. 3 correspond to the optical Fourier transform of high resolution images of WS_2 catalysts prepared at 673 K. These diffraction patterns are calibrated using lattice planes of gold. From the pattern in Fig. 3a, interplanar spacings of 0.23 nm and 0.25 nm are measured which correspond to (103) and (102) planes, respectively. The appearance of three different orientations of (102) planes in the pattern is due to the presence, in the original micrograph, of several domains with the same lattice distance (0.25 nm) but misoriented with respect to each other. The pattern in Fig. 3b shows several arcs instead of spots, indicating that domains with the same crystallinity are misoriented with respect to each other by a few degrees. Such results suggest that several microdomains with the right stacking order appear in the micrograph (at least two layers keeping the hexagonal sequence must be present to form (102) and (103) planes) but slightly misoriented with respect to each other. The fact that these domains, belonging to the same family of planes, are misoriented by only a few degrees also suggests that they may be reoriented to give a uniform structure with only a little additional energy.

High resolution micrographs of the two WS_2 catalysts (673 K and 1073 K) are shown in Fig. 4, from which the periodicities of gray intensities

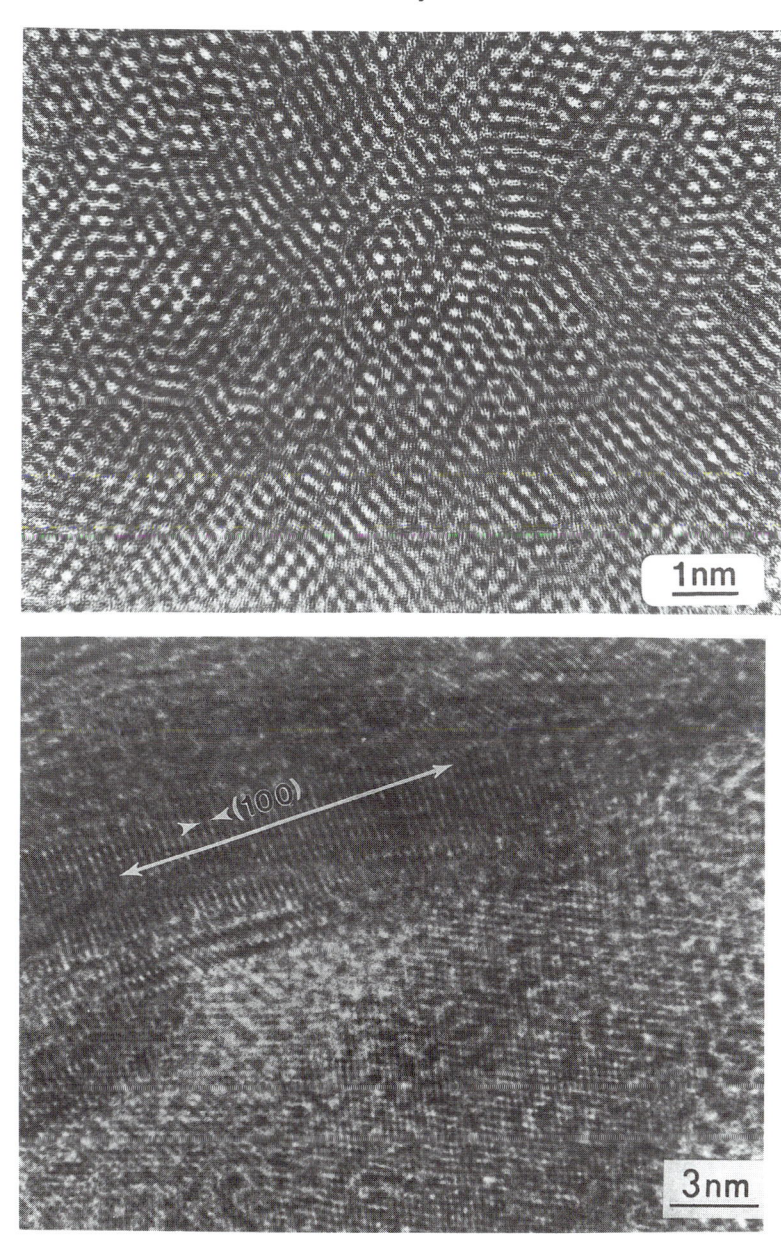

Microstructural Characterization of Catalysts

Fig. 4 High resolution micrographs of WS_2 catalysts prepared at

a) 673 K, and b) 1073 K. The line running perpendicular

to the (100) planes in (b) indicates the scanned zone.

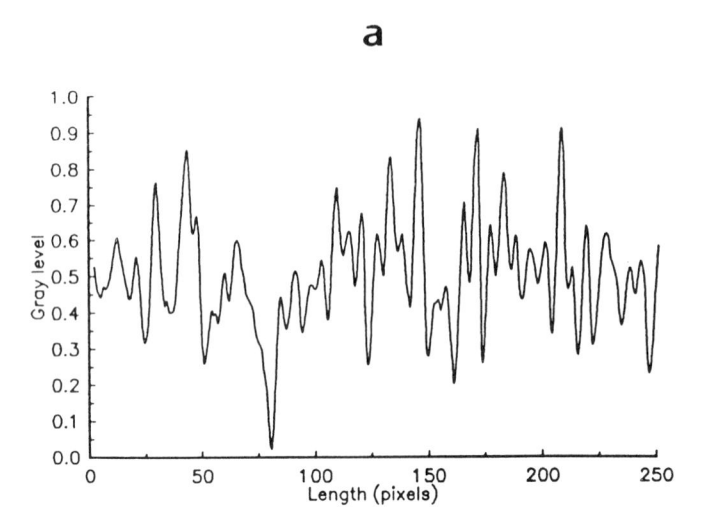

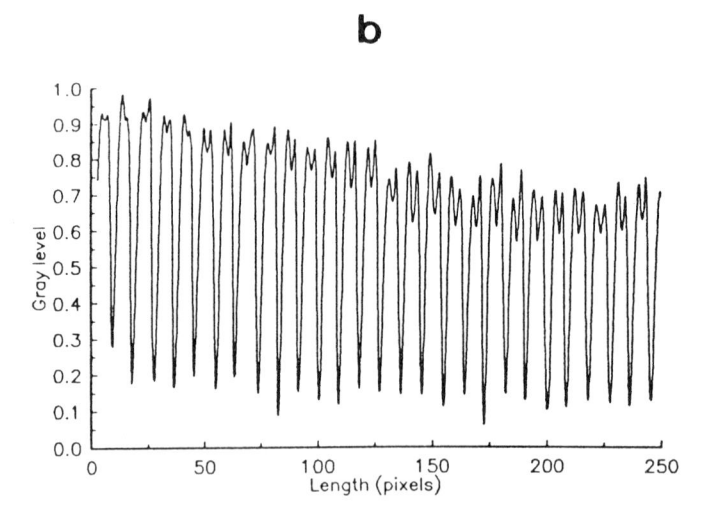

Fig. 5 Periodicity of gray intensities in HREM micrographs of WS$_2$ prepared at a) 673 K, and b) 1073 K. The spacing between the peaks is proportional to the lattice spacing.

Microstructural Characterization of Catalysts

presented in Fig. 5 were derived according to the procedure described by Yacamán et al. (26). For the first sample, Fig. 5a presents different periodicities which are maintained only for short distances, suggesting that the zone (considered representative of the overall structure in Fig. 4a) has poor crystallinity. In contrast, the plot in Fig. 5b suggests that periodicity holds constant over several nanometers, indicating that the zone (the well ordered central region of the structure in Fig. 4b) has good crystallinity. The lattice spacing in this scheme measured 0.27 nm, corresponding to the (100) planes. These results show that by increasing the temperature to 1073 K, the poorly crystalline structure is reconstructed and long range order becomes evident.

Fig. 6a shows a high resolution micrograph of the WS_2 catalyst treated at 673 K. Disordered material is the prevalent feature of the micrograph, although a few areas are better ordered than the rest. To improve visibility, the micrograph was digitized, filtered in the Fourier space and ultimately reconstructed to selectively enhance the features of the original micrograph. Enhancement of internal spots gives the image in Fig. 6b, where (002) planes exhibit a stacking fault which induces

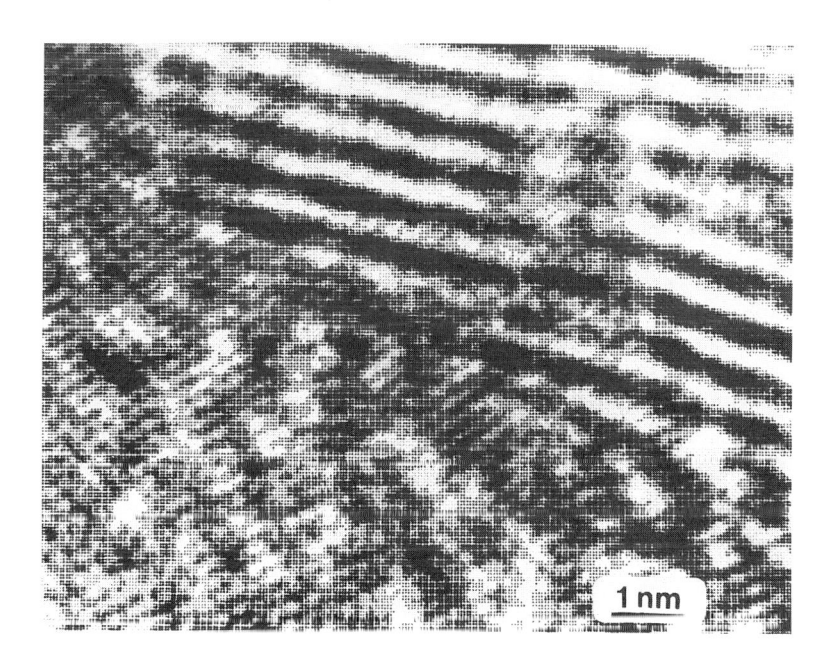

Fig. 6a High resolution micrograph of WS_2 prepared at 673 K,

before computer processing.

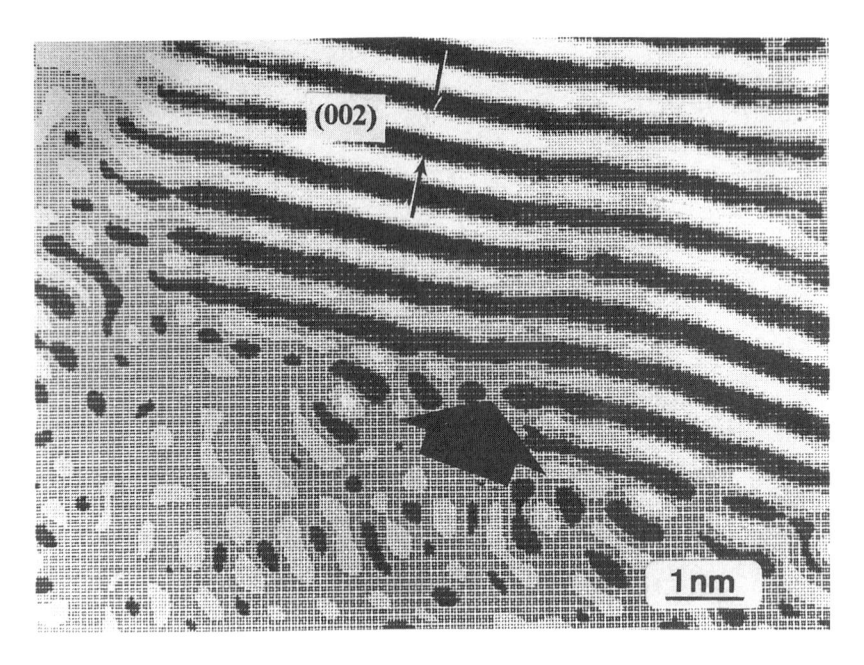

Fig. 6b Computer-processed HREM micrograph of WS_2 prepared at 673 K, using the Fourier transform method. The micrograph shows an improved image of (002) planes.

a change in direction of up to 10 layers. Enhancement of the external spots of the Fourier pattern gives the image in Fig. 6c, where the outstanding features are a zone in which (101) and (01$\bar{3}$) planes form a 73° angle, and another where (102) and (01$\bar{3}$) planes form a 53° angle; their corresponding zone axes being [$\bar{1}$31] and [010], respectively. This confirms that a number of domains of the same family of planes are misoriented by small angles.

Figs. 7 and 8 offer images of WS_2 catalysts (sulfided at 673 K and 1073K, respectively) that have been computer-processed as described above. The irregular crystallite shown in Fig. 7 is 9 X 6 nm in size, with lattice periodicities of 0.27 nm forming a 60° angle. The lattice planes are the (100) and the (010), indicating that

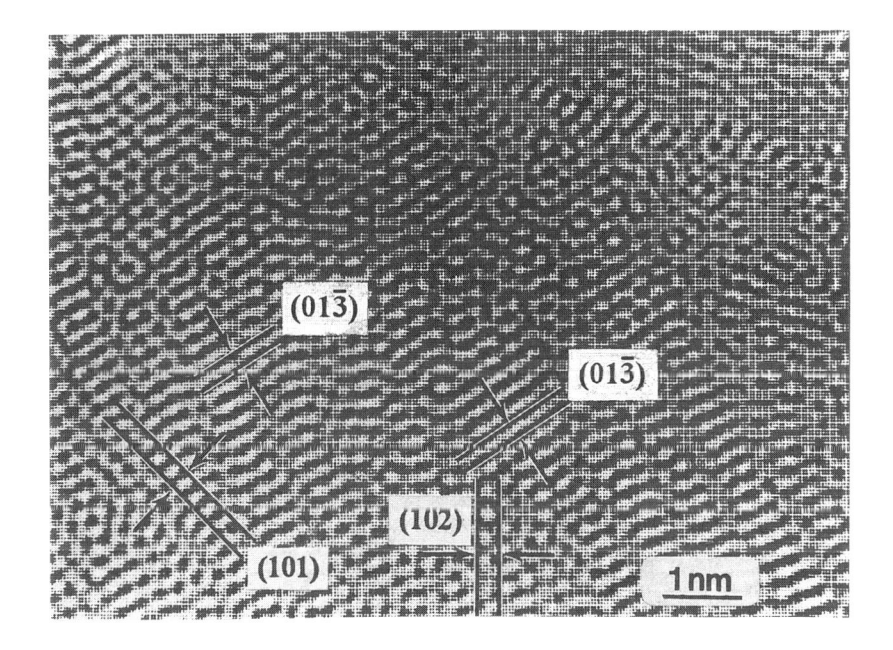

Microstructural Characterization of Catalysts

Fig. 6c Computer-processed HREM micrographs of WS_2 prepared at 673 K, using the Fourier transform method. The micrograph shows an improved image of planes other than (002).

the image corresponds to the basal plane. On the border of the crystallite edge planes, defects such as kinks or steps are present which contribute to the irregular shape of the crystallite. Fig. 8 corresponds to a region of the catalyst where channels appear between crystalline zones. These channels may be evidence of the microporosity formed in the catalyst as it eliminates gases (NH_3 and H_2S) during the decomposition of ATT. The typical surface area of WS_2 catalysts is about 50 m^2/g, with an important contribution coming from micropores (less than 0.25 nm in size). Fig. 9 is an image of WS_2 sulfided at 1073 K, where a better crystallinity of the sample (more organized regions) with respect to the 673 K catalyst is observed. The net formation of (01$\bar{3}$) and (101) planes indicates that intracrystallite rearrangement has taken place.

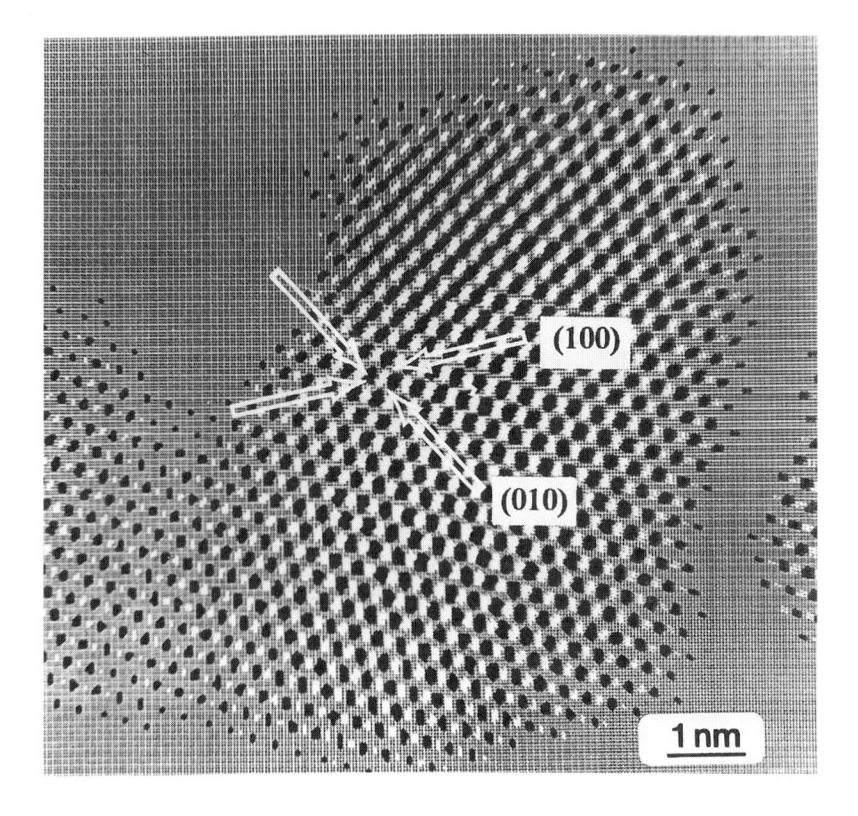

Fig. 7 Computer-processed HREM micrograph of a WS$_2$ catalyst particle (673 K), offering a top view of basal planes resulting from intersection of (100) and (010) planes at a 60° angle.

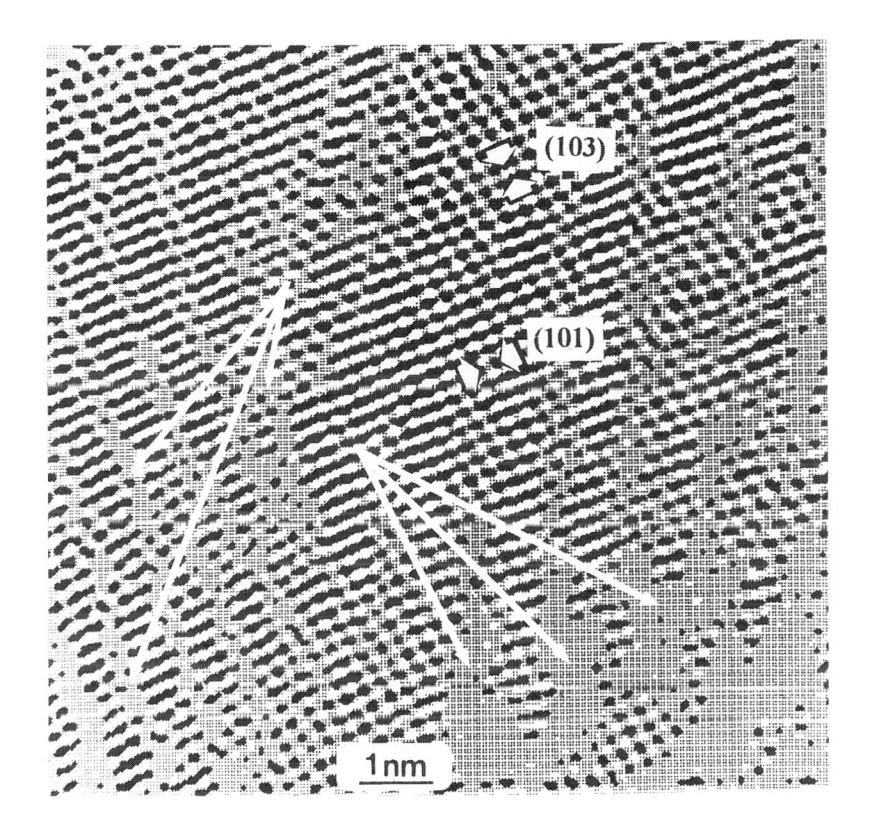

Microstructural Characterization of Catalysts

Fig. 8 Computer-processed HREM micrograph of a WS$_2$ catalyst prepared at 1073 K, as seen along the [$\bar{1}$31] zone axis. The arrows indicate channels between crystalline zones.

These results show that important information is provided by high resolution micrographs of $\{100\}$, $\{101\}$, $\{102\}$ and $\{103\}$ planes. The disorder generally observed in $\{002\}$ planes of MoS$_2$ (WS$_2$) catalysts also extends to these planes. This disorder consists of extended basal-like surfaces where poorly crystalline microdomains are spread througout the catalyst. Defects such as kinks, steps and microporosity are present as well. As expected, increasing the preparation temperature significantly reduces the amount of defects in the material, leading to a less disordered structure. The restructuring of the catalyst is favored by the fact that microdomains with the same orientation are very close to each other.

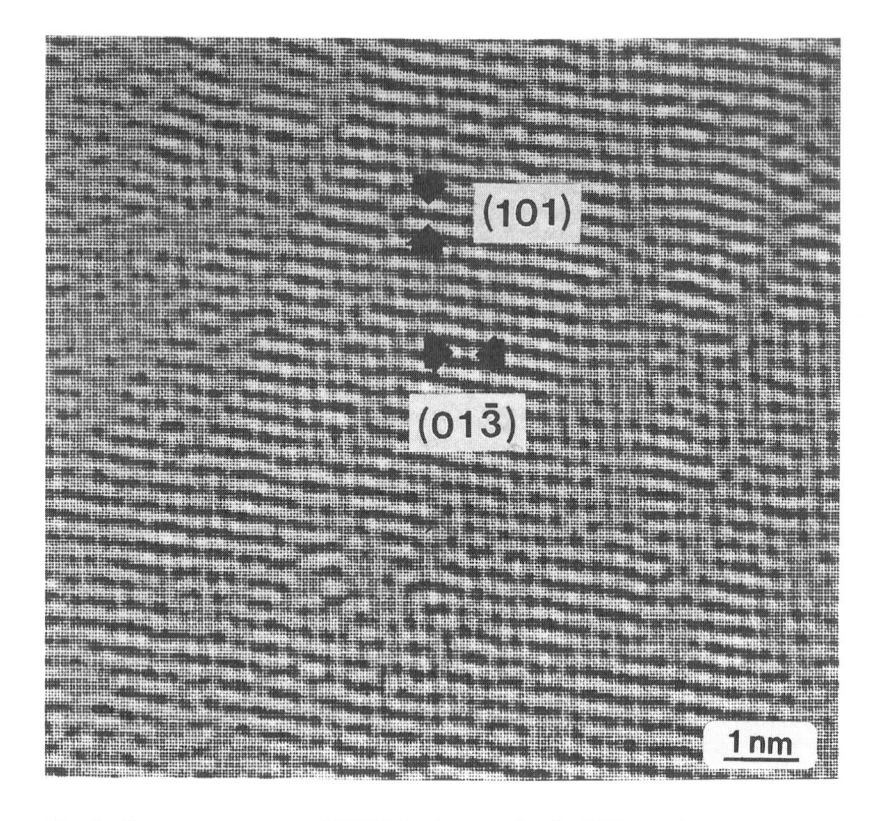

Fig. 9 Computer-processed HREM micrograph of a WS$_2$ catalyst prepared at 1073 K, showing ordered domains of (101) and (01$\bar{3}$) planes in a $[\bar{1}31]$ zone axis.

The characteristic "poorly crystalline" structure of MoS$_2$, consisting of small crystallites composed in all directions, has been shown in Fig. 2. Statistical measurements have shown that, typically, these crystallites have an average stacking (N) between 4-6 layers and an average length between 3-6 nm. These dimensions tend to be smaller when MoS$_2$ is supported on Al$_2$O$_3$ or TiO$_2$ (27, 28). Another structural feature of MoS$_2$ catalysts is the formation of stacking faults, bent and shifted planes, and dislocations. The temperature, as well as the reductive atmosphere (H$_2$S/H$_2$), used during the decomposition-activation step also contributes strongly to the disorder observed in MoS_2 catalysts. At 673 K, the transition from amorphous MoS_3 to poorly crystalline MoS_2 takes place (29), while the presence of hydrogen creates sulfur vacancies (30).

The changes in surface area, composition and catalytic activity for a MoS_2 catalyst after the test reaction are shown in Table 1. All three variables are modified by the reaction, indicating a high instability of the catalyst under the reaction environment. Both the surface area and the catalytic activity decrease by about the same factor, suggesting that the activity loss may be related to loss in surface area, in agreement with previous results (31, 32). Although sulfur elimination may also be a cause of deactivation by itself (33), we believe that in this case it is only a secondary factor, since resulfiding the sample under 15% H_2S/H_2 for two hours only recovers 20% of the initial catalytic rate. The main cause of activity is therefore assumed to be a reduction in the amount of accesible surface.

The selectivity of the MoS_2 catalyst with time on stream is indicated in Table 2. During the reaction, butane selectivity decreases, while that for 1-butene remains constant and both the cis- and trans-2-butene selectivities increase. Such variations indicate that the hydrogenation, isomerization, and HDS reactions that occur during the conversion of thiophene take place on different sites of the surface, in agreement with prevalent models (34, 35). The reduction in butane

Table 1. Surface area, S/Mo atomic ratio and reaction rates for the MoS_2 catalyst before and after the HDS reaction at 623K.

	Catalyst condition	
	fresh	spent
Specific surface area (m^2/g)	41	8
S/Mo ratio (% at)	2.19	1.81
Reaction rate ($X10^7 mol\ s^{-1}\ g^{-1}$)	5.25*	1.06**

* After reacting for 1 h
**After reacting for 15 h

selectivity suggests that hydrogenation sites disappear more rapidly than does the surface area, while the constant selectivity for 1-butene and the increased selectivity for 2-butenes suggest that HDS sites remain constant while the isomerization sites increase.

Several authors have studied the nature of the active sites in unsupported MoS_2 catalysts for HDS (36), hydrogenation and isomerization (34, 37, 38). Siegel (37) has proposed that hydrogenation reactions occur on three-coordinatively unsaturated (CUS) molybdenum-ions, whereas isomerization occurs on 2-CUS Mo-ions.

Kasztelan et al. (34) have shown similar results and they have proposed a geometrical model where CUS Mo-ions are located on (100) planes of MoS_2 crystallites. According to both models, the loss of a sulfur atom (2-CUS to 3-CUS) would decrease isomerization and increase hydrogenation. This, however, runs contrary to the observed change in the selectivity of the catalyst during the reaction.

Fig. 10 presents the statistical results on the length and stacking of MoS_2 layers (fringes) in catalyst crystallites, as determined by HREM, where frequency denotes the number of particles. According to Fig. 10a, a slight shift in the number of stacking layers that correspond to the maximum frequency takes place in the course of the reaction test, while the average number of layers (derived from the data in Fig. 2a) increases from 5.5 to 7.0. Fig. 10b shows a similar shift, and the average crystallite length increases from 3.0 to 6.2 nm by the end of the catalytic test, suggesting that the crystallites tend to sinter. The favored growth mechanism would thus seem to be a sintering of crystallites (involving border planes) rather than an increase of crystallite stacking (involving basal planes). Such a growth

Table 2. Selectivity of the MoS_2 catalyst in the HDS of thiophene at 623 K, as a function of time on stream.

	Selectivity (%)	
	1h	15 h
n-butane	32	23
1-butene	24	24
cis-2-butene	15	19
trans-2-butene	25	31
$C_1 - iC_4$	04	03

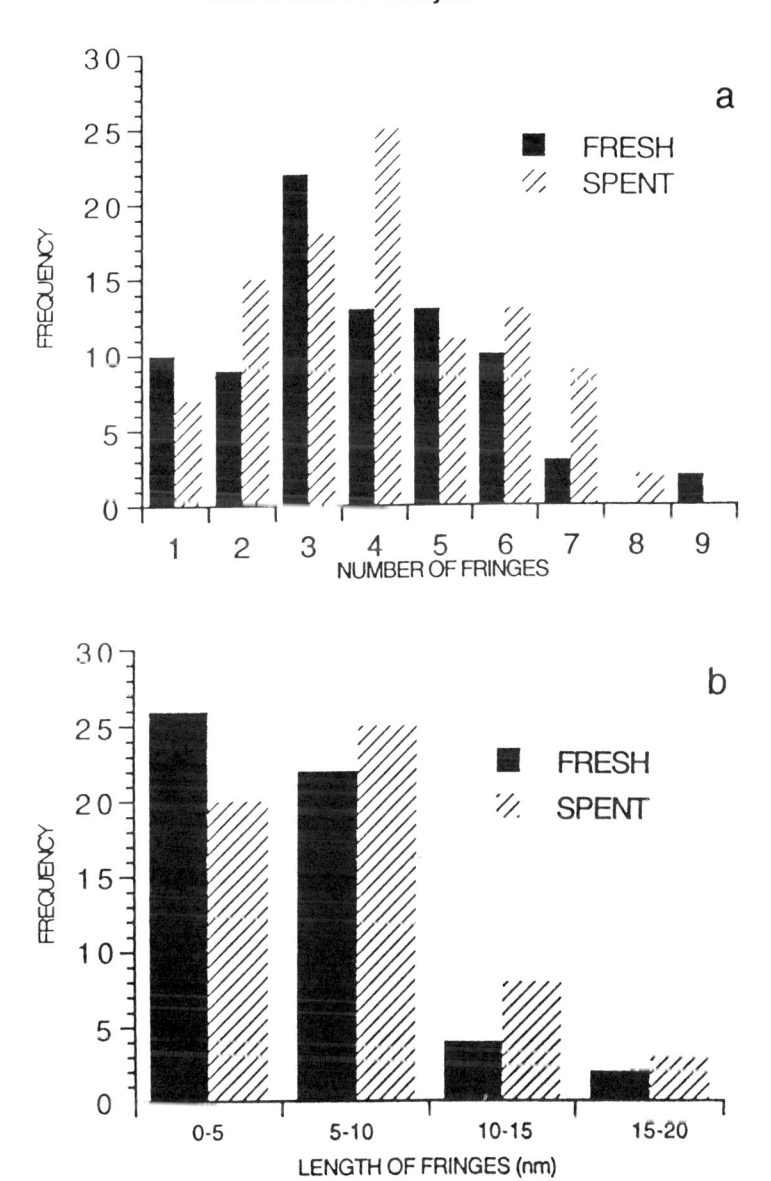

Fig. 10 Statistical distribution of MoS$_2$ catalyst microcrystals according to the a) number, and b) length of fringes (layers), measured before and after the reaction test.

mechanism is esentially the same as the active-sites nucleation model for radio frequency sputtered molybdenum disulfide films discussed by Hilton and Fleischauer (39), and is a consequence of the much greater chemical reactivity of edge planes over that of basal planes.

Since sulfur elimination mechanisms alone cannot explain the variations in catalytic activity, selectivity and surface area with time on stream, alternate mechanisms are required. Thus, these variations may be understood in light of the results of electron microscopy concerning the growth phenomenon of MoS_2 crystallites and assuming, as proposed by Daage et al. (35), that HDS takes place on the edge surfaces and hydrogenation on rim surfaces. The correlation between surface area and activity losses may be explained considering that microporosity occurring between MoS_2 crystallites is mostly responsible for the surface area; this microporosity is formed by the gases (NH_3 and H_2S) evolved during the decomposition of ATM. Micropores are then eliminated when MoS_2 crystallites grow via edge sintering, and so the catalytic activity decays at about the same rate as the surface area.

Crystallite growth due to layer stacking is considered to have a low influence on the surface area, since only a minor variation in stacking takes place during the reaction. Edge surfaces decrease at almost the same rate as the surface area because edge sintering eliminates both edge planes and microporosity, so that HDS selectivity remains constant with time on stream. Hydrogenation (butane selectivity) sites located on rims are affected by both the layer stacking and edge sintering mechanisms, causing the reaction to decay more rapidly than the surface area. Finally, the increase in selectivity of isomerization may be explained by assuming that isomerization sites can occur on imperfect basal planes, where layer stacking decreases the number of sites at a lower rate than the surface area so that the relative selectivity increases.

IV. SUMMARY

Theoretical and experimental HREM images of MoS_2 (WS_2) have been obtained. Catalysts prepared by decomposition of thiosalts at 673 K have a very disordered microstructure, similar to the poorly crystalline structure reported by Liang et al. (3). The disorder in structure is apparent in (100) and (10*l*) edge-like planes, as well as in the better known (002) basal-like planes. Besides the stacking faults, bent planes and dislocations which are the typical defects observed in (002) planes, other such as kinks, steps and microporosity appear in (100) and (10*l*) planes.

The characterization by HREM of layered dichalcogenide catalysts suggests that the mechanism of steady layer growth typical of these materials is absent during the preparation and use of these catalysts. Instead, a growth mechanism involving transient conditions with the participation of active impurities (such as H_2S and sulfur) is occuring. Vicinal and defective surfaces are formed when crystal growth takes place out of equilibrium.

Observed defects in these crystallites, such as stacking faults, dislocations, bent and shifted planes, are mainly related to the very weak bonding energy between layers in the c direction. Other defects, such as kinks and steps (observed by TEM) as well as vacancies (not readily observable by TEM) are related to the mechanism of crystal growth. Under transient conditions far from equilibrium, as is the case in the preparation of catalysts, the number of these defects is expected to be very high. On the other hand, the formation of channels between crystallites is mainly related to the evolution of gases during the decomposition of the precursor.

HREM has also provided information about the modification in the stacking and length of layers in MoS_2 catalysts as a function of the reaction test of thiophene HDS. The observed variations are believed to occur mainly by sintering of particles through edge-like planes, although layer growth of basal-like planes is also observed. With the sintering hipothesis, and assuming that active sites are located on different planes (HDS on edge sites, hydrogenation on rim surfaces and isomerization in basal sites), the variation of selectivity with time on stream is explained. Microporosity is believed to play an important effect in the catalytic properties of MoS_2 by allowing access to inner edge-surfaces. As microporosity diminishes due to sintering, a parallel decrease in activity with surface area is observed.

ACKNOWLEDGMENTS

We thank DGAPA-UNAM for funding this research under Project No. IN102692. We also thank L. Rendón for valuable technical assistance. M. Del Valle acknowledges fellowship support from CONACYT (Reg. No. 052621).

REFERENCES

1.- O. Weisser and S. Landa, Sulfide Catalysts: Their Properties and Applications, Pergamon, Oxford, 1973.
2.- R. Chianelli in Solid State Chemistry in Catalysis, A.C.S. Symp. Series 279, 13 (1985) p. 221.
3.- K.S. Liang, R.R. Chianelli, F.Z. Chien and S.C. Moss, J. Non-Cryst. Sol. 79:251 (1986).
4.- J.V. Sanders and K.C. Pratt, J. Catal. 67:331 (1981).
5.- F. Delannay, Appl. Catal. 16:135 (1985).
6.- S. Kasztelan, H. Toulhoat, J. Grimblot and J.P. Bonnelle, Appl. Catal. 13:127 (1984).
7.- S. Eijsbouts, J.J.L. Heinerman and H.W.J. Elzerman, Appl. Catal. A, 105:53 (1993).
8.- J. Cruz-Reyes, M. Avalos-Borja, M. Farías and S. Fuentes, J. Catal. 137:232 (1992).
9.- J.R. Gunther, O. Marks, T. Koranyi and Z. Paal, Appl. Catal. 39:285 (1988).
10.- T.F. Hayden and J.A. Dumesic, J. Catal. 103:366 (1987).
11.- S. Srinivasan, A.K. Datye and C.H.F. Peden, J. Catal. 137:513 (1992).
12.- A. Sachdev, J. Lindner, J. Schwank and M.A. Villa García, J. Sol. State Chem. 87:378 (1990).
13.- O. Sorensen, B.S. Clausen, R. Candia and H. Topsoe, Appl. Catal. 16:135 (1985).
14.- W. Elzner, M. Breysse, M. Lacroix, C. Leclercq and M. Vrinat, Polyhedron 7:2405 (1988).
15.- R.A. Kemp, R.C. Ryan and J.A. Smegal in Proceedings of the 9th International Congress on Catalysis, Calgary, 1988, Vol. 1, (M.J. Phillips and M. Ternan, eds.) The Chemical Institute of Canada, Ottawa, 1988, p. 128.
16.- M. Salmeron, G.A. Somorjai, A. Wold, R.R. Chianelli and K.S. Liang, Chem. Phys. Letts. 90:105 (1983).
17.- K. Tanaka and T. Okuhara in Proceedings of the IIIrd International Conference on "The Chemistry and Uses of Molybdenum", Climax Molybdenum 1979, p. 170.
18.- R.W. Phillips and A.A. Fote, J. Catal 41:168 (1976).
19.- R.J.H. Voorhoeve, S.C.M. Stuiver, J. Catal. 23:243 (1971).
20.- A.L. Farragher and P. Cossee in Proceedings of the 5th. International Congress on Catalysis, Palm Beach, 1972, (J.W. Hightower, ed.), North Holland, Amsterdam, 1973, p. 1301.
21.- H. Topsoe, B.S. Clausen, R. Candia, C. Wivel, and S. Morley, J. Catal. 68:433 (1981).

Microstructural Characterization of Catalysts

22.- S. Fuentes, M. Avalos-Borja, D. Acosta, F. Pedraza and J. Cruz in New Frontiers in Catalysis, Proc. 10th. International Congr. on Catalysis (L. Guczi, ed.), Elsevier Sci. Publ., Budapest, 1993, p. 611.
23.- S. Fuentes, G. Díaz, F. Pedraza, H. Rojas and N. Rosas, J. Catal. 113:535 (1988).
24.- K. Ramanathan and S. Weller, J. Catal. 95:249 (1984).
25.- Mc Tempas, a program for the Simulation of HREM Images, Total Resolution, Berkeley, California.
26.- M.J. Yacamán, R. Herrera, S. Tehuacanero, A. Gómez and L. Beltrán del Río, Ultramicroscopy 33:133 (1990).
27.- K.C. Pratt, J.V. Sanders and V.J. Christov, J. Catal. 124:416 (1990).
28.- J. Ramirez, S. Fuentes, G. Díaz, M. Vrinat, M. Breysse and M. Lacroix, Appl. Catal. 52:211 (1989).
29.- E. Diemann and A. Muller, Coord. Chem. Rev. 10:79 (1973).
30.- B. Scheffer, N.J.J. Dekker, P.J. Mangnus and J.A. Moulijn, J. Catal. 121:31 (1990)
31.- R. Frety, M. Breysse, M. Lacroix and M. Vrinat, Bull. Soc. Chim. Belg. 93:663 (1984).
32.- B.S. Clausen, H. Topsoe, B. Lengeler and R. Candia, in EXAFS and Near Edge Structure III (K.O. Hodgson, B. Hedman and J. E. Penner-Hahn, eds.), Springer Proceedings in Physics 2, Springer-Verlag, Berlin, Heidelberg, 1984, p. 181.
33.- M. Ledoux, G. Maire, S. Hantzer and O. Michaux in Proceedings of the IX IX International Congress on Catalysis, Calgary, 1988, Vol. 1, (M.J. Phillips and M. Ternan, eds.), The Chemical Institute of Canada, Otawa, 1988, p. 74.
34.- A. Wambeke, L. Jalowiecky, S. Kasztelan, J.P. Grimblot and J.P. Bonelle, J. Catal. 109:320 (1988).
35.- M. Daage, R.R. Chianelli and A.F. Ruppert, in Proceedings of the 10th International Congress on Catalysis, July 19-24, 1992, Budapest, Hungary, (L. Guczi ed.), Hungarian Academy of Sciences, 1992, p. 177
36.- C.B. Roxlo, M. Daage, A.F. Rupert and R.R. Chianelli, J. Catal. 100:176 (1986).
37.- S. Siegel, J. Catal. 30:139 (1973).
38.- Y. Okamoto, A. Maezawa and T. Imanaka, J. Catal. 120:29 (1989).
39.- M.R. Hilton and P.D. Fleischauer, J. Mater. Res. 5:406 (1990).

STM and AFM Studies of Transition Metal Chalcogenides

S. P. Kelty, A. F. Ruppert and R. R. Chianelli[*]
Exxon Research and Engineering Co.
Annandale, NJ 08801-0998

J. Ren and M. -H. Whangbo
Dept. of Chemistry
North Carolina State University
Raleigh, NC 27695-8204

Abstract

We have used atomic force microscopy (AFM), and scanning tunneling microscopy (STM) to characterize the local surface electronic structure of several transition metal chalcogenides. The observed AFM and STM images are interpreted in terms of the structural and electronic properties of the surface and by calculating the total, $\rho(r_0)$, and partial electron density distribution, $\rho(r_0, e_f)$, respectively. The influence of local surface defects and edge planes on the observed surface electronic structure is also described. The possible implications of these findings on the chemical properties of these materials will be discussed.

I. Introduction

There is long standing interest in the surface properties of transition metal sulfides (TMS) within the petrochemical industry due to their commercial importance in such catalytic hydrotreating processes as hydrogenation, hydrodesulfurization (HDS), and hydrodenitrogenation (HDN). As a result, a extensive knowledge base has been generated concerning the bulk structural, electronic and optical properties of TMSs.[1,2,3] Further development in hydrotreating catalysts could benefit from a detailed understanding of the local properties of the catalytic site. In the early 1980's, a new generation of microscopic techniques were developed which provide, for the first time, a direct probe of the electronic and structural properties of solid surfaces at atomic resolution. These techniques show great promise in complementing other surface sensitive techniques in helping to resolve long-standing controversies in hydrotreating catalysts.

Over the past ~75 years, a number of new hydrotreating catalysts have been introduced. In general, these catalysts are based on treated (promoted)

[*]*Current affiliaton*: University of Texas at El Paso, El Paso, Texas

layered TMSs, particularly MoS$_2$. Several different structures and models have been proposed as to the nature of the active sites in these promoted catalysts.[3] It is now recognized that the active sites of these layered materials are localized at edges and corners.[4] However, many crucial questions remain unanswered as to the elemental composition, atomic structure and electronic configuration of the active site. For example, both ReS$_2$ and (unpromoted) MoS$_2$ are layered TMSs that exhibit HDS catalytic activity. Of the two, ReS$_2$ has about an order of magnitude higher activity. The activity of MoS$_2$ can be greatly improved or "promoted" by co-doping an amount of Co or Ni. The precise composition of the Co(or Ni) in the catalyst and how it enhances the activity of MoS$_2$ are not well understood. It is known that the Co or Ni promoter is not homogeneously distributed throughout the crystal but is localized at edge sites as has been demonstrated in depth profiling studies of Co$_x$MoS$_2$.[5] It is also known that ReS$_2$ is isostructural and isoelectronic with CoMo$_2$S$_4$. There also are open questions concerning what role (if any) defect states may play in catalysis. It is clear that the underlying surface electronic properties of the material play a fundamental role in determining chemical properties of the surface such as heterogeneous catalysis. It is therefore essential to have a fundamental understanding of how bulk electronic properties, which are generally well established, are represented at the surface. STM and AFM are direct probes of the electronic states of the surface atoms, and as such, are uniquely suited techniques for studying important surface properties of heterogeneous catalysts. In this paper we begin by briefly describing the operation of STM and AFM. As an example of the capabilities of these techniques, we then focus on recent studies of the basal plane of ReS$_2$. Finally, we describe some STM studies of edges and defect sites.

II. Instrumentation

A. STM

To achieve atomic resolution in STM, an atomically sharp metal tip is brought within a few angstroms of a conductive sample surface and a bias applied between the sample and tip.[6] When the tip and sample are sufficiently close that their respective wavefunctions overlap in the classically forbidden region (gap), quantum mechanical tunneling may occur from a filled level of one electrode to an unfilled level of the other (Figure 1). In a one dimensional tunneling junction approximation, the tunneling current shows an exponential dependence on the tip-sample separation s

$$I \approx V_b \rho_s e^{-1.025 s \sqrt{\Phi}}$$

STM and AFM Studies of Transition Metal Chalcogenides

where Φ is the average barrier height of the junction $((\Phi_s+\Phi_t)/2)$ and V_b is the applied bias voltage. The wavefunction overlap, and thus the observed

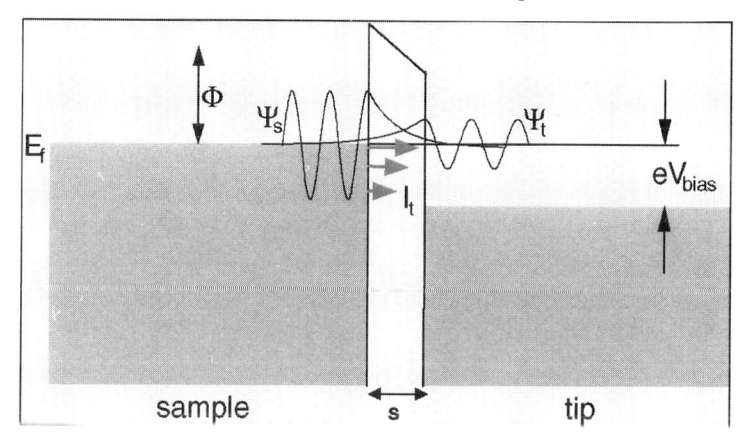

Figure 1. 1D tunnel junction schematic

tunneling current, is a function of the applied bias V_b, the tip-sample separation s, and the density of electronic states of the sample, ρ_s.[6b,c] An image is acquired serially by scanning the tip over the surface and measuring the tunneling current as a function of tip position (Figure 2).

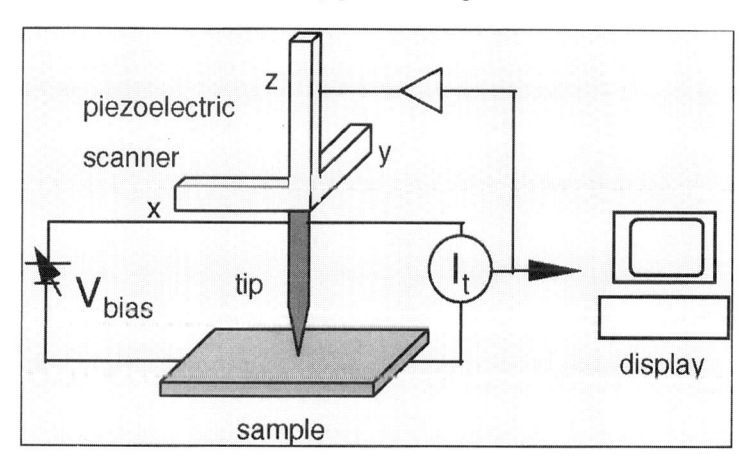

Figure 2. STM schematic

STM derives its sensitivity from the exponential decay of the wavefunction overlap with tip-sample separation. Since the applied bias is usually limited to within a few eV of the Fermi energy, STM is most sensitive to the electronic states associated with highest occupied and lowest unoccupied electronic levels.

The STM image represents a contour of the local electronic state distribution of the energy levels associated with chemical bonding. It has been demonstrated that STM images can be accurately modeled using an Extended Huckel Tight Binding (EHTB) computational method based on partial density of states $\rho(r_0, e_f)$.[7] We used a commercially available instrument (Nanoscope II, Digital Instruments) for these studies operated in ambient air. To aid in image analysis a 2-Dimensional Fast Fourier Transform (2DFFT) method was use to filter the data.

B. AFM

In AFM, a rigid probe tip in physical contact with the surface is scanned over the surface in a fashion similar to STM. The tip is attached to a soft cantilever (Figure 3). The deflection of the cantilever in response to the

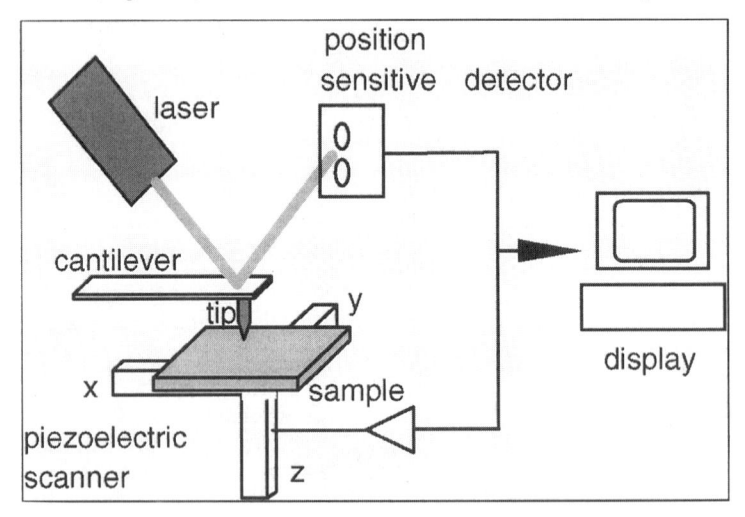

Figure 3. AFM schematic

surface topography is measured as a function of lateral tip position to provide a force image. The deflection is measured as a displacement of a laser beam reflected from the back (top) of the cantilever. Typical forces measured in an image are on the order 10^{-10}-10^{-8} N. The AFM images shown herein are taken in constant height mode, in which the z-drive of the piezo scanner is modulated to achieve a constant preset deflection on the cantilever. The measured forces are dominated by the core level electronic states and thus the images are usually dominated by the local atomic structure. Specifically, it has been found that AFM images may be accurately modeled as images of the total density of states $\rho(r_0)$.[7b] Used in combination with STM, AFM can provide a detailed understanding of the relationship between local crystal and electronic structure. It is expected that the unique electronic and structural properties of the active site in these may play a key role in determining the activity and specificity of the catalysts.

III. SPM Studies

A. Basal Plane of ReS$_2$

As an example of the utility of SPM in the characterization of catalyst surfaces, we describe a recent study of the basal plane of ReS$_2$. This material has a layered structure in which the Re atoms lie in a plane sandwiched between two bounding S atom planes (Figure 4). The Re(IV) (d^3) atoms

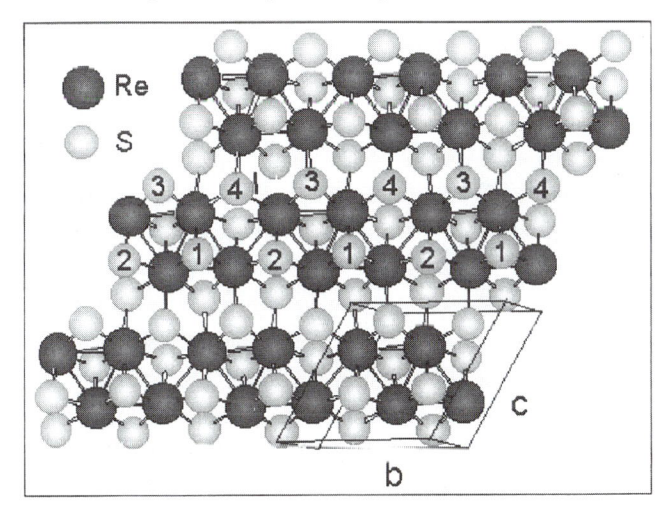

Figure 4. ReS$_2$ basal plane. Numbers indicate relative height of S atoms

occupy a distorted octahedral (CdI$_2$-type) coordination environment consisting of Re$_4$ tetramer-chains. The S layers are distorted or "puckered" due to the underlying Re plane structure such that four distinct S atom sites can be identified. These S atom sites are distinct in both their crystal geometry and in their local electronic density of states (see below). The study involves characterizing the surface S-atom layer as to its electronic and structural topography and interpreting these results in terms of calculated ReS_2 surfaces. The information obtained helps provide a basis for understanding the local details of the basal plane structure.

The ReS_2 is a semiconducting layered dichalcogenide with a 1.23 eV optical band gap.[8] Details of the crystal structure are a = 6.417 Å; b = 6.377 Å; c = 6.461 Å; α = 119.06°; β = 91.62°; γ = 105.12°.[9] Samples were prepared by cleaving the material along the basal plane using adhesive tape. STM images were collected in the constant current imaging mode although similar results were obtained in constant height mode. AFM images shown herein were obtained using the repulsive contact force imaging mode.

1. AFM and STM Data

Figure 5 is a AFM image of the ReS_2 sulfur basal plane in which the spot intensity scales with the local contact force. The image indicates the presence of parallel alternating high and low intensity rows. In the image we can resolve individual spots in these rows. The 2D spot pattern observed is consistent with the crystal structure of the surface unit cell (b ~ c = 6.4 ± .3 Å, α = 120°± 1°). In figure 6 is shown a typical STM image obtained in the

Figure 5. AFM image of ReS_2 (2DFFT filtered)

surface-to-tip imaging mode of the ReS$_2$ basal plane surface. This image also shows an alternating row structure similar to the AFM image. The STM and

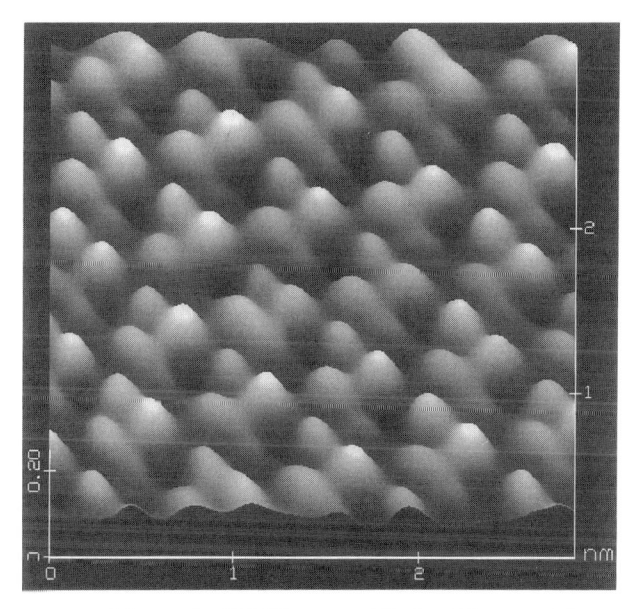

Figure 6. STM image of ReS$_2$; V$_{bias}$ = -1.8 eV

AFM images were collected in areas free from defects and so correspond to images of the bulk basal plane crystal structure. Defect sites can significantly alter the local electronic structure, as described in a later section.

2. Model Calculations

We noted above that STM and AFM image data corresponds to a convolution of the height of the features (s) and total and partial density of states for AFM and STM respectively. In this section we address the relative contribution of s and $\rho(r_0)$ (and $\rho(r_0,e_f)$) to the AFM (and STM) images using a model calculation an ReS$_2$ surface. The model calculations correspond to surface projected local DOS. Figure 7a shows a calculated AFM image of an S-Re-S single layer slab obtained using the EHTB method. The calculated AFM image also indicates alternating rows of high and low atom sites. The spots are centered on atomic positions and indicate only the top S atom state density. The Re and bottom S layer atoms are not observed in the calculated images. The relative intensity of the spots in the calculated image follows the relative height of the S atoms positions: the brightest spots are those most projected from the surface (closer to the AFM tip).

The STM data may be similarly modeled except here we distinguish between tip-to-surface (conduction band edge) images and surface-to-tip (valance band edge) images. In the former case we calculate $\rho(r_0, e_f)$ within a narrow energy interval around the conduction band edge while in the latter we calculate $\rho(r_0, e_f)$ around the valance band edge. These are shown in figures 7b and 7c respectively. We observe significant differences between these two band edges as a result of the different origins of the conduction and valance band orbital contributions. The surface projected valance band edge DOS is dominated by $S(p_z)$ orbitals whereas the conduction band edge is dominated by $S(p_x, p_y)$ orbitals. These variations are manifest in the intensity and location of the HED spot patterns. The p_z orbitals (valance edge) are centered over the S atom positions and are oriented outward (toward the tip). The p_x and p_y orbitals (conduction edge) are directed in-plane and have a lower height profile. In fact one S atom (farthest from the tip) is not predicted to be observed in the conduction band image.

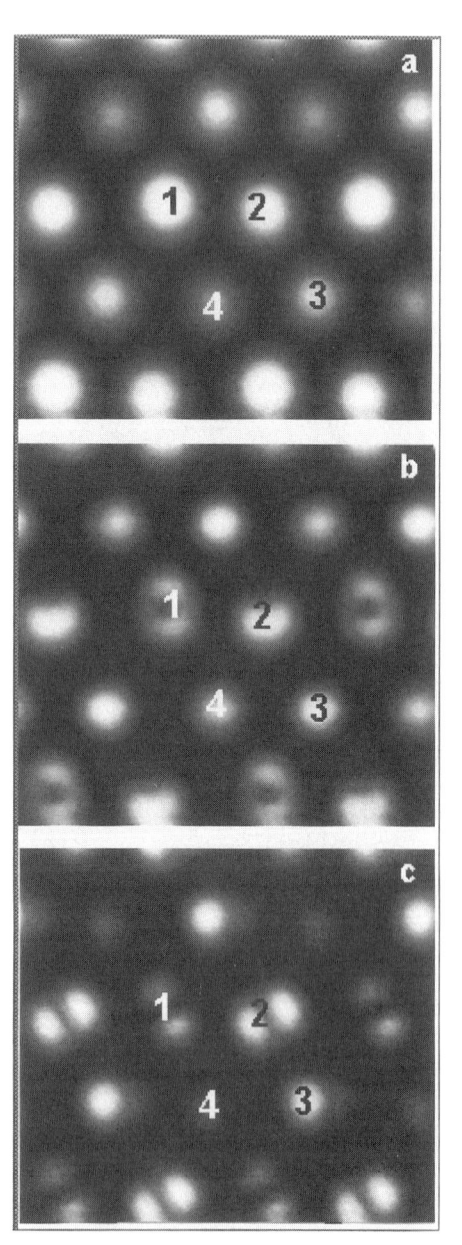

Figure 7. Calculated ReS_2 images of; a) $\rho(r_0)$, b) $\rho(r_0, e_f)$ (VB), c) $\rho(r_0, e_f)$ (CB)

STM and AFM Studies of Transition Metal Chalcogenides

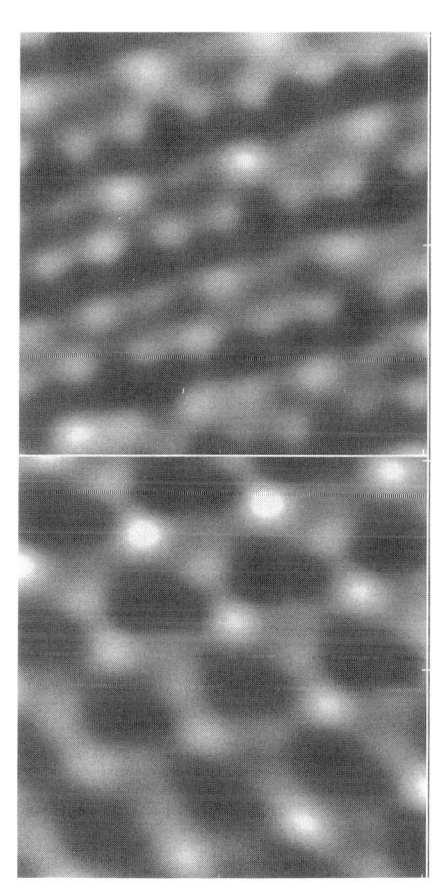

Figure 8. Simultaneous STM images of ReS$_2$ at V$_{bias}$ = -1.6 eV (top), and V$_{bias}$ = +2.7 eV (bottom). 2nm x 2nm.

Figure 8 shows two STM images of ReS$_2$ which were simultaneously obtained at (Figure 8, top -1.6 eV (valance band) and (Figure 8, bottom) +2.4 eV (conduction band). The VB image shows four distinct atomic sites per surface unit cell while the CB image shows only three (the weakest atomic feature in Figure 8a missing). Also, the CB features are displaced from those of the VB image. These observations are consistent with the predicted STM images shown in Figure 7b,c. The calculated images predict several image features which would not have been expected from a simple interpretation of the bulk band structure. In fact, the bulk band structure would not predict that the S atoms be observed at all since the Re atom d-orbitals dominate the valance and conduction band state density.[7] It is interesting that the relative intensity of the features in the STM image does not rigorously follow the relative height of the S-atom positions. The order of the observed height is S(2),S(3) > S(1),S(4) rather than the S(1)>S(2)>S(3)>S(4) pattern expected from the crystal structure. These results demonstrate that properly modeling STM and AFM data is essential for reliable interpretations. They also demonstrate the importance of these techniques in obtaining a understanding of the local environment of the catalyst surface.

B. Defects, Edges & Steps

In the previous section we described how local (atomic) information can be obtained and interpreted of a catalytic material. It is widely accepted that the active sites on layered transition metal chalcogenides are localized at the basal plane edge where the crystal may exhibit unique electrical and structural properties. Accordingly, further advances in the development of heterogeneous catalysts is currently focused on obtaining a better understanding of the local properties of the crystal edges. If one could obtain the same level of detailed information about the edge planes as we described above for the basal plane using SPM, this would allow a fundamentally new level of understanding of the catalytic activity of these materials. Due to experimental considerations of the probe tip, it is not generally feasible to obtain reliable data from a direct study of the edge planes; however, one can obtain local information about how the bulk crystal properties vary as a function of edge or defect proximity. For example, Figure 9 shows how these properties vary in a prototypical layered

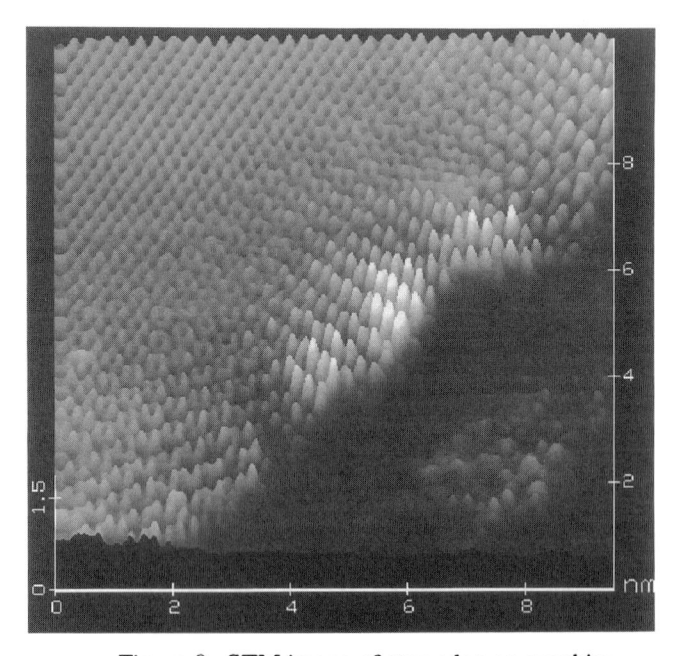

Figure 9. STM image of step edge on graphite.

material (graphite). The bulk (1x1) periodicity (upper left) is seen to undergo perturbations (pairing, height modulations, etc.) caused by the presence of the step edge (lower right). It is desireable to obtain similar information on the surface of a catalyst such as MoS_2.

Figure 10 shows a low resolution STM image of a step edge in MoS_2. Because of the large variation in DOS moving from semiconducting bulk to

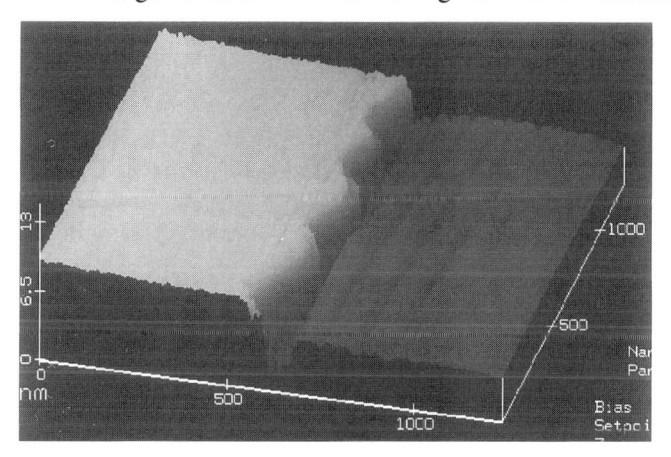

Figure 10. STM image of MoS_2 step edge.

matallic-like near the edge, it difficult to obtain stable images. The difficulty can be overcome by selecting only the edge states for our sample. This is acomplished by instead of imaging bulk MoS_2, we image only a small particle of MoS_2 which is effectively comprized of only edge regions. These particles are then deposited onto a metallic surface such as graphite of other single crystal substrate. An example of such a particle is shown in Figure 11. This MoS_2 particle on graphite retains its layered bulk yet it can be imaged at relatively low bias voltages (near the graphite Fermi level).

The crystal structure of TMSs is sensitive to the oxidation state of the metal. In the edge region, we expect that this oxidation state may be different than in the bulk due to the edge being sulfur rich, sulfur poor or oxidized. STM images should reflect these local variations in structure and provide some new insight about the unique properties of the edge and defect sites.

It is also possible to images of adsorbates on surfaces. By imaging changes to the edge and defect states caused by surface adsorbates, it may be possible to use STM as a direct in situ probe of the interactions between absorbates and the active site.

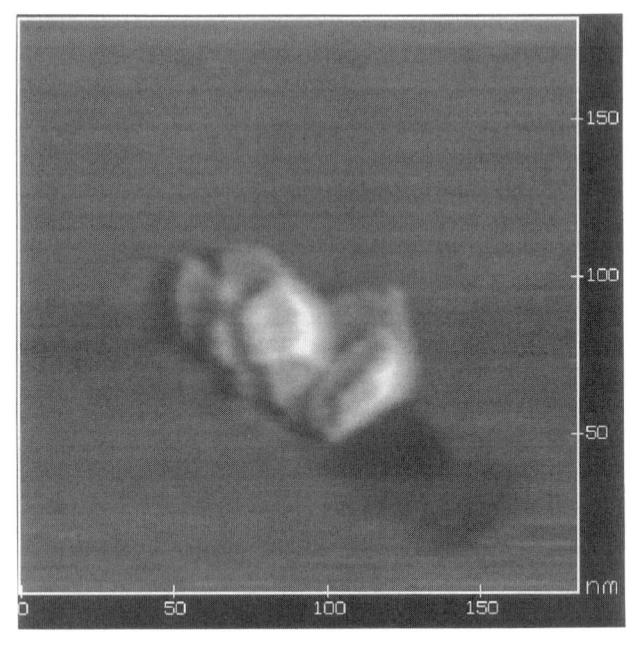

Figure 11. STM image of MoS_2 particle on graphite substrate.

We have shown that STM and AFM can be useful tools for helping to elucidate the local details of materials which are of current interest in heterogeneous catalysis. Other high resolution microscopic techniques such as HRTEM has begun to provide a better understanding of the local structure of edge and defect site. When properly interpreted, STM data can yield a wealth of new and complementary information regarding unique electronic properties of the active site as well.

References

1. W. Jaegermann, H. Tributsch. *Prog. Surf. Sci.*, 29:1-167. (1988)
2. J. A. Wilson, A. D. Yoffe. *Adv. Phys.*, 18:193 (1969)
3. R. R. Chianelli, M. Daage, M. J. Ledoux. *Adv. Catal.*, 40:177-232 (1994)
4. M. B. Salmeron, G. A. Somerjai, A. Wold, R. R. Chianelli, K. S. Liang. *Chem. Phys. Lett.*, 90:105-107 (1982)
5. R. R. Chianelli, A. F. Ruppert, S. K. Behal, A. Wold, R. Kershaw. *J. Cat.*, 92:56 (1985)
6. (a) G. Binnig, H. Rohrer. *Rev. Mod. Phys.* 59(3):615-625 (1987), (b) J. Tersoff, D. R. Hamman. *Phys. Rev. B*, 31:805 (1985). (c) J. Tersoff, *Phys. Rev. Lett.*, 57:440 (1986)
7. (a) M.-H. Whangbo, J. Ren, E. Canadell, D. Louder, B. A. Parkinson, H. Bengel, S. N. Magonov. *J. Am. Chem. Soc.*, 115:3760 (1993), (b) S. P. Kelty, A. F. Ruppert, R. R. Chianelli, J. Ren, M. H. Whangbo. *J. Am. Chem. Soc.*, 116:7857-7863 (1994)
8. J. V. Marzik, R. Kershaw, K. Dwight, A. Wold. *J. Sol. St. Chem.*, 51:170-175 (1984)
9. H. H. Murray, S. P. Kelty, R. R. Chianelli, C. S. Day. *Inorg. Chem.*, 33:4418-4420 (1994)

HREM Characterization of MoS$_2$, MoS$_2$:Co, and MoS$_2$:Fe

A. Vázquez-Zavala and M. José Yacamán
Instituto de Física UNAM. Apdo Postal 20-364 México 01000 D.F. México

Abstract

HREM has recently given new information about layered transition metal sulfide (LTMS) catalytic materials. It provides a direct way to measure the lattice parameters of the chemical crystalline phases present in the sample.

HREM techniques are beginning to elucidate the structure of the catalytically edge active sites and to reveal the atomic structure of these edge terminations. Also, information can be obtained about the highly folded planes that have not a simple explanation for this kind of crystalline structure. Finally, HREM is yielding insight into the very complex "promotion" problem. Some results in MoS$_2$:Co and MoS$_2$:Fe are presented in this work.

Introduction

Molybdenum and tungsten sulfides have played a crucial role in hydrotreating processes. Among the industrial applications of these sulfides are hydrogenation of olefines, ketones and aromatics, hydrodesulfurization (HDS), hydrodenitrogenation (HDN), hydrodemetallization (HDM), dealkilation and ring opening of aromatics (1). Also due to the layered structure of the MoS$_2$, it is widely used as a lubricant at high temperatures.

In comercial applications of the LTMS (Layered Transition Metal Sulfides), they almost always occur as alumina supported, extruded pellets with Mo or W present in the range of 10 to 20% by weight of the metal. They are usually "promoted" by Co or Ni which is present in the range of 20 to 30 atomic per cent with respect to the Mo or W.

The metals are sulfided and activated prior to use by "in situ" sulfiding techniques at temperatures in the range of 300 to 400° C. H_2S is the commonly using sulfiding agent but other agents are also used.

In spite of their wide and longstanding use, an explanation of the active sites in these catalysts needs a complete formulation. Fifteen years ago Massoth reviewed the field of Mo and W based hydrotreating catalysts (2). At the time of the Massoth article, it was not clear what role the alumina support played in the activity and selectivity of these catalysts originates in the sulfide phase. Since the availability of probes such as EXAFS (Extended X-ray Fine Structure), it has become clear that the alumina's primary function is to disperse and stabilize the MoS_2 and WS_2 hydrotreating catalysts (3). The role of the Co and Ni promoter phase has been extensively studied with much progress being made in elucidating its role. In past years, three models have been proposed to explain the action of the promoter in the Mo and W sulfides. In the first model (the intercalation model) (4) it is proposed that the Co or Ni atoms are intercalated into octahedral sites at the edges between MoS_2 or WS_2 slabs. The second model or synergistic model proposes that the promoter phase (Ni_3S_2 or Co_9S_8) exist as a separated phase. Finally, the Topsoe's model or CoMoS, NiMoS phase where the promoter atoms as in the intercalation model are located on in octahedral sites but in the same plane as the Mo atoms.

One of the techniques that provides direct information about the crystalline structure and morphology of these LTMS is High Resolution Electron Microscipy. This technique has played, and will continue to play, a key role in developing an understanding of the relation between the catalytic activity and selectivity of LTMS. In fact, it is precisely the atomic level structural information provided by HREM, which is needed for developing further insight into the origin of the catalytic activity and selectivity of these catalytic materials.

In our study of LTMS we divide this work into three parts, the first one dedicated to MoS_2 and the other two when Co and Fe are added to MoS_2.

Experimental

HREM was performed with a JEOL 4000EX electron microscope with a resolution limit of 17nm. The sulfides were dispersed with ultrasound in n-heptane. A drop of the solution was deposited in a holey carbon covered copper grid of 3mm diameter. The grids with the drop of solution were dried at room temperature (296 K) in air.

Results

Currently the atomic structure of the edge planes which terminate the two dimensional layers is unknown. It is the purpose of this paper to investigate this.

Metals at the edge of MoS_2 are thought to be at the origin of the promotion effect which is central to the catalytic importance of layered transition metal sulfides. The interaction of Co and Ni at the edges of the LTMS appears to be quite strong, resulting in charge transfer from the Co or Ni to the Mo and W. Due to the importance of this edges, direct images of these structures are now being obtained. Fig.1 shows MoS_2 crystals with hexagonal shape showing smooth basal planes terminated by edges planes. Fig.2 is an image of MoS_2 edges, several areas shows lattice images corresponding to 6.15 Å MoS_2 planes.

The MoS_2:Co System

One of the main questions asked in the case of MoS_2 or WS_2 promoted catalysts is where are the promoter atoms located. As mentioned previously there

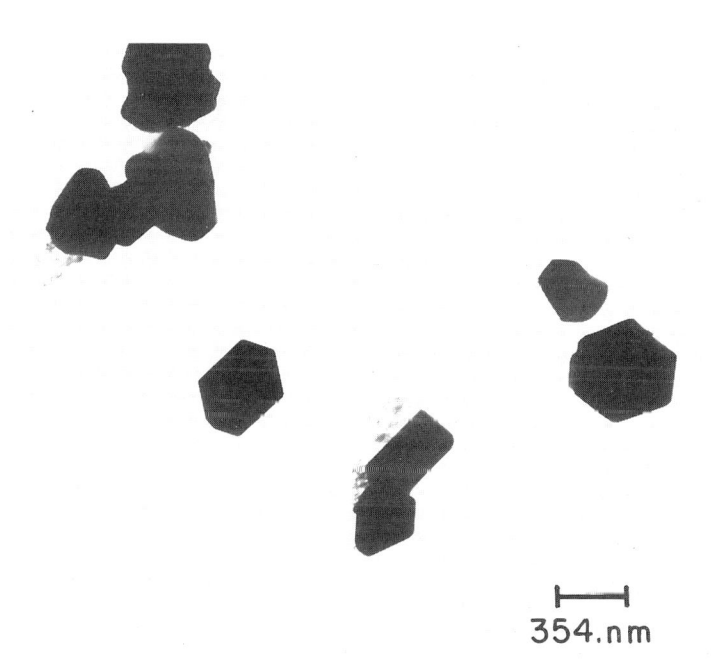

Figure 1. MoS_2 Crystals presenting hexagonal shape.

exist three models to explain the increase in activity of these catalysts. Voorhoeve and Stuiver (4) and Farragher and Cosee (5) from Shell proposed a model derived from the structure of sulfides called intercalated solids. In these types of compounds, the group VIII transition metals of the first row of the periodic table (Fe, Co and Ni) can penetrate the Van der Waals layers of the lamellar tungsten sulfide to occupy the octahedral symmetrical voids. However, for MoS_2 in the presence of Co or Ni, these intercalated compounds can only be prepared at temperatures greater than 800° C resulting in surface areas too low for catalytic measurements. Furthermore, it was difficult to prove that intercalates existed under catalytic conditions, and the author of the model had to make the assumption that only the sites located on the edges of the MoS crystallites were occupied by the Co and Ni ions under catalytic conditions. They called this model the "pseudo-intercalation" model to incorporate this idea.

In the "contact synergy" model proposed by Delmon (6), it is proposed that the MoS_2 and Co_9S_8 act together by being in close contact. The HDS reaction takes place at the interface between the two sulfides with each phase "helping".

The concept of a mixed phase containing both cobalt and molybdenum

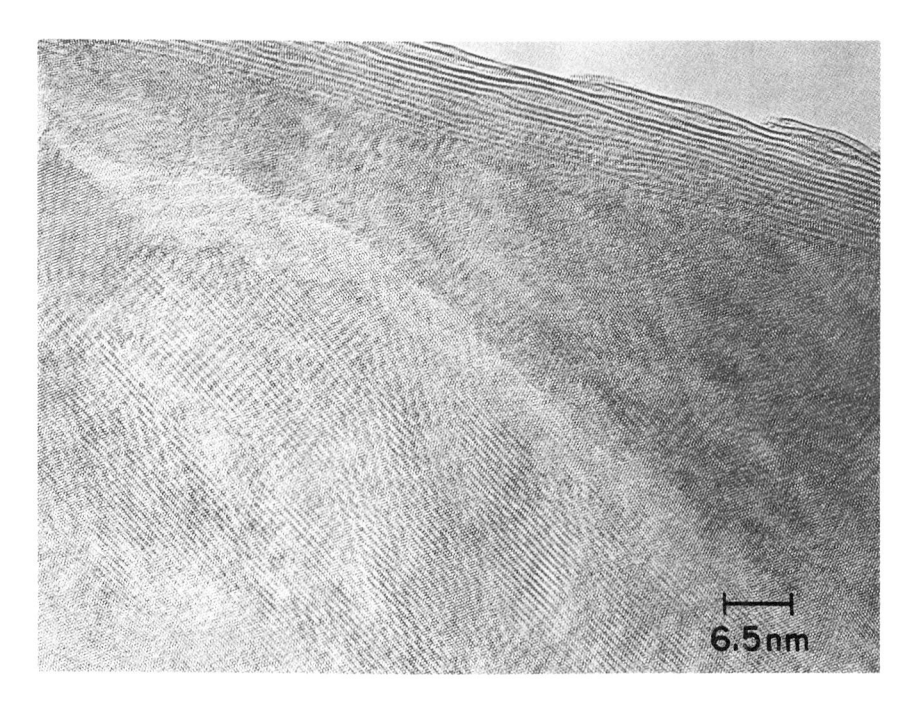

Figure 2. MoS_2 crystal edges.

which was responsible for the synergy was considered seriously, but the first physical proof of a specific Co environment was present by Topsoe et.al. They located the cobalt inside or on the edges of MoS_2 crystallites and called this the "CoMoS" phase (7).

Only microanalysis of MoS_2 crystallites, containing a small amount of Co, showed convincingly that the Co atoms were located on the edges of the MoS_2 platelets (8), leading to a model very close to the pseudo-intercalation model with similar limitations. In a paper peresnted by Phillips and Fote (9) they take the view that the promoted systems may be view as mixtures of two immiscible phases (Co_9S_8 and MoS_2) which interact by forming surface phases or surface complexes. This behavior has been termed "symmetrical synergy" (10).

Due to problems in the contrast interpretation related to interference of the alumina with highly dispersed sulfides, electron microscopy has failed to give a clear picture of the CoMoS or NiMoS phase. But some researchers among them Prat and Sanders (11) for bulk NiMo sulfide and Vrinat and Demourgues (12) for bulk CoMo sulfides were able to produce pictures from electron microscopes, showing a high dispersive effect of cobalt and nickel added to molybdenum sulfide.

Figure 3. MoS_2:Co crystal edges.

However, Ledoux and coworkers (13) were able to measure the activity of catalysts, highly dispersed on carbon support, and at the same time to observe the size and shape of the active particles by high resolution electron microscopy, since the carbon support is almost transparent to the electron beam.

Model catalysts are beginning to yield valuable insight because they allow simplification of the complex catalyst system. In Fig. 3 it is shown a HREM micrograph of MoS_2 doped with Co. The edge structure of this crystal is clearly seen in the micrograph with possible areas of insertion between the layers of MoS_2. Fig. 4 shows atomic resolution where the atomic chains appear to be "entangled" as in a DNA helix. At the present, there does not exist a model to

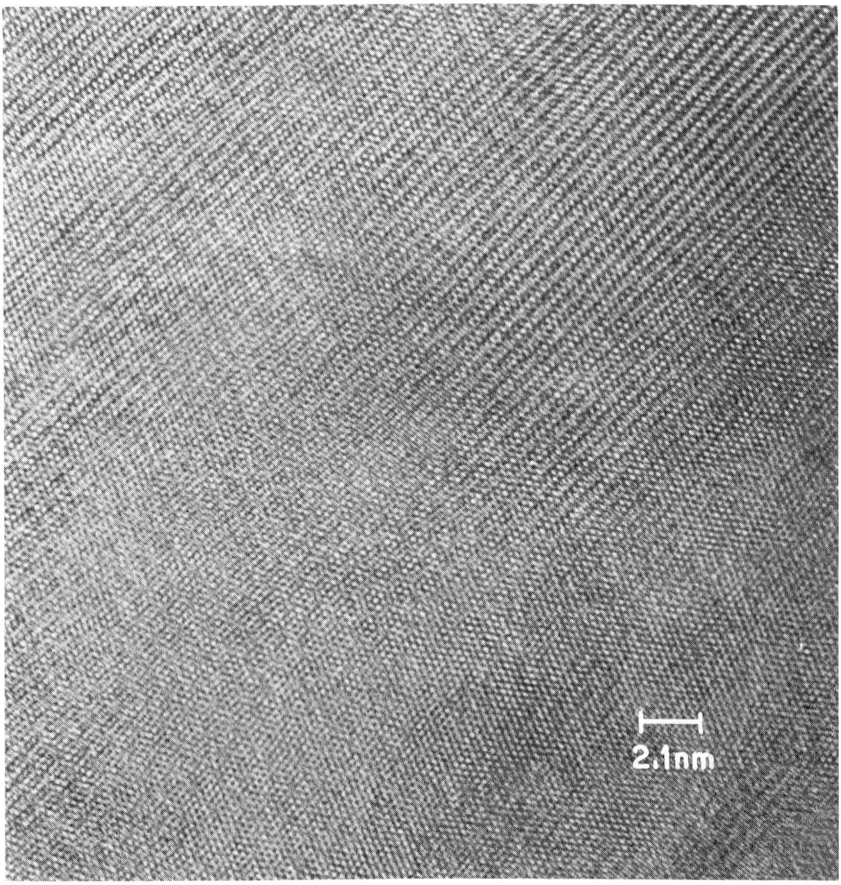

Figure 4. Atomic Resolution of MoS_2 crystals where atomic chains appear "entangled".

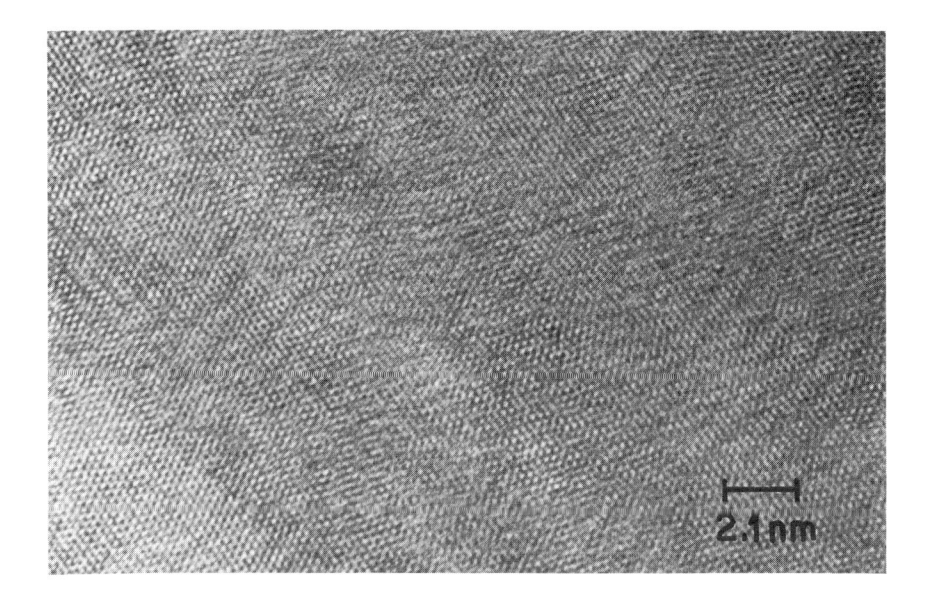

HREM Characterization of MoS_2, MoS_2:Co, and MoS_2:Fe 91

Figure 5. Hexagonal arrengements typical of MoS_2 crystalline structures.

explain this behavior and further study is necesary. Fig. 5 shows a superstructure of atomic resolution observed with typical hexagonal arrangements of the MoS_2 crystalline structure.

MoS_2 with Fe

In this part of the work we will show several images that corresponds to MoS_2:Fe. In Fig. 6a-c it is shown a typical MoS_2 edge with the sulfide stacks. In this case a tilting sequence around the (001) axis was performed. Surprisingly as we increased the angle of tilt the stacks started to disappear until in the last image they have completely vanished. This experiment confirms that there are no such things as "bookends" and the stacks are located in the basal plane of MoS_2 sheets and is because of the imaging conditions that these stacks appear in the microscope images.

Conclusions

As we see from our observations HREM is a direct technique for the observation of the layered transition metal sulfides (LTMS) It provides information about the complex problem of the promotional effect and about the sulfides edge terminations where the active sites are located.

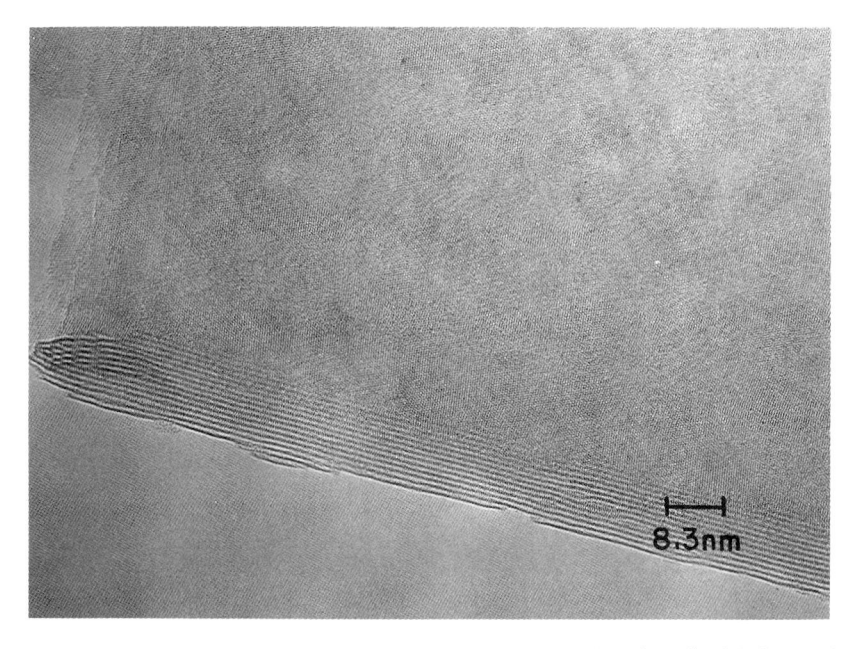

Figure 6a. Tilting sequence of MoS_2:Fe crystals showing that the MoS_2 stacks disappear with the tilting angle.

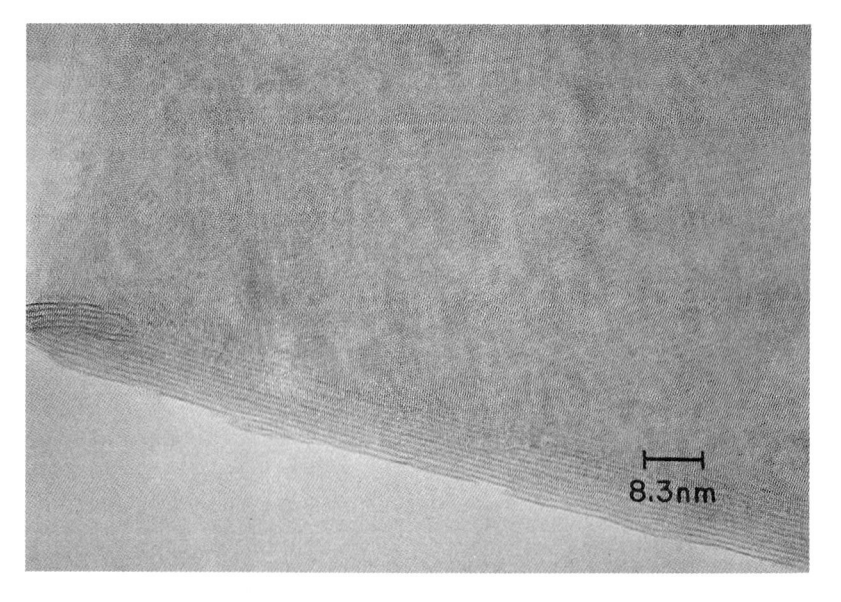

Figure 6b. Tilting sequence of MoS_2:Fe crystals showing that the MoS_2 stacks disappear with the tilting angle.

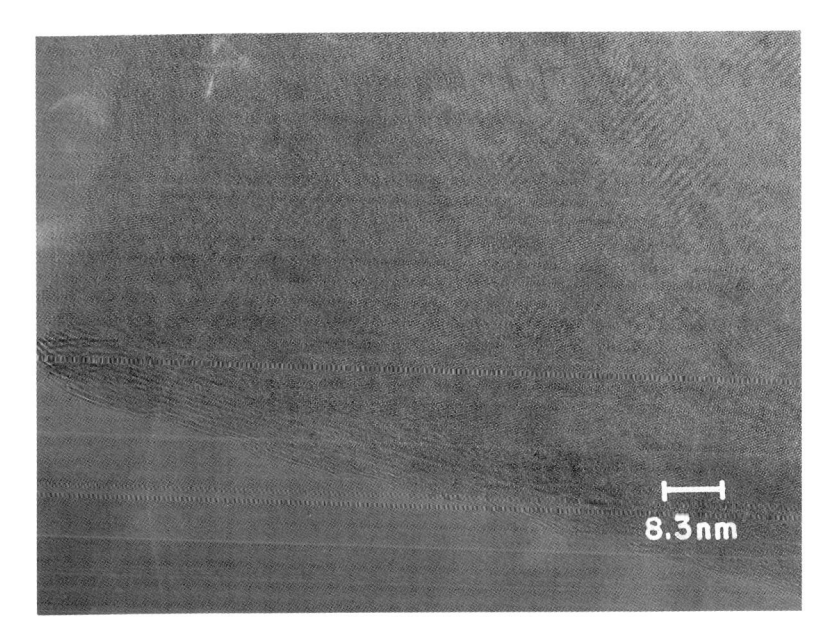

HREM Characterization of MoS$_2$, MoS$_2$:Co, and MoS$_2$:Fe 93

Figure 6c. Tilting sequence of MoS$_2$:Fe crystals showing that the MoS$_2$ stacks disappear with the tilting angle.

Acknowledgement

Thanks are due to Mr. Luis Rendón for preparation of several of the HREM photographs.

References

1.- O. Weisser and S. Landa, in Sulphide Catalysts, Their properties and Applications, Pergamon Press, New York, 1973.

2.- F.E.Massoth, Adv. Catal. 27.265 (1978).

3.- R.R. Chianelli, M. Daage and M.J. Ledoux, Adv. Catal. 1994 in press.

4.- R.J.H. Voorhoeve and J.C.M Stuiver. J. Catal. 23:228 (1971).

5.- A. L. Farragher and P. Cosee, Proc.5th ICC, Miami 1972, (1973) p.1301.

6.- G. Hagenbach, P. Courty and B. Delmon. J. Catal. 23:295 (1971).

7.- H. Topsoe, B.S. Clausen, R. Candia, C. Wivel and S. Morup. J. Catal. 68:433 (1981).

8.- O. Sorensen, B.S. Clausen, R. Candia and H. Topsoe. Appl. Catal. 13:363 (1985).

9.- R.W. Phillips and A.A. Fote. J. Catal. 41:168 (1976).

10.- K.C. Pratt and J.V. Sanders, Proc. 7th ICC, Tokyo 1980, (1981) p.1420.

11.- M.L. Vrinat and L. De Mourgues. Appl. Catal. 5:43 (1983).

12.- M.J. Ledoux, G. Maire, S. Hantzer and O. Michaux, Proc.9th ICC Calgary 1988, (1988) p.74.

On the Identification of "Co-Sulfide" Species in Sulfided Supported Co and CoMo Catalysts

M.W.J. Crajé, V.H.J. de Beer[1], J.A.R. van Veen[2], and A.M. van der Kraan

Interfacultair Reactor Instituut, Delft University of Technology, Mekelweg 15, 2629 JB Delft, The Netherlands
[1]Schuit Institute of Catalysis, Eindhoven University of Technology, P.O. Box 513, 5600 MB Eindhoven, The Netherlands
[2]Koninklijke/Shell-Laboratorium (Shell Research B.V.), Badhuisweg 3, 1031 CM Amsterdam, The Netherlands

Abstract

Mössbauer emission spectroscopy and X-ray absorption spectroscopy were applied for the identification of 'Co-sulfide' species formed during sulfidation of supported Co and Co-Mo hydrotreating catalysts. In the sulfided catalysts a 'Co sulfide' species is formed which is characterized by a doublet with a varying Q.S. value, depending on e.g. the presence of Mo, sulfidation temperature and Co-loading. Combining the variation in Q.S. values observed by MES and the coordination parameters deduced from EXAFS, the 'Co-sulfide' species turns out to be a particle characterized by its *size* and *ordering*. In Co/C highly dispersed Co_9S_8-like species are formed in which no ordering can be discerned beyond 4 Å, while in uncalcined Co/Al_2O_3 these 'Co-sulfide' particles are substantially larger. In CoMo catalysts the highly dispersed 'Co-sulfide' species are smaller than in the Co/C catalysts and are most likely located at the edge of the MoS_2 crystallites that function as a secondary support. By varying the Co/Mo ratio the number of Co atoms associated with a 'Co-sulfide' particle is varied continuously over a large range, with the lower and upper limit being the structure proposed in the 'Co-Mo-S' and contact synergy model, respectively. In the case of commercially applied Co/Mo ratios 'Co-sulfide' particles will be involved rather than the atomically dispersed Co of the 'Co-Mo-S' structure.

I. Introduction

For several decades many attempts have been made to find a relation between the structure of sulfided CoMo hydrodesulfurisation (HDS) catalysts and the synergetic behaviour of Co and Mo in these catalysts. Nevertheless, so far no consensus has been reached about the *local structure* of the 'Co-sulfide' species in these catalysts. In the beginning, the investigations leading to the first structural models for these catalysts [1-4] were of limited validity due to the application of ex-situ characterization techniques. With the introduction of Mössbauer emission spectroscopy (MES) by the Topsøe group [5,6] an important step forward was made. MES makes use of 14.4. keV γ-rays and can be applied as an in-situ technique. In the MES spectra of sulfidic CoMo catalysts a signal was observed that was not present in the spectra of sulfided Co catalysts (without Mo) or in the spectrum of any known crystalline Co(-Mo)-sulfide [7]. Topsøe et al. [7] assigned this spectral contribution to a species containing Co+Mo+S and they called this species the "Co-Mo-S" phase. Wivel et al. [8] reported a linear relation between the amount of Co present as "Co-Mo-S" and the thiophene HDS activity. These two findings form the basis of the "Co-Mo-S" model.

In this now most widely used structural model the Co-atoms are located at *certain* edge positions of MoS_2 slabs. However, CoMo catalysts show some features that cannot be explained in terms of the "Co-Mo-S" model. In fact, results of characterization studies of CoMo catalysts by different research groups often appear to be conflicting [9,10]. Still, all these studies are dealing with sulfidic CoMo catalysts and thus all these results should be interrelated. It is therefore reasonable to suppose that the apparent discrepancies are the consequence of different experimental conditions. It is shown in the literature that experimental details (like e.g. sulfidation procedure, catalyst preparation etc.) strongly influence the catalyst structure [11,12] and its properties [13,14], but in spite of this, experimental details are often missing in papers that deal with the subject. Hence, due to the lack of information and systematics in the characterization studies of CoMo catalysts it is not possible to develop a model that can interpret all the structural information obtained.

The emphasis in the present study has been on a systematic approach such that series of catalysts were measured and only *one* parameter was changed in

Identification of "Co-Sulfide" Species in Catalysts

between successive measurements. The observed changes between successive measurements could then be related to the variation of that particular parameter. By changing the sulfidation temperature (stepwise sulfidation treatment), the Co-loading, the presence/absence of Mo, the support material (carbon or alumina) and (in case of alumina) the application of a calcination procedure, a framework of results was created. In it various trends could be discerned which have been published before [15-19] but which will be discussed together in this paper. (The experimental details are presented before [15-19,23,31,32] and will not been included here. As a catalyst notation is used: Co(x)Mo(y)/C(T) with x and y the Co- and Mo-loading in wt%, respectively, C the carbon support and T the temperature at which the catalyst is sulfided in a H_2S/H_2 gas mixture.) It is found that basically all results can be considered to reflect the particle size of the 'Co-sulfide' species formed in the catalysts. Finally, the proposed structural model of the active phase and the implications of this model, and of the applied catalyst investigation method will be discussed.

II. On the similarity of Co/C and CoMo/C, according to MES

Fig. 1 shows the MES spectra recorded for a series of CoMo/C catalysts sulfided up to 673K, and the MES parameters of the sulfidic doublets obtained are summarized in Table 1 for those cases where the high-spin 2^+ doublet forms a minor spectral contribution. In the sulfided CoMo/C catalysts [15,16] species are formed which are characterized by a MES spectrum that differs from the MES

Table 1. Q.S. values in mm/s (±0.03 mm/s) of the 'Co-sulfide' doublet in the MES spectra obtained during the stepwise sulfidation of a set of CoMo/C catalysts. Unless otherwise stated, the spectral contribution of the 'Co-sulfide' doublet is 100%.

	300 K	373 K	473 K	573 K	673 K
Co(0.04)Mo(6.84)/C	*	1.26***	1.20	1.12	1.31
Co(0.08)Mo(6.84)/C	**	1.28***	1.07	1.08	1.24
Co(2.25)Mo(6.84)/C	1.22****	1.20	0.90	0.85	0.88
Co(4.3)Mo(7.0)/C	1.24	1.24	0.72	0.69	0.64

* Spectral contribution high-spin 2+ doublet 40±5%.
** Spectral contribution high-spin 2+ doublet 60±5%.
*** Spectral contribution of the 'Co-sulfide' doublet 85±5%, high-spin 2+ doublet 15±5%.
**** Spectral contribution of the 'Co-sulfide' doublet 80±5%, high-spin 2+ doublet 20±5%.

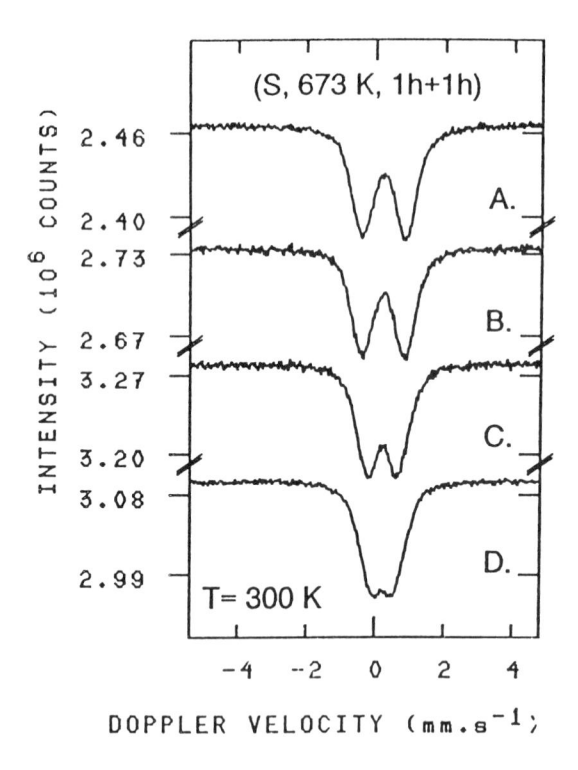

Fig. 1. MES spectra obtained at 300K with four CoMo/C catalysts with different Co(wt%)/Mo(wt%) loading after stepwise sulfidation up to 673K. The notation (S, 673K, 1h+1h) indicates that the sample is linearly heated to 673K within 1h and kept at that temperature for 1h before it is cooled to room temperature.
A: Co(0.04)Mo(6.84)/C;
B: Co(0.08)Mo(6.84)/C;
C: Co(2.25)Mo(6.84)/C;
D: Co(4.3)Mo(7.0)/C.

spectra of crystalline Co(-Mo)-sulfides.. From such an observation on sulfided CoMo/Al$_2$O$_3$ catalysts, Topsøe et al. [7] concluded already before that apparently a sulfidic phase containing both Mo and Co was formed in these catalysts and designated this phase "Co-Mo-S". Following Topsøe's definition of the "Co-Mo-S" phase, we concluded from our results that the Q.S. value of Co in such a phase can vary over a much larger range (at least 0.64-1.31 mm/s) than the one indicated by Topsøe et al. [20] (1.0-1.3 mm/s). In addition we have found that under certain conditions (sulfidation at low temperature) the so-called

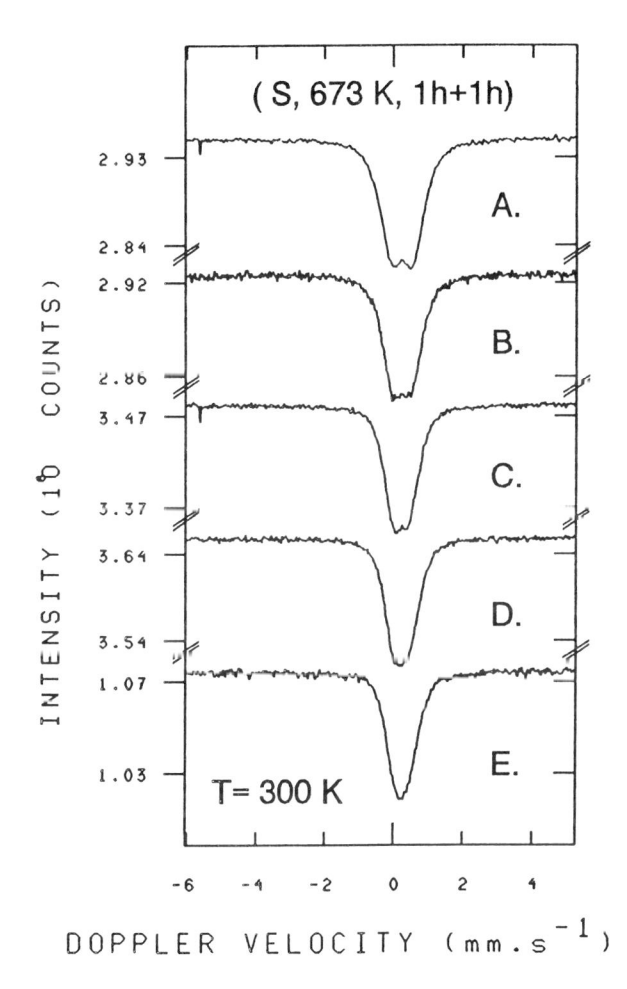

Identification of "Co-Sulfide" Species in Catalysts

Fig. 2. MES spectra obtained at 300K with five Co/C catalysts with different Co(wt%) loading after stepwise sulfidation up to 673K. The notation (S, 673K, 1h+1h) indicates that the sample is linearly heated to 673K within 1h and kept at that temperature for 1h before it is cooled to room temperature.
A: Co(0.04)/C; B: Co(0.08)/C;
C: Co(0.4)/C; D: Co(2.45)/C;
E: Co(4.3)/C.

"Co-Mo-S" MES spectrum can also be observed in sulfided Co/C catalysts (in the absence of Mo). At increasing sulfidation temperature Q.S. value in these sulfided catalysts strongly decreased. Although, however, after sulfidation at 673K the thermodynamically stable Co_9S_8 phase is expected to result, the Q.S. value remains much larger than the Q.S. value of 0.26 mm/s for crystalline Co_9S_8. Fig. 2 shows the MES spectra of a series of Co/C catalysts sulfided at 673K while the results of the analyses of the stepwise sulfided catalysts are presented in Table 2. In fact the observed parameters cannot be assigned to any crystalline Co-sulfide [7]. However, according to X-ray Photoelectron Spectroscopy (XPS) [21] and Extended X-ray Absorption Fine Structure (EXAFS) measurements [11] a highly dispersed Co_9S_8-like species ('Co-sulfide' species showing certain Co_9S_8 features) has been formed in the sulfided Co/C catalysts. This result can only be related to our MES results by assuming that the Q.S. value of highly dispersed Co_9S_8 deviates markedly from that of the crystalline compound. Such an effect has been found before by Van der Kraan [22] for Fe_2O_3. He has shown that the Q.S. value of surface atoms in Fe_2O_3 deviates strongly from that of bulk atoms causing an increase of the Q.S. value with decreasing particle size. So, we believe that also MES spectra of crystalline sulfides cannot be considered as proper reference spectra for the highly dispersed sulfidic phases in Co(-Mo)/C catalysts as the MES parameters of the latter will be much different from those of the former. Furthermore, the results obtained with the Co/C catalysts [15,16,18,24] indicate a correlation between the Q.S.

Table 2. Q.S. values in mm/s (±0.03 mm/s) of the 'Co-sulfide' doublet in the MES spectra obtained during the stepwise sulfidation of a set of Co/C catalysts. Unless otherwise stated, the spectral contribution of the 'Co-sulfide' doublet is 100%.

	S, 300 K	S, 373 K	S, 473 K	S, 573 K	S, 673 K
Co(0.04)/C	*	*	1.36	0.50[***]	0.62
Co(0.08)/C	*	*	1.31	0.53	0.54
Co(0.4)/C	1.32[**]	1.27	1.19	0.46	0.46
Co(2.45)/C	1.24[**]	1.19	0.64	0.45	0.44
Co(4.3)/C	1.11	1.08	0.53	0.42	0.39

[*] Spectrum dominated by high-spin 2+ doublet.
[**] Spectral contribution of the 'Co-sulfide' doublet 90±5%, 10±5% high-spin 2+ doublet.
[***] This Q.S. value does not fit properly in the sequences mentioned in the text.

Identification of "Co-Sulfide" Species in Catalysts 101

value observed by MES and the particle size of the 'Co-sulfide' species present in these catalysts. So, it has been demonstrated by us before [15,16,18,23] that the original "Co-Mo-S" model does not satisfy the MES results of the sulfided CoMo/C and Co/C catalysts. Consequently, the question arises to which species the observed MES spectra of the sulfided CoMo/C catalysts have to be assigned, and to what extent the species giving rise to a "Co-Mo-S" doublet in sulfided Co/C and CoMo/C catalysts are similar. Next, therefore, the structural similarities between Co/C and CoMo/C catalysts will be discussed on the basis of their MES spectra. In addition, further attention will be paid to the possible relationship between the Q.S. value and the particle size and ordering of the sulfided cobalt ('Co-sulfide') species.

A. *Temperature dependence of the MES parameters of the "Co-Mo-S" doublet in Co/C*

To start the discussion on the similarity of the sulfidic species in sulfided Co/C and CoMo/C catalysts, first the temperature dependence of the Q.S. value of the so-called "Co-Mo-S" doublet which is observed in sulfided Co/C catalysts will be considered. From the temperature dependence of $\ln(RAA/RAA_0)$ of this "Co-Mo-S" doublet in Co(0.04)/C(473) a Debije temperature θ_D of about 170 K has been derived [17]. This θ_D is in excellent agreement with the value of 175 K for the "Co-Mo-S" phase in Co-Mo/C reported by Topsøe et al. [20]. From a comparison between the temperature dependence of the Q.S. value of "Co-Mo-S" in Co(0.04)/C(473) and that reported by Topsøe et al. [7] it follows that it is the same as that of "Co-Mo-S" in CoMo/Al$_2$O$_3$.

Hence, it appears that the "Co-Mo-S" doublet in Co/C corresponds to a similar Co-species as the "Co-Mo-S" doublet in Co-Mo/Al$_2$O$_3$, which Topsøe et al. [7] assigned to the Co, Mo and S containing phase that supposedly is the main active phase in Co-Mo sulfide catalysts. Thus, apparently a local Co-S configuration might be obtained in the absence of MoS$_2$ that is the same or at least very similar to that of Co in the "Co-Mo-S" phase in supported Co-Mo catalysts (i.e. Co decorating the edge positions of MoS$_2$ slabs etc.).

From the above, it is concluded that the 'Co-sulfide' species giving rise to the "Co-Mo-S" doublets in sulfided Co/C and CoMo/C catalysts cannot be distinguished by MES. Therefore, we will now discuss in more detail whether there are more similarities between the sulfided Co/C and the CoMo/C catalysts and an attempt will be made to explain the observed behaviour of all sulfided Co/C and CoMo/C catalysts in terms of a correlation between the Q.S. value and the particle size and ordering of the 'Co-sulfide' phases formed.

B. *Sulfidability of Co/C and CoMo/C catalysts*

It is shown by Crajé [23] that a relationship exists between the Co-loading of Co/C catalysts and their sulfidability. In Table 2 it can be seen that a 100%

"Co-Mo-S" doublet is observed in the catalyst with a high Co-loading (4.3 wt%) after sulfidation at room temperature, whereas sulfidation up to 473 K is necessary to observe a 100% "Co-Mo-S" doublet in Co/C catalysts with low Co-loadings (0.04 and 0.08 wt%). The local concentration of the Co-oxide phases in the fresh catalysts is the only parameter that may be different in these catalysts. So, the observation that the sulfidation of the Co/C catalysts takes place at lower temperatures if the Co-loading is higher points to a possible relationship between the local concentration of the Co-oxide in the fresh catalyst and its sulfidability. In Table 2 it can also been seen that the Q.S. value of the "Co-Mo-S" doublet is smaller when the temperature at which 100% "Co-Mo-S" is formed is lower. As pointed out by Crajé et al. [18] all this can be taken to mean that a higher local concentration (lower 'dispersion') of the Co-oxide species in the fresh catalyst sulfides more easily and leads to larger 'Co-sulfide' particles (characterized by a lower Q.S.; see further below)

In Table 1 it can be seen that also the CoMo/C catalyst with the highest Co-loading (and also highest Co/Mo ratio because all catalysts have the same Mo-loading) is sulfided to 100% "Co-Mo-S" at room temperature like the Co(4.3)/C catalyst. The CoMo/C catalysts with lower Co-loadings have to be sulfided at higher temperatures in order to obtain a 100% "Co-Mo-S" MES spectrum. If a comparison is made between the Co/C and CoMo/C catalysts with similar Co-loading, it is seen that the temperatures at which a 100% "Co-Mo-S" doublet is observed are quite similar for both catalysts. Nevertheless, the Mo influences the sulfidation behaviour of the Co, as is apparent from (i) the substantial "Co-Mo-S" contribution to the MES spectra of Co(0.04)Mo(6.84)/C [15] and Co(0.08)Mo(6.84)/C [16] after sulfidation at room temperature, while such a contribution is not observed in their Co/C counterparts, and (ii) a decrease in Q.S. value of the "Co-Mo-S" doublet with increasing Co loading found for Co/C catalysts sulfided at 300-473K whereas for comparable CoMo/C catalysts the Q.S. value remains constant.

The observed differences between the Co/C and the CoMo/C catalysts during sulfidation at room temperature suggest that Co and Mo are already associated immediately after the preparation of the catalysts. This may occur whenever Co is preferentially attached to the Mo during the preparation procedure. In this respect it is of interest to note that Kasztelan et al. [24] studied Mo/Al$_2$O$_3$ and CoMo/Al$_2$O$_3$ catalysts by XPS and from the shadowing effect of Co on the Mo they concluded that Co covers the Mo in the fresh catalysts. Ramselaar et al. [25,26] extensively studied Fe/C and FeMo/C catalysts by Mössbauer absorption spectroscopy at 4.2 K. From their results, it followed that the Mo strongly prevents the Fe from sintering during a similarly mild drying treatment in H$_2$ as we have applied to remove nitrate groups from our catalysts. This supports the indication that Fe (Co) is attached to the Mo during the preparation process. Due to the presence of Mo, the sulfidation of the Co in the CoMo/C

Identification of "Co-Sulfide" Species in Catalysts 103

catalysts even with low Co-loading *starts* already at room temperature. A possible reason for this behaviour may be that the Mo catalyzes the sulfidation of the Co in the CoMo/C catalysts, but this is only a reasonable assumption as long as the Mo itself becomes sulfided during exposure to the H_2S/H_2 gas mixture at room temperature. However, from Temperature Programmed Sulfidation (TPS) studies by Mangnus [27], EXAFS measurements [28] and our recent XPS measurements [29] it is found that this is not the case. On the contrary, it has been reported that the Co catalyzes the sulfidation of the Mo. Therefore, it is very unlikely that the Mo catalyzes the sulfidation of the Co at room temperature. Another possibility is that the local concentration of Co in the fresh CoMo/C catalysts is higher because the Co is preferentially attached to the Mo. Due to such higher local concentration of Co it sulfides more easily as we already concluded from the results on the Co/C catalysts [15,16,18,23]. Although we cannot really distinguish between the two possibilities, a higher local concentration (lower 'dispersion') of Co due to preferential adsorption on Mo seems to be a more reasonable explanation.

C. *Relation between the Q.S. value and the particle size of the formed 'Co-sulfide' species*

The question is now, whether we can in fact understand the trends in the MES spectra of the sulfided Co/C and the CoMo/C catalysts in terms of a correlation between the Q.S. value and the particle size and ordering of the 'Co-sulfide' species. The answer can straightforwardly be given. With one exception (Co(0.04)/C(573)) the behaviour of the Co/C catalysts can be understood quite well on this basis. First of all, the observation that for none of the Co/C(673) catalysts the spectrum of crystalline Co_9S_8 is observed and that the Q.S. value of the characteristic doublet depends on the Co-loading already led to the assumption that the Q.S. value might be correlated with the particle size of the 'Co-sulfide' species. Furthermore, it can be seen from Table 2 that an increase of the Co-loading and/or increase of the sulfidation temperature, which favour an increase of the particle size, indeed lead to a decrease of the observed Q.S. value.

The behaviour of the CoMo/C catalysts is more complicated. If we examine the Q.S. value of the "Co-Mo-S" doublet recorded at the lowest sulfidation temperature at which no other spectral contributions are present (Table 1) we find with decreasing Co-loading the following sequence of Q.S. values 1.24(300 K)-1.20(373 K)-1.07(473 K)-1.20(473 K) mm/s (in parentheses the sulfidation temperatures are given). No correlation can be observed between these Q.S. values and the Co/Mo ratio. Furthermore, if we examine the relationship between the Q.S. value and the sulfidation temperature, only for the Co(4.3)Mo(7.0)/C catalyst a continuous decrease of the Q.S. value with increasing sulfidation temperature is observed. The other catalysts show a decrease of the Q.S. value during a sulfidation treatment up to 573 K, but sulfidation at 673 K leads to an increase of the Q.S. value. So, at first sight no relationship between the Q.S. value and the particle size of the 'Co-sulfide' species seems to exist for sulfided CoMo/C catalysts. However, if we examine only the CoMo/C(673) catalysts then we find a relationship between the Co/Mo ratio and the Q.S. value. Such a relationship cannot be understood in terms of the "Co-Mo-S" model and might instead point to a correlation between the Q.S. value and the particle size of the 'Co-sulfide' species.

If we compare the Q.S. values observed for the CoMo/C(673) catalysts with those of the so-called "Co-Mo-S" doublet observed in the spectra of the Co/C catalysts recorded at the lowest sulfidation temperature at which no other spectral contributions are present, we see a remarkable agreement. For CoMo/C(673) the sequence of Q.S. values with increasing Co-loading is: 1.31-1.24-0.88-0.64 mm/s and for Co/C: 1.36(473 K)-1.31(473 K)-1.27(373 K)-1.19(373 K)-1.11(300 K) mm/s, respectively (in parenthesis the sulfidation temperatures are given). In the Co/C catalysts this correlation between the Co-loading and the Q.S. value could be understood in terms of a particle size effect. As is demonstrated above (***Temperature depend...***), the 'Co-sulfide' species giving rise to these "Co-Mo-S" doublets in Co/C and CoMo/C cannot be distinguished by MES. It is reasonable to assume, therefore, that for the CoMo/C catalysts sulfided at 673 K the same trend can *also* be interpreted in terms of a particle size effect. This implies that "Co-Mo-S" in CoMo/C catalysts is very highly dispersed (small particles) 'Co-sulfide'. Such a conclusion is also supported by the observation that the Q.S. value for Co(4.3)Mo(7.0)/C(673) resembles that for Co(0.04)/C(673), and by the similar behaviour of the Co(4.3)Mo(7.0)/C and the Co(4.3)/C catalysts. Both the latter catalysts become fully sulfided at room temperature and the Q.S. value decreases, i.e. the particle size of the 'Co-sulfide' species increases, during sulfidation treatment at higher temperatures. From Tables 1 and 2 it is clear that the Q.S. value of the doublet characterizing the Mo containing catalysts after sulfidation up to 573 and 673 K is systematically larger than that of the catalyst with the same Co-loading without Mo. This can be understood on the basis of the previous suggestion that Co is preferentially attached to the Mo during the preparation of the catalysts; apparently Mo stabilizes the 'Co-sulfide' species in a higher dispersed state.

So, by assuming a correlation between the Q.S. value and the particle size of the 'Co-sulfide' species we can interpret the MES spectra of the CoMo/C(673) catalysts, and the sulfidation behaviour of the Co(4.3)Mo(7.0)/C catalyst. However, in the previous paragraph it was argued that a clear relationship between the Co/Mo ratio and the Q.S. value of the 'Co-sulfide' species formed at low sulfidation temperatures (up to 473 K depending of the Co/Mo ratio) does not exist. This feature can be explained in terms of a local concentration effect: as Co is preferentially attached to the Mo during the

Identification of "Co-Sulfide" Species in Catalysts

preparation, its 'apparent loading' (local concentration) is higher in the presence of Mo than in its absence. Finally, sintering can be the cause that the Q.S. value of the 'Co-sulfide' species in the CoMo/C catalyst decreases during sulfidation at temperatures up to 573 K, see Table 1. Hence, only the increase of the Q.S. value during the sulfidation at 673 K of the CoMo/C cannot be explained in this way. But, from TPS it is known [27] that the sulfidation of the Mo takes place via a MoS_3 phase which is transformed into MoS_2 at about 550 K. This result is confirmed by recent XPS measurements [29] performed on a Co(4.3)Mo(7.0)/C catalyst, and by EXAFS measurements of silica-supported MoO_3 catalysts by de Boer et al. [28]. So, the increase of the Q.S. value can be qualitatively explained as follows. A highly dispersed 'Co-sulfide' species is formed at low sulfidation temperatures. This 'Co-sulfide' species will sinter gradually during the sulfidation treatments at higher temperatures and this is reflected in a decrease of the Q.S. value. After the sulfidation treatment at 573 K the Q.S. is minimal for all CoMo/C catalysts (except Co(4.3)Mo(7.0)/C). At this sulfidation temperature the MoS_2 formation is presumably completed. Toulhoat and Kasztelan [30] showed that it is energetically favourable that the Co decorates the edges of the MoS_2 crystallites. Therefore, it is possible that during the sulfidation at 673 K the 'Co-sulfide' species will settle at the edge of the MoS_2. The observed increase of the Q.S. value indicates an accompanying redispersion or disordering of the 'Co-sulfide' species. Thus, after sulfidation at 673 K, the CoMo/C catalysts should consist of small MoS_2 crystallites with small 'Co-sulfide' particles at the edges.

According to this model all Co is located at the edges of the MoS_2 crystallites, which is supported by the results of the Co/C and CoMo/C catalysts sulfided at 573 and 673 K. If the relevant data from Tables 1 and 2 are plotted as in Fig. 3, it can be seen that the Q.S. value of highly dispersed Co_9S_8 (Co/C(573) and Co/C(673)) is in the range 0.4-0.55 mm/s over a large Co-loading range. Furthermore, the Q.S. value of very highly dispersed 'Co-sulfide' as formed in CoMo/C(373) is approximately 1.25 mm/s, which value also is observed for Co(0.04)Mo(6.84)/C(673) and Co(0.08)Mo(6.84)/C(673). So, *if* part of the Co would not have an interaction with Mo, it should be expected that the spectra of CoMo/C(673) catalysts are composed of a weighted summation of a doublet with Q.S.=1.25 and one with Q.S.=0.4-0.55 mm/s (due to Co_9S_8 separately from Mo on the support). However, this is not the case. For example in the spectrum of Co(2.25)Mo(6.84)/C(673) none of both doublets can be fitted. Thus, no 'Co-sulfide' segregated on the support is present and consequently all 'Co-sulfide' should be located at the MoS_2 crystallites.

D. *Conclusions*

The 'Co-sulfide' species formed in Co/C and CoMo/C catalysts are essentially

★CoMo/C(373) + CoMo/C(673) ♦ Co/C(673)

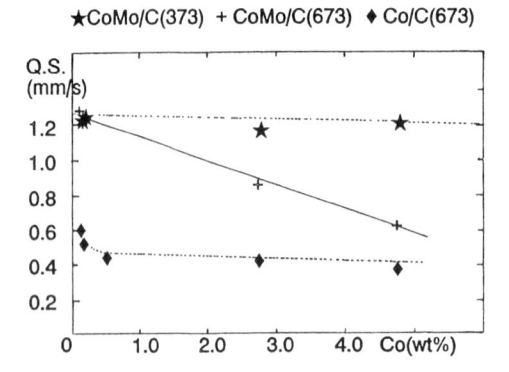

Fig. 3. Relationship between the Q.S. value and the Co-loading for Co/C and CoMo/C after successive treatments.

the same, only the particle size and/or ordering differs. In Co/C catalysts at low sulfidation temperatures a highly dispersed 'Co-sulfide' species is formed that by MES cannot be distinguished from Co in the "Co-Mo-S" phase. Upon sulfidation at higher temperatures the 'Co-sulfide' species sinter and after sulfidation up to 673 K highly dispersed Co_9S_8-like phases are formed.

In the CoMo/C catalysts the Co is attached to the Mo during the preparation of the catalyst. Therefore, the local Co concentration ('apparent loading') is originally higher than in the case of Co/C catalysts, resulting in the formation of 'Co-sulfide' species already at room temperature in CoMo/C catalysts (even in catalysts with a very low Co-loading). During sulfidation up to 573 K the 'Co-sulfide' species sinter but they remain attached to the Mo-sulfide species. After sulfidation at 573 K the formation of small MoS_2 crystallites presumably is completed. During sulfidation at 673 K the 'Co-sulfide' species settles at the edge of the MoS_2 crystallites. After sulfidation up to 673 K, the CoMo/C catalysts consist of small MoS_2 crystallites with small 'Co-sulfide' particles at the edges.

III. On the influence of the support

We have found [31] that the trends in sulfidation behaviour of the $CoMo/Al_2O_3$ catalysts are similar to those of their carbon-supported counterparts. The spectra of the $CoMo/Al_2O_3$ catalysts after sulfidation up to 673 K also show a relationship between the Q.S. value of the sulfidic doublet and the Co/Mo ratio as found for CoMo/C. Similarly, the Q.S. value of the 'Co-sulfide' doublet in the spectra of the $CoMo/Al_2O_3$ catalysts continuously decreases with increasing sulfidation temperature up to 573 K whereas it increases during sulfidation at

673 K. Finally, we found that the temperature at which the high-spin 2+ doublet has disappeared completely from the spectrum decreased with increasing Co/Mo ratio. The only difference found is that the CoMo/Al$_2$O$_3$ catalyst with a rather high Co/Mo ratio (0.61 at/at) become completely sulfidic at slightly higher temperatures than the CoMo/C catalysts with a rather high Co/Mo ratio (>0.53 at/at). Apparently the sulfidation of the alumina-supported catalysts is hampered due to a stronger particle support interaction.

We have also found [31] the same trends in the MES parameters during the sulfidation of Co/Al$_2$O$_3$ catalysts with different Co-loadings, as in the case of Co/C catalysts. During sulfidation treatment up to 473 K, 'Co-sulfide' species are formed which exhibit the "Co-Mo-S" doublet. The sulfidation temperature required to form this doublet decreased with increasing Co-loading and so did the Q.S. value of the "Co-Mo-S" doublet. At elevated sulfidation temperatures the "Co-Mo-S" doublet collapsed and a sulfidic phase exhibiting a doublet with a Q.S. value that decreased with increasing sulfidation temperatures was formed. In addition, also the Q.S. value of the sulfidic doublet observed after sulfidation at 673 K decreased with increasing Co-loading. However, as expected, the behaviour of uncalcined Co/Al$_2$O$_3$ catalysts differed from that of the carbon-supported counterparts in one respect. In the case of the Co/Al$_2$O$_3$ catalysts a stronger Co-support interaction was observed from the behaviour at low sulfidation temperatures and from the contribution of Co:Al$_2$O$_3$ in the spectra of the fully sulfided catalysts.

As outlined above, the trends observed in the CoMo/Al$_2$O$_3$ and Co/Al$_2$O$_3$ catalysts are essentially the same as those observed in the CoMo/C and Co/C catalysts. Therefore, we have assumed that the MES spectra of the Al$_2$O$_3$-supported catalysts may be interpreted in the same way as those of their C-supported counterparts. From this assumption it follows that also the 'Co-sulfide' species present in Co/Al$_2$O$_3$ and CoMo/Al$_2$O$_3$ catalysts are essentially similar and only the particle size and/or ordering of the 'Co-sulfide' species differs. Although a stronger Co-support interaction is observed in case of the Al$_2$O$_3$-supported catalysts, it is found from a comparison of the spectra of Co(4.3)/C(673) and Co(0.7)/Al$_2$O$_3$(673), that despite the lower Co-loading (0.25 at/nm^2 versus 0.4 at/nm^2), the Q.S. value of the 'Co-sulfide' species formed on the alumina support is much smaller than that on carbon. Apparently the 'Co-sulfide' species can sinter less easily on the carbon support which most probably is connected to the pore structure of the activated-carbon. Sulfidation at higher temperatures leads to growth of the 'Co-sulfide' species. In Co/Al$_2$O$_3$ catalysts with rather high Co-loading this leads to the formation of a 'Co-sulfide' species exhibiting the MES spectrum of crystalline Co$_9$S$_8$. In all MES spectra of Co/Al$_2$O$_3$ catalysts with low Co-loadings a subspectrum of Co:Al$_2$O$_3$ is observed.

IV. Structural model

Based on the various trends observed in the systematic study of the supported Co and CoMo catalysts, as presented above and before [15-19,24,31], we propose a structural model which applies to both carbon- and alumina-supported catalysts. In the sulfided catalysts a 'Co-sulfide' species is formed which is characterized by a doublet with a varying Q.S. value, depending on e.g. the presence of Mo, sulfidation temperature, Co-loading [18]. Combining the variation in Q.S. values observed by MES and the coordination parameters deduced from EXAFS measurements, the 'Co-sulfide' species turns out to be a particle characterized by a *particle size* and an *ordering*. Here, the *particle size* indicates the number of Co atoms that is associated with a 'Co-sulfide' particle, while the 'Co-sulfide' is said to be more *ordered* the better its coordination parameters agree with those of Co_9S_8 (Co-S$^{(1)}$: N=4.22, R= 2.21 Å; Co-Co$^{(1)}$: N=3.0, R= 2.51 Å; Co-Co$^{(2)}$: N= 6, R= 3.5 Å). Within a 'Co-sulfide' particle both four-fold and six-fold sulfur-coordinated Co atoms may be present.

Using this model, it is found that during the stepwise sulfidation up to 673 K highly dispersed Co_9S_8-like species are formed in Co/C catalysts. The EXAFS results reveal that in these Co_9S_8-like species no ordering can be discerned beyond 4 Å. Therefore, their particle size most probably does not exceed 10 Å. In uncalcined Co/Al_2O_3 catalysts 'Co-sulfide' particles are formed that exhibit the MES parameters of crystalline Co_9S_8, indicating that their particle sizes are substantially larger than for their carbon-supported counterparts. In supported CoMo catalysts very highly dispersed 'Co-sulfide' species are formed. These species will be considerably smaller than the Co_9S_8-like species formed in the Co/C catalysts and are most likely located at the edge of the MoS_2 crystallites that function as a secondary support. By varying the Co/Mo ratio of the CoMo catalysts sulfided at 673 K, the number of Co atoms associated with a 'Co-sulfide' particle can be varied continuously over a large range. In the limit of a small Co/Mo ratio, the number of Co atoms involved in one 'particle' becomes one and this is the case of the "Co-Mo-S" model as presented by Topsøe et al. [7]. In the limit of an extremely high Co/Mo ratio, the number of Co atoms involved in one 'Co-sulfide' particle becomes so large that it approaches crystalline Co_9S_8. Then we will end up with the *structure* of the catalysts as proposed in the contact synergy model [4].

Applying this model on Co/C catalysts with extremely low Co-loading and on CoMo/C catalysts, and relating the obtained 'Co-sulfide' particle sizes to the thiophene HDS activity of the catalysts shows that no clear relation between the 'Co-sulfide' particle size and the intrinsic activity of the Co atoms exists [32].

V. Implications

It is interesting to discuss some literature results in the light of the proposed model. Firstly, the broad line width that is observed in the MES spectra of

Identification of "Co-Sulfide" Species in Catalysts

CoMo catalysts [7,15,16] indicates a distribution in local surroundings of the Co atoms. This does not fit with the structural models proposed recently by Topsøe et al. [33] and Bouwens et al. [11]. Both very detailed edge-decoration models are based on the "Co-Mo-S" model, the Co atoms being thought to be present at well-defined edge positions of small MoS_2 slabs (7 to 8 Mo atoms, [34]), even the Miller indices of the MoS_2 plane are given. However, such well-defined Co positions should lead to a much narrower line width than those observed by Topsøe et al. [7] and by ourselves. The large line width, however, can be understood in terms of our model in which both 4- and 6-fold coordinated Co atoms are present within one and the same very small Co_9S_8-like particle. This particle, which most likely contains only a few Co atoms, is located somewhere on the MoS_2 crystallites. In agreement with previous results [11,35] it is most probably located at the edge because the EXAFS results yield a Co-Mo coordination distance of 2.9 Å which cannot be assigned to Co on the basal plane of the MoS_2 slabs.

Secondly, the variation in Q.S. values for "Co-Mo-S" will be discussed. In terms of the above mentioned promoter site structures proposed by Topsøe et al. [33] and Bouwens et al. [11] the variation in Q.S. values that was reported for "Co-Mo-S" by Topsøe et al. [10], and which according to our results [15,16,18] is even much larger, cannot be understood. Again, *if* the Co is located at such well-defined positions it should exhibit a fixed Q.S. value. In our model the variation in Q.S. value is understood in terms of a changing size and/or ordering of the very small 'Co-sulfide' species (including changing atomic distances and coordination numbers).

Thirdly, it is interesting to consider the relation between the MES spectrum of the catalyst and its activity. From our combined MES and thiophene HDS activity studies [32] we have concluded that no relationship exists between the Q.S. value and the *intrinsic* activity (QTOF) of the 'Co-sulfide' species. This result is not unique. It was already reported by Candia et al.[13] that high temperature sulfidation (873 K) of calcined $CoMo/Al_2O_3$ catalysts leads to the formation of a species exhibiting a normal "Co-Mo-S" MES spectrum but having twice the activity of "Co-Mo-S" in $CoMo/Al_2O_3$ catalysts sulfided at 673 K. Also van Veen et al. [14] combined MES and thiophene HDS activity tests of supported CoMo catalysts and they found that even a third species, supported on carbon instead of alumina or silica, exists which exhibits the normal "Co-Mo-S" MES spectrum but twice the thiophene HDS activity of "Co-Mo-S" II. This leads to the conclusion that a simple relation between the amount of "Co-Mo-S" and the activity of the catalysts cannot be established. Such a relation is only observed within a group of catalysts. So, one cannot deduce the activity of a catalyst only from its MES spectrum. Additional knowledge is required about the preparation and sulfidation of the catalyst under study, thus underlining the previous remark that it is very important to provide experimental details together with the results of characterization studies.

Finally, the question is addressed whether the structural model proposed in this investigation applies to other hydrotreating catalysts as well. Therefore, we will consider the Mössbauer absorption spectroscopy (MAS) studies of Ramselaar et al. [26,27,36] of Fe and FeMo catalysts supported on both carbon and alumina. These authors in fact were the first to apply a stepwise sulfidation procedure. Many of the results they obtained for the Fe-based catalysts apply to the Co-based catalysts too. A "Fe-Mo-S" doublet is observed after sulfidation of Fe catalysts (without Mo) at intermediate sulfidation temperatures. The same *trend* between the Q.S. value and the particle size is observed. Also the influence of Mo on the behaviour of the Fe resembles that of the influence of Mo on the Co catalysts. Thus, the results of the Fe-based catalysts lead to the same structural model. From 4.2 K in-situ measurements the proposed relation between the Q.S. value and the size of the Fe(Co)-sulfide particles could be established [27]. Nevertheless, in some aspects the behaviour of the Fe catalysts differs essentially from that of their Co counterparts. First of all, if the relation between the Fe-loading in Fe/C and the mean particle size of the Fe-species (oxidic as well as sulfidic) is examined, a minimum in the mean particle size is observed at a Fe-loading of about 2 wt% Fe/C. At both higher and lower Fe-loadings a lower dispersion is observed. This behaviour clearly differs from that of the Co/C catalysts where the dispersion of the Co-species continuously increases with decreasing Co-loading down to Co-loadings in the range 10^{-4} wt% [32]. Another important difference between the Fe and the Co catalysts is that Fe shows a much stronger tendency to diffuse into the alumina support material in both catalysts with and without Mo. If one keeps these differences in mind, the Fe system is very suitable to gather structural information of the group VIII-sulfide species in hydrotreating catalysts. In this connection, it may be mentioned that both ^{57}Fe and ^{57}Co do have certain advantages with respect to each other. The (non radioactive) ^{57}Fe can be applied easier both at room temperature and at 4.2 K. Furthermore, the assignment of subspectra in ^{57}Fe MAS experiments is not masked by problems with after-effects which may occur in the MES measurements. Finally, ^{57}Fe is much cheaper than ^{57}Co. An advantage of Co over Fe is that the spectrum of the thermodynamically stable Co-sulfide (Co_9S_8 under all applied conditions) differs considerably from that of the fresh catalyst, whereas the spectra of the Fe-sulfides (FeS_2 and $Fe_{1-x}S$ depending on the stage of the process) are rather similar to those of the fresh Fe(III)-oxyde samples, which hampers a straightforward interpretation of the MAS spectra. Room temperature spectra of Co catalysts are therefore in general easier to interpret than those of Fe catalysts. The thermodynamically stable Co-sulfide Co_9S_8 is not magnetic, so, (difficult to perform) 4.2 K measurements of Co-based catalysts are less relevant than such measurements of Fe-based catalysts. As Co is generally applied in commercial hydrotreating catalysts, characterization studies of Co-based catalysts attract more attention than those of their Fe-based counterparts.

Also characterization studies of Ni-based catalysts, which are especially used for hydrotreatment of feeds containing much nitrogen attract more attention. Therefore, the same systematic method used in this investigation has been applied for a preliminary investigation of Ni-based hydrotreating catalysts. As Ni has no suitable Mössbauer isotope, we added ^{57}Co at a Co/Ni ratio of about 10^{-4} at/at as a probe for the Ni. It turned out [37] that interesting information concerning the local structure of the Ni atoms can be derived from *trends* in the MES spectra. These trends can also be interpreted in terms of the structural model we proposed. Such 'doping' experiments on Ni(-Mo) catalysts have also been applied with ^{57}Fe dope, now the ^{57}Fe/Ni ratio was about 0.05 at/at. As ^{57}Fe is experimentally easier to use and is relatively cheap, more extensive studies could be performed. The results agreed very well with those obtained with ^{57}Co [37].

In conclusion, results of previous characterization studies can be understood in terms of the structural model proposed. Furthermore, in all cases it is of great importance to use *trends* in characterization studies of catalysts. Results of characterization studies of catalysts are of limited value if the experimental details necessary to repeat the experiments so as to compare those results are not provided.

References

[1] R.J.H. Voorhoeve and J.C.M. Stuiver, J. Catal., 23:243 (1971).
[2] A.L. Farragher and P. Cossee, in Proc. 5th Intern. Conf. on Catal., (Palm Beach, 1972), (J.W. Hightower, ed.), North-Holland, Amsterdam, 1973, p. 301.
[3] B.C. Gates, J.R. Katzer and G.C.A. Schuit, in Chemistry in Catalytic Processes, McGraw-Hill, New-York, 1979, p. 411.
[4] B. Delmon, in Proc. 3rd Conf. on the Chem. and Uses of Molybdenum, (Ann Arbor, 1979), (H.F. Berry and P.C.H. Mitchell, eds.), Climax Molybdenum Company, Ann Arbor, 1979, p. 73.
[5] H. Topsøe and S. Mørup, in Proc. Int. Conf. Mössbauer Spectr, (Cracow, Poland, 1975), (A.Z. Hrynkiewicz and J.A. Sawicki, eds), Cracow, 1975, vol. 1, p. 305.
[6] B.S. Clausen, S. Mørup, H. Topsøe and R. Candia, J. Phys., 37:C6-249 (1976).
[7] H. Topsøe, B.S. Clausen, R. Candia, C. Wivel and S. Mørup, J. Catal., 68:433 (1981).

[8] C. Wivel, R. Candia, B.S. Clausen and H. Topsøe, J. Catal., 68:453 (1981).
[9] R. Prins, V.H.J. de Beer and G.A. Somorjai, Catal. Rev.-Sci. Eng., 31:1 (1989).
[10] H. Topsøe and B.S. Clausen, Catal.Rev.- Sci.Eng., 26:395 (1984).
[11] S.M.A.M. Bouwens, J.A.R. van Veen, D.C. Koningsberger, V.H.J. de Beer and R. Prins, J. Phys. Chem., 95:123 (1991).
[12] W.L.T.M. Ramselaar, M.W.J. Crajé, E. Gerkema, V.H.J. de Beer and A.M. van der Kraan, Bull. Soc. Chim. Belg., 96:931 (1987).
[13] R. Candia, H. Topsøe and B.S. Clausen, in Proc. 9th Iberoamerican Symp. on Catalysis, (Lisbon, 1984), Lisbon, 1984, p. 211.
[14] J.A.R. van Veen, E. Gerkema. A.M. van der Kraan and A. Knoester, J. Chem. Soc., Chem. Commun., :1684 (1987).
[15] A.M. van der Kraan, M.W.J. Crajé, E. Gerkema, W.L.T.M. Ramselaar and V.H.J. de Beer, Hyp. Int., 46:567 (1989).
[16] M.W.J. Crajé, E. Gerkema, V.H.J. de Beer and A.M. van der Kraan, in Studies in Surf. Science and Catalysis 50, (M.L. Occelli and R.G. Anthony, eds.), Elsevier, Amsterdam, 1989, p. 165.
[17] M.W.J. Crajé, V.H.J. de Beer and A.M. van der Kraan, Appl. Catal., 70:L7 (1991).
[18] M.W.J. Crajé, V.H.J. de Beer and A.M. van der Kraan, Bull. Soc. Chim. Belg., 100:953 (1991).
[19] M.W.J. Crajé, S.P.A. Louwers, V.H.J. de Beer, R. Prins and A.M. van der Kraan, J. Phys. Chem., 96:5445 (1992).
[20] H. Topsøe, R. Candia, N.-Y. Topsøe and B.S. Clausen, Bull. Soc. Chim. Belg., 93:783 (1984).
[21] J.P.R. Vissers, Ph.D. thesis, Eindhoven University of Technology, Eindhoven (1985).
[22] A.M. van der Kraan, Phys. Stat. Sol. (a), 18:215 (1973).
[23] M.W.J. Crajé, Ph.D. thesis, Delft University of Technology, (I.S.B.N. 90-73861-08-X), Delft, (1992).
[24] S. Kasztelan, J. Grimblot and J.P. Bonelle, J. Phys. Chem., 91:1503 (1981).
[25] W.L.T.M. Ramselaar, R.H. Hadders, E. Gerkema, V.H.J. de Beer, E.M. van Oers and A.M. van der Kraan, Appl. Catal., 51:263 (1989).
[26] W.L.T.M. Ramselaar, M.W.J. Crajé, E. Gerkema, V.H.J. de Beer and A.M. van der Kraan, Appl. Catal., 54:217 (1989).
[27] P.J. Mangnus, Ph.D. thesis, University of Amsterdam, Amsterdam, (1991).
[28] M. de Boer, Ph.D. thesis, Utrecht University, Utrecht, (1992).
[29] M.W.J. Crajé, A.A. de Jong, J.W. Niemantsverdriet, V.H.J. de Beer and A.M. van der Kraan, to be published.
[30] H. Toulhoat and S. Kasztelan, in Proc. 9th Int. Congr. on Catal., (Calgary 1988), (M.J. Phillips and M. Ternan, eds.), Calgary, 1988, p. 152.

[31] M.W.J. Crajé, V.H.J. de Beer, J.A.R. van Veen and A.M. van der Kraan, J. Catal., 143:601 (1993).
[32] M.W.J. Crajé, V.H.J. de Beer, J.A.R. van Veen and A.M. van der Kraan, Appl. Catal. A: General, 100:97 (1993).
[33] H. Topsøe, B.S. Clausen, N.-Y. Topsøe and P. Zeuthen, Stud. Surf. Sci. Catal., 53:77 (1990).
[34] S.M.A.M. Bouwens, R. Prins, V.H.J. de Beer and D.C. Koningsberger, J. Phys. Chem., 94:3711 (1990).
[35] N.-Y. Topsøe and H. Topsøe, J. Catal., 84:386 (1983).
[36] W.L.T.M. Ramselaar, M.W.J. Crajé, R.H. Hadders, E. Gerkema, V.H.J. de Beer and A.M. van der Kraan, Appl. Catal., 65:69 (1990).
[37] M.W.J. Crajé, V.H.J. de Beer and A.M. van der Kraan, Catal. Today, 10:337 (1991).

Properties of MoS$_2$-Based Hydrotreating Catalysts Modified by Electron Donor or Electron Acceptor Additives (P, Na, Cs, F)

R.Hubaut [1], O.Poulet [1,3], S.Kasztelan [2], E.Payen [1] and J.Grimblot [1]

1-*Université des Sciences et Technologies de Lille*
Laboratoire de catalyse hétérogène et homogène
URA CNRS 402 Bât. C3
59650 Villeneuve d'Ascq Cédex, France

2-*Institut Français du Pétrole*
1-4 -Avenue de Bois-Préau, BP 311
92506 Rueil-Malmaison Cédex, France

3-*Present adress: Centre de recherches Total*
Gonfreville-l'Orcher, 76700 Harfleur, France

ABSTRACT

A conventional MoS$_2$/γ-Al$_2$O$_3$ hydroprocessing catalyst has been tested in hydrogenation of conjugated dienes and of toluene and in hydrodenitrogenation of pyridine to probe the influence of elements like phosphorus, fluorine or alkaline ions added into the catalyst formulation. The activity depends on the nature and concentration of the additive and also on the temperature (T_R) at which the catalysts have been submitted to hydrogen after sulfidation but prior the tests for creating coordinative unsaturated sites

(CUS). The results suggest that the additives directly interact with the MoS_2 supported phase by poisoning the sites and/or by blocking the labile sulfur species which are progressively removed during the activation pretreatment under H_2. As a consequence, the studied additives influence both the local structure of the sites of the MoS_2 phase as well as their number.

1.INTRODUCTION

The hydrotreating reactions performed under hydrogen pressure play a very important role in petroleum refining for producing clean fuels well adapted to the market and to environmental legislations. They consist in the hydrogenation (HYD) of unsaturated compounds and in the removal of heteroatoms chemically bound to the hydrocarbon molecules, namely sulfur (hydrodesulfurization-HDS), nitrogen (hydrodenitrogenation-HDN), oxygen (hydrodeoxygenation-HDO) or metals (hydrodemetalization-HDM).Due to the high bond energy between carbon and the heteroatom on one hand and the aromatic character of the molecules to be transformed on the other hand, the hydrotreating reactions have to be performed under severe conditions (i.e. high hydrogen pressure and temperature ranging from 570 to 770K) with adapted catalysts. The standard catalyst formulations contain the MoS_2 or WS_2 active phase promoted by Co or Ni and supported on transition alumina. It is now admitted that a strong interaction exists between the promoter and the transition metal disulfide to form in optimized conditions the so-called CoMoS or NiMoS phases (1). The activity of the resulting solid may be simultaneously due, among numerous reasons, to an adapted bond energy between S and the transition metal by reference to the Bond Energy Model proposed by Topsøe and coworkers (2) and to the capacity of the catalyst to activate hydrogen and transport it by spillover to the active centers in relation to the Remote Control Model of Delmon (3).In recent catalysts formulations, other elements like phosphorous (4-14), alkaline ions (15-17) or fluorine (18-22) are added with the objective to improve activity. However, in some cases, their presence is rather negative. Therefore, their role is still matter of debate, particularly because their loading, the conditions of their incorporation, the calcination and sulfidation steps may have major importance.

The aim of the present work is to study the role of additives considered, at first, as electron donors or electron acceptors on the activity and selectivity for transforming modele molecules such as isoprene , pentadiene or toluene (HYD) and pyridine (HDN) of a $MoS_2/\gamma\text{-}Al_2O_3$ reference catalyst without classical promotor which could introduce difficulties in interpretation. The catalysts are pretreated under hydrogen at various temperatures (T_R). This procedure allows to create coordinative unsaturated sites (CUS) (23-24) which nature and number depend on the Mo-S bond energy (in relation with the measured S/Mo ratio) and on the rate of hydrogen spillover.

II. EXPERIMENTAL

IIA. Catalyst preparation

The $MoO_3/\gamma-Al_2O_3$ (14 wt % Mo) catalyst was prepared by the pore filling impregnation method using $\gamma-Al_2O_3$ extrudates (252 $m^2.g^{-1}$; 0.6 $cm^3.g^{-1}$) and a $(NH_4)_6Mo_7O_{24}, 4H_2O$ solution of natural pH (~5.5). Calcination was performed in air at 773K for 4h. The Mo-P/Al_2O_3 catalyst was obtained by simultaneous impregnation of a H_3PO_4 and ammonium heptamolybdate solution at a pH of ~2. The Mo-F, Mo-Na, and Mo-Cs /Al_2O_3 catalysts were obtained after a second impregnation step of the previous MoO_3 /Al_2O_3 sample using respectively HF, $NaNO_3$ and $CsNO_3$ solutions. In all the cases, the final calcination was always performed in air at 773K for 4h. The amount of P and F were respectively 2.8 and 2.0 wt % whereas loadings up to 5 wt % were used for Na and Cs. The oxidic samples were sulfided at 623K for 4h with a H_2/H_2S (90/10) mixture and treated with purified H_2 at various temperatures (T_R), from 623 to 973K, for 12 h before the catalytic tests without any intermediate exposure to air.

IIB. Catalytic tests

Dienes (isoprene or cis penta-1,3-diene) HYD tests were carried out at 323K in an all glass set-up at atmospheric pressure with a H_2 flow rate of 30 $l.h^{-1}$ and a diene partial pressure of 20 torrs. Toluene HYD and pyridine HDN were performed in a high pressure catalytic flow reactor with a sulfur-free feed at respectively 6 Mpa, 623K and a LHSV of 1.8 h^{-1} for toluene in hexane (70/30 v/v) and 3.3 Mpa, 573K and a LHSV of 2 h^{-1} for pyridine in decaline (90/10 v/v). The reactants were supplied by Fluka (grade > 99%) and H_2 and H_2S by Air Liquide. The products were analysed by gas chromatography with a FID.

IIC. Physical characterizations

IIC1. X-Ray Photoelectron Spectroscopy. X Ray photoelectron spectra were obtained by using a AEI ES 200 B spectrometer equipped with an Al X-ray source working at 300W. Binding energy (BE) was referenced against the Al 2p level at 74.8eV, a value generally encountered for $\gamma-Al_2O_3$.

IIC2. High Resolution Electron Microscopy. The sulfided powder was ultrasonically dispersed in hexane for the preparation of the electron microscope grid covered with a thin carbon holed film. The instrument was a JEOL 100 CX microscope with a double side entry tilt which provides an instrument resolution over a continuous range of periodicities down to 0.3nm.

III. RESULTS

IIIA. Chemical analyses and XPS results of the sulfided catalysts

It is important to determine whether the considered elements added to the catalyst formulation influence the extent of sulfidation of the molybdenum oxo-species initially present on the support after calcination and/or the amount of labile sulfur removed during H_2 treatment when the MoS_2 phase has been formed. The initial S/Mo ratio can be determined from chemical analysis of the sulfided catalytst and it is controlled by XPS analysis which gives elemental composition from the ratio between the S 2p peak over the Mo 3d doublet (with only the Mo^{IV}/MoS_2 component).The second aspect of analysis is given by the S/Mo ratio obtained after the H_2 treament before tests and according to a procedure already described (24).

Table 1: S/Mo atomic ratio

Catalyst	S/Mo (a)	S/Mo (b)	% S removed
Mo	2.3	1.2	48
Mo-P	1.55	0.8	48
Mo-Na(1) (c)	2.2	1.3	41
Mo-Cs(5) (c)	2.4	1.7	30
Mo-F	2.0	1.5	25

(a) Analyses performed after sulfidation
(b) Analyses performed after treament by H_2 at 973K
(c) (1) and (5) indicate the Na and Cs loadings (wt%)

Table 1 shows that, after sulfidation, the reference solid MoO_3/Al_2O_3 has a S/Mo ratio of about 2.3 which is in agreement with the presence of small MoS_2 crystallites on the alumina support. This ratio does not change very much when Na or Cs are present whereas F (S/Mo=2.0) and P (S/Mo=1.55) noticeably influence sulfidation of the Mo-oxo species deposited on the support. Treatment under H_2 at 973K removes from 25 (Mo-F) to 48% (Mo and Mo-P) of the amount of S present after the sulfidation step. This S removal, as H_2S, provokes the formation of anionic vacancies, the number of which is obviously depending on the nature of the additive introduced in the catalyst formulation. The effect is more pronounced with P than with F and appears moderate with the alcaline ions.

Figure 1 presents typical photoelectron spectra of the Mo species present on the catalysts after sulfidation. The Mo^{VI} remaining species have typical binding energies at 235.5 (Mo $3d_{3/2}$) and 232.5eV (Mo $3d_{5/2}$) whereas the peaks of the Mo^{IV} entities in the MoS_2 structure are shifted to 232.0 and 229.0eV respectively. Presence of the S 2s component is detected on the low BE side. It is clear that a large part of the Mo-oxo species are transformed into Mo^{IV} in a sulfidic environment whatever the catalyst; the proportion of remaining Mo^{VI} varies from 15% in the reference solid (no additive) to 22%

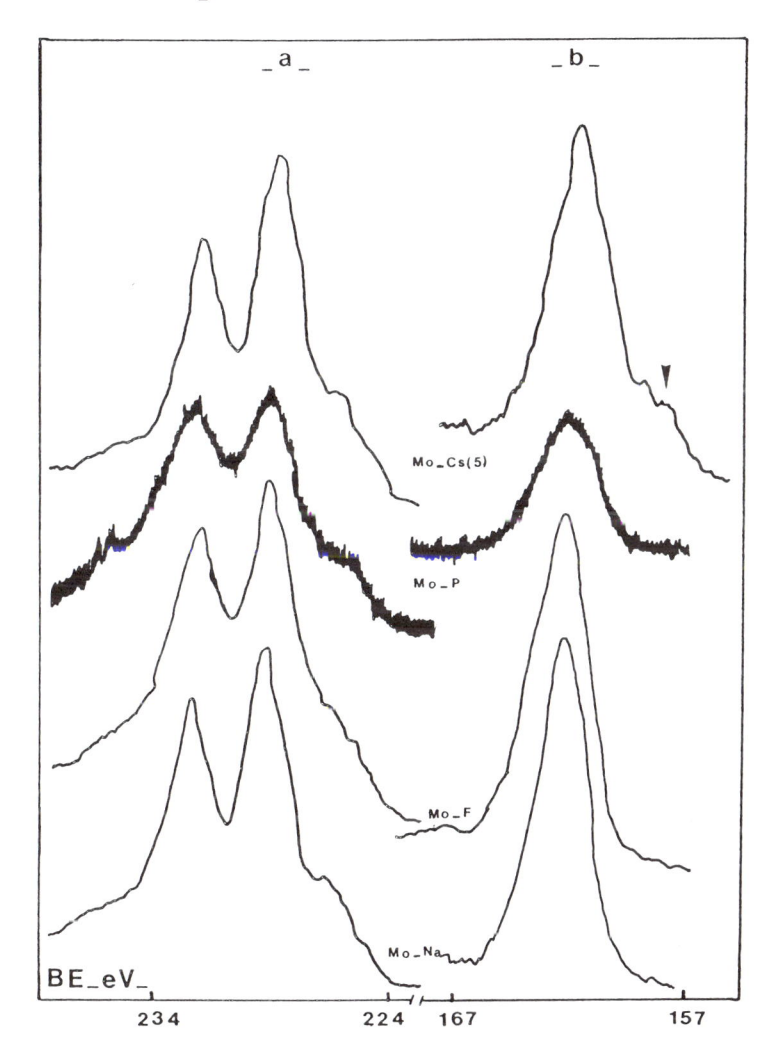

Figure 1-XPS of (a) the Mo $3d_{3/2-5/2}$ doublet and the S 2s component on the low BE side and (b) the S 2p level of the sulfided catalysts.

in the Mo-P catalyst. This observation has been already reported and was assigned to the presence of residual (Al_2O_3)-P-O-Mo^{VI}-(MoS_2) bonds (14). Such a direct interaction does not seem to occur between MoS_2 and the alkaline ions. However, when rather large amounts of Cs are present (5 wt %,sample Mo-Cs(5)), a new S 2p peak at 159.5eV is detected with the classical peak of MoS_2 at162.0eV. This new species is assigned to the presence of Cs sulfides in which the strong electronegativity of Cs makes the electron density on S higher than that found on more covalent sulfides (i.e. MoS_2). In presence of F, the residual amount of Mo^{VI} is higher than in the Mo reference catalyst which suggests incomplete sulfidation of Mo; this in agreement with the results of table 1 where the ratio S/Mo is 2.0 instead of 2.3. Nevertheless, the S 2p photoelectron peak remains unique at162.0eV, indicating that F does not electonically influence the sulfur species but rather takes the place of sulfur.

IIIB. Morphology of the supported MoS_2 phase

HREM permits the detection of the small MoS_2 particles supported on alumina. Based on the detection of more than 400 crystallites on several micrographs of each sample, repartition of the length L of the elemental layers as well as their stacking N can be obtained (25). The results reported in table 2 concern the mean $\bar{L}$ and $\bar{N}$ values deduced from the micrographs. The most striking observation concerns the very weak, if any, influence of the presence of the additives on the mean length $\bar{L}$ of MoS_2 (32 to 38Å). On the contrary, the presence of the additives induces a more marked stacking and sometimes N values higher than 5 layers are observed; the mean values $\bar{N}$ vary between 1.4 for the reference sample to 2.7 when the additive is F. The sequence is:

$$\bar{N}_{Mo} < \bar{N}_{Mo-Na} < \bar{N}_{Mo-Cs} < \bar{N}_{Mo-P} < \bar{N}_{Mo-F}$$

This effect is, among other reasons, probably related to weaker interactions between the alumina support and the supported MoS_2 patches. Consequently, the additives have also an influence on the support properties. Such a stacking could also modify the sulfur removal during the treatment with.H_2.

IIIC. Catalytic performances

The presence of additives to Mo in the catalytic formulation induces changes in the catalytic results as expected from the physical

Table 2: Morphology evolution of the MoS_2 deposited phase as a function of the nature of the added element. $\bar{L}$ and $\bar{N}$ refer to the mean values of length and number of layers respectively.

catalyst	N	L
Mo	1.4	35
Mo-P	2.5	32
Mo-Na(1)	1.8	38
Mo-Cs(5)	2.4	34
Mo-F	2.7	33

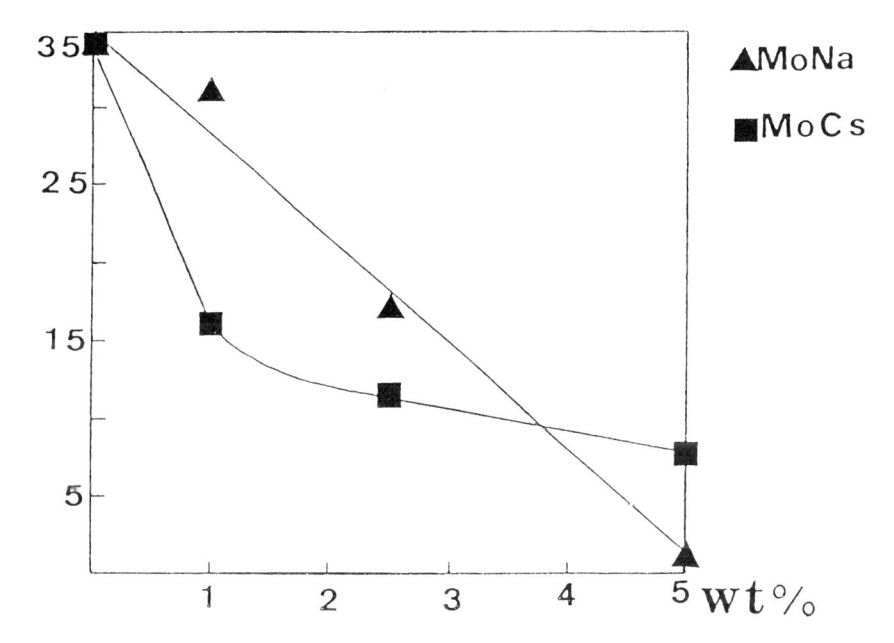

Figure 2 : Cis penta-1,3-diene hydrogenation activity (in 10^3 mol.g^{-1}.h^{-1}) of MoS$_2$- based catalysts treated under H$_2$ at 773K versus the amount of added alkaline ion in wt % of the element.

characterizations reported in sections IIIA and IIIB; they play an important role in the structure and number of active sites. This influence depends on the nature and electronic properties of the added element, on the type of reaction considered and on the loading. In particular, HYD activity of cis penta-1,3-diene decreases with the amount of added alkaline ions; this decrease is initially more pronounced with Cs, very electropositive, than with Na (fig. 2).These results are in agreement with the evolutions shown by Muralhidar et al.(26). However, the influence of the presence of additives is more significant if the catalytic data examined with comparable amounts of alkaline ions defined by the number of atoms per surface area (nm^2) of support. Data for such a comparison have been reported on table 3. Isoprene HYD activity exhibits larger differences than toluene HYD and pyridine HDN.

Phosphorous induces a decrease of diene HYD and pyridine HDN activities whereas its presence enhances toluene Hyd activity by a factor of ~2 by reference to the Mo standard catalyst. Such a result, associated with a more pronounced stacking of the MoS$_2$ layers in Mo-P (table 2) has been ascribed to some ordering of the layers by the presence of remaining P O Mo bridges bound to alumina (14). Adsorption of planar toluene at the edges of the MoS$_2$ platelets, or more selectively at the top edges if the rim-edge model

Table 3 : Influence of the presence of P, Na, Cs or F additives in a Mo/Al$_2$O$_3$ catalyst on the activity (in mol.g^{-1}.h^{-1}) for various reactions

catalyst	isopreneHYD (a)	tolueneHYD (b)	pyridineHDN (c)
Mo 8wt % 2.31 at.nm^{-2}	23 x 10^{-3}	0.8 x 10^{-7}	9 x 10^{-7}
Mo-P(2.8 wt%) 2.18 P At.nm^{-2}	15 x 10^{-3}	1.7 x 10^{-7}	6 x 10^{-7}
Mo-Na(1wt%) 1.03 Na At.nm^{-2}	22 x10^{-3}	0.7 x 10^{-7}	7 x 10^{-7}
Mo-Cs(5wt%) 1.0 Cs At.nm^{-2}	4 x 10^{-3}	0.5 x 10^{-7}	2 x 10^{-7}
Mo-F(2wt%) 2.5 F At.nm^{-2}	2 x 10^{-3}	0.8 x 10^{-7}	5 x 10^{-7}

The temperature of treatment under H$_2$ of the sulfided catalysts is **(a)** : 820K, **(b)** : 623K, **(c)** :623K

of Daage and Chianelli (27) is considered, through a π-complex intermediate is thus facilitated and deserves some analogy with thiophenic intermediates created during HDS (3,28).

In presence of any of the other tested additives, activity is always smaller than the one obtained with the reference MoS$_2$ supported catalyst whatever the reaction. For the Mo-Na catalyst, isoprene HYD is hardly affected and the product distribution is quite similar to the one obtained with the Na-free sample: isopentane is always the major product. On the contrary, in presence of Cs, HYD activity of isoprene decreases with changes in selectivity: 2-methyl but-1-ene, the primary product following Lebedev'rule is dominant. On the other hand, isomerization of trans penta-1,3-diene into the cis isomer, which needs a Lewis acid site, is less important when alkaline ions are present in the catalyst. Fluorine also causes a drastic decrease of the isoprene HYD activity but does not greatly affects toluene HYD. This sharp decrease is probably the result of a rapid and important poisoning coming from cracking or more certainly from polymerized dienes near the active sites. Such a poisoning is far less important with molecules like pyridine or toluene of lower reactivity.

IV.DISCUSSION

The results presented in this paper reveal that the additives under investigation (P, F, Na, Cs) have generally several effects,both in catalytic activity and selectivity and in the MoS$_2$ morphology and reactivity (lability of sulfur) when incorporated to a conventional Mo/Al$_2$O$_3$ catalyst. These effects can be combined altogether and it is not excluded that some of them are related to modifications induced on the support (i.e. stacking of MoS$_2$, diene polymerization ...).

Properties of MoS₂-Based Hydrotreating Catalysts

The activity level and the product distribution during HYD and isomerization at low temperature of cis penta-1,3-diene show a strong dependence with the extent of sulfur unsaturation occuring at the MoS$_2$ edge planes and controlled by treating under H$_2$ the sulfided catalysts at different temperatures T$_R$. Higher T$_R$ is, lower the S/Mo stoichiometry is and more lacunar the sites are. Schematically, as the S/Mo decreases, and considering the relative Mo-S bond energies at the ($10\bar{1}0$) and ($10\bar{1}\bar{0}$) edge planes, 2-CUS are first generated at the ($10\bar{1}0$) edge plane followed by formation of 3-CUS and 4-CUS. A satisfactory fit has been found between the maximum of activity of penta-1,3-diene isomerization and the statistical optimum of the presence of 2-CUS whereas maximum HYD activity corresponds to the optimum in the presence of 3-CUS (29). Even if the definition of active sites is over-simple (only S unsaturation are considered), it is obvious that sulfur mobility has a primary effect in relation with the results presented here. Therefore, the presence of additives which may have a real influence on the Mo-S bond energy leads to different repartitions of anionic vacancies for a given T$_R$ and, consequently, to different activities and selectivities. This argument implies that the additives are localized very close to the MoS$_2$ phase. A previous study (30) on HYD of isoprene with a standard MoS$_2$/Al$_2$O$_3$ catalyst has shown, through the product distribution (-1,2- ;-1,3- and -1,4- addition within monohydrogenation), that an ensemble of at least 2 adjacent Mo atoms with coordinative unsaturations has rather to be considered as the active sites. Presence of additives may, as well, modify the structure of these ensembles and change the repartition of the sulfur vacancies.

The presented results can be compared with the «Bond Energy Model» for sulfides proposed by Topsøe et al. (2) in which the bond energy between sulfur and the metallic center, and consequently the number of anionic vacancies around it, are depending on the nature of the transition metal. Indeed, HDS activities are related to the differences in the CUS concentration. Changing the nature of the transition metal corresponds, by analogy, to the change, for given Mo-S bonds, of temperature T$_R$. As a matter of fact, a typical Balandin-type volcano curve, noticeably for isoprene HYD activity versus the pretreatment temperature T$_R$ has been obtained (24). Moreover, for a given temperature T$_R$, if activity and selectivity patterns are modified by presence of additives, it is inferred that the distribution of active sites will be modified.

The number of active sites is also an important aspect to take into account, particularly when the number of stacked layers of MoS$_2$ in the crystallites increases. It is expected that sites at the edges of MoS$_2$ layers inside the crystallites are less accessible to reactants and such a situation leads to a decrease of the catalytic activity. This aspect is similar to the «Rim-Edge» model of Daage et al. (27) where the most accessible MoS$_2$ layers provide the effective active sites. Stacking of MoS$_2$ in Mo-P, Mo-F and Mo-Cs systems, as revealed by HREM, is clearly related to the isoprene HYD activity decrease. Lower activity with Mo-Cs is due to the presence of Cs sulfides near the active sites, whereas the results with F possibly indicate replacement of sulfide by F, leading to a decrease of anionic vacancies around Mo.

Besides the needs of an optimum number of anionic vacancies constituting the site with the ensemble of Mo atoms, it is necessary to provide hydrogen species for hydrogenating the unsaturated molecules through, at least partly, hydrogen spill-over according to the «Remote Control Model» proposed by Delmon (3).It was already shown that, in the case of MoS_2/γ-Al_2O_3 catalysts, the support plays a role in the storage of reactive hydrogen (31) and consequently has an indirect, but significative, influence on the catalytic data. Therefore, any additive present on the support could also modify the catalytic behaviors of a MoS_2-based catalyst. Presence of additive acting on the alumina support properties is evidenced, in the light of the results presented in this paper, for F and P. the polymerization and/or cracking of conjugated diene, which drastically influences HYD activity was found when F is incorporated in the catalyst formulation. F on the alumina support modifies its acidic properties by changing the acid site density and also probably their strength. Changes in the MoS_2 layer stacking, and more particularly the detection of residual $P-O-Mo^{VI}$ bridges (14) after sulfidation of the Mo-P catalyst show that P is also directly interacting with the alumina support.

In table 4 are schematically summarized the modifications induced by the presence of additives in the solid formulation and the consequences they induced on the catalytic properties.

The final aspect we wish to consider in this discussion is relative to the mode of adsorption of some probe molecules on the CUS and to the mechanistic aspects of the molecule transformation. Only the molecule having an aromatic character -toluene and pyridine - will be considered. On a conventional MoS_2 supported catalyst, toluene HYD activity decreases very slowly when T_R increases (32). At the maximum activity, the ratio S/Mo is about 1.9 which statistically corresponds to a 2-CUS. With such a low sulfur deficient site, it is expected that the reactive toluene intermediate is adsorbed through the distorted ring having partly lost its aromaticity with formation of a π-surface complex like in orgonometallic compounds modelling HDS of thiophene where the heterocyclic adsorption mode is η^4 (33).this mode requires at least two sulfur vacancies. In such a view, an additive like P which favors formation of 2-CUS to the detriment of 3-CUS has a beneficial effect on toluene HYD (see table 3). A dual mechanism has been proved for pyridine HDN (34). For low T_R, which removes only small amount of sulfur, the reaction network is similar to the one described by Mc Ilvried (35) whereas a second route with no desorption of piperidine occurs at higher T_R. As the level of sulfur unsaturation is different, the mode of pyridine adsorption on the catalyst treated at low or high T_R will change. However, as

Properties of MoS₂-Based Hydrotreating Catalysts

Table 4: Schematic evaluation of the effects induced by the presence of P, Na, Cs and F on the properties of a MoS$_2$/γ-Al$_2$O$_3$ catalyst.

additive	effect on support acidity (a)	MoS$_2$ stacking (b)	active sites (c)	poison effect (d)	toluene HYD	pyridine HDN
P	yes ↗	yes ↗	P-MoS$_2$ bonds	no	+ effect	– effect small
Na	yes	small effect	no effect	no effect	no effect	– effect small
Cs	yes	yes ↗	bulk Cs sulfide present	yes	– effect small	– effect high
F	yes very high	yes ↗	S lability ↘	yes	no effect	– effect

(a) The additives may directly interact with the alumina support and therefore modify the acidic properties and change the density of surface hydroxyls. Na and Cs have a neutralizing effect.
(b) Results obtained by HREM.
(c) Some additives may directly act on the MoS$_2$ active sites through electronic effects; indirect influence is given by considering S mobility which modifies the number of coordinative unsaturation of the sites.
(d) Poisoning of the active phase may be due to coke deposition (with F) or to site blocking by formation of bulk compounds (bulk Cs sulfide).

one transformation route is progressively replaced by the other one, the effects of the presence of the additives are rather complex (table 3).

V. CONCLUSION

Presence of additives of different nature like phosphorous, alkaline ions or fluorine in a MoS$_2$/γ-Al$_2$O$_3$ reference catalyst reveals more or less pronounced effects on the catalytic performances with model molecules. The origins of this influence are likely combined and complex: the additives may directly interact with the supported phase, with the active sites or modify the surface properties of the support. the amount of additive is also an important parameter to take into account. Na ions appear to be the additive which has the less influence on either the MoS$_2$ morphology, the sulfur lability or the catalytic performances.

VI.REFERENCES

1. H.Topsøe, B.S.Clausen, N.Y.Topsøe and P.Zeuthen, in Studies in Surface Science and Catalysis (D.L.Trimm et al, eds.), Elsevier, Amsterdam, 1990, Vol.53, 77.
2. J.K.Nørskov, B.S.Clausen and H.Topsøe, Catal. Lett., 13:1 (1992). H.Topsøe, B.S.Clausen, N.Y.Topsøe, J.Hyldtoft and J.K.Norskov, ACS,38(3):638 (1993).
3. B.Delmon, in Studies in Surface Science and Catalysis (D.L.Trimm et al, eds.),Elsevier, Amsterdam, 1990, Vol.53,1.
4. J.N.Heresnape and J.E.Morris, British Patent, 701,217, (1953).
5. J.D.Colgan and V.Chomitz, U.S. Patent 3,287,280, (1966).
6. C.W.Fitz and H.F.Rase, Ind.Eng.Chem.Prod.Res.Dev., 22:40 (1983).
7. K.Gishti, A.Iannibello, S.Marengo, G.Morelli and P.Titarelli, Appl. Catal.,12:381 (1984).
8. M.McMillan, J.S.Brinen and G.Haller, J.Catal.,97:243 (1986).
9. A.Moralès and M.M.Ramirez de Agudelo, Appl.Catal.,23:23 (1986).
10. S.M.A.M.Bouwens, J.P.R.Vissers, V.H.J.DeBeer and R.Prins, J.Catal., 112:401 (1988).
11. P.Atanasova and T.Halachev, Appl.Catal., 48;295 (1989).
12. R.Lopez-Cordero, S.Lopez-Guerra, J.L.G.Fierro and A.Lopez-Agudo, J.Catal., 126:8 (1990).
13. P.J.Mangnus, V.H.J.DeBeer and J.A.Moulijn, Appl.Catal., 67:119 (1990).
14. O.Poulet, R.Hubaut, S.Kasztelan and J.Grimblot, Bull.Soc.Chim.Belg., 100:857 (1991) and references therein.
15. S.J.Decanio, M.C.Cataldo, E.C.Decanio and D.A.Storm, J.Catal., 119:256 (1989).
16. A.R.Saini, B.G.Johnson and F.E.Massoth, Appl.Catal. 40:157 (1988).
17. J.F.Kelly and M.Ternan, Can.J.Chem.Eng., 57:726 (1979).
18. L.G.Tejuca, C.H.Rochester, A.Lopez Agudo and J.L.G.Fierro, J.Chem.Soc.Faraday Trans.I, 79:2543 (1983).
19. J.L.G.Fierro, A.Lopez Agudo, L.G.Tejuca and C.H.Rochester, J.Chem.Soc.Faraday Trans.I, 81:1203 (1985).
20. P.M.Boorman, R.A.Kydd, Z.Sarbak and A.Somogyvari, J.Catal., 96:115 (1985).
21. P.M.Boorman, R.A.Kydd, Z.Sarbak and A.Somogyvari, J.Catal., 100:287 (1986).
22. H.Matralis,C.Papadopoulou and A.Lycourghiotis, Appl.Catal., 116:221 (1994) and references therein.
23. D.G.Kalthod and S.W.Weller, J.Catal., 98:572 (1986).
24. A.Wambeke, L.Jalowiecki, S.Kasztelan, J.Grimblot and J.P.Bonnelle, J.Catal., 109:32 (1988).

25. E.Payen, R.Hubaut, S.Kasztelan, O.Poulet and J.Grimblot, J.Catal., 147:123 (1994).
26. G.Muralidhar, F.E.Massot and J.Shabtai, J.Catal., 85:44 (1984).
27. M.Daage and R.R.Chianelli, J.Catal., 149:414 (1994).
28. H.Kwart, G.C.A.Schuit and B.C.Gates, J.Catal., 61:128 (1980).
29. S.Kasztelan, L.Jalowiecki, A.Wambeke, J.Grimblot and J.P Bonnelle, Bull.Soc.Chim.Belg., 96:1003 (1987).
30. S.Kasztelan, A.Wambeke, L.Jalowiecki, J.Grimblot and J.P.Bonnelle, J.Catal., 124:12 (1990).
31. L.Jalowiecki, J.Grimblot and J.P.Bonnelle, J.Catal., 126:101 (1990).
32. J.P.Bonnelle, A.Wambeke, A.Kherbeche, R.Hubaut, L.Jalowiecki, S.Kasztelan and J.Grimblot, in Advances in Hydrotreating Catalysts (M.L.Ocelli and R.G.Anthony Eds.), Elsevier, Amsterdam, 1989, p;123.
33. R.J.Angelici, Bull.Soc.Chim.Belg., in press (1995).
34. A.Kherbeche, R.Hubaut, J.P.Bonnelle and J.Grimblot, J.Catal. 131:204 (1991).
35. M.G.McIlvried, Ind.Eng.Chem.Proc.Des.Dev., 10:125 (1971).

Modeling HDS Catalysis with Sulfido Bimetallic Clusters

M. David Curtis
Department of Chemistry
The University of Michigan
Ann Arbor, MI 48109-1055

I. Introduction

A. **HDS Catalysts.** Hydrodesulfurization (HDS) of fossil fuels continues to attract a great deal of research attention [1]. The HDS reaction, eq. 1, is arguably one of the largest scale catalytic reactions practiced world wide in the petroleum industry [2]. During HDS, C-S bonds in organic sulfur compounds are cleaved and the sulfur is then removed from the catalyst surface as H_2S. In addition to C-S bond hydrogenolysis, hydrogenation of unsaturated molecules and hydrocracking of larger molecules into smaller ones can also occur. The catalyst normally used in an industrial setting consists of a sulfided mixture of Mo and Co on a high surface area support, usually γ-alumina, although other combinations, e.g. W/Ni, are also used. The late transition metal (TM), Co or Ni, is referred to as the "promoter", although Mo can "promote" Co as well [1]. The nature of the sulfided catalyst has been the subject of research and controversy for over 30 years.

About 17 years ago, Massoth reviewed the state of knowledge of these metal sulfide catalysts [3]. At that time, even the structure of the catalyst surface was not well established and essentially nothing was known about the actual reaction mechanism(s) responsible for the catalytic action in the HDS reaction. The 1980's witnessed the application of a battery of new spectroscopic techniques to the study of the promoted MoS_2-based hydrotreating catalysts. These new methods, including EXAFS, HREM, PES, and Co-emission Mössbauer spectroscopy (MES), along with more traditional techniques, e.g. FTIR, Raman, and chemisorption studies, eventually led to a model of the catalyst surface [4,5]. The model that currently enjoys a rather wide acceptance is the 'CoMoS' phase model, developed by Tøpsoe et al..

According to this model, the surface of the Al_2O_3 is covered by small crystallites that resemble the layered structure of MoS_2. The Co is found in several phases, the relative proportions of which depend, inter alia, on the absolute Co-loading, the Co/Mo ratio, and the calcination temperature. Very low Co loadings and high calcination temperatures promote the insertion of Co into vacancies in the alumina lattice; this Co is apparently inactive in the HDS reaction. As the Co loading increases a new "CoMoS" phase, identified by its MES signature, builds up. At even higher Co loading, the Co_9S_8 phase becomes an important Co-containing phase, and this pentlandite structure has low activity for HDS catalysis [4].

According to Tøpsoe et al., the CoMoS phase consists of Co-atoms decorating the edges of the basal planes of the MoS_2 crystallites. Although there is no direct evidence that the Co actually resides on the edges of the MoS2 crystallites on the high surface area supports, Chianelli et al. have shown unequivocally that the edges of the basal planes are the preferred loci of the Co-atoms adsorbed onto bulk MoS_2 crystals [7,8], and NMR evidence also strongly supports the edge location [9].

Even if Topsøe's "CoMoS" model accurately depicts the physical nature of the catalyst surface, the exact role of the promoter ion is still unknown. Several hypotheses have been advanced for the role of Co. Chianelli et al. [8] find that the HDS activity of various binary metal sulfides correlates with the $\Delta H_f°$ of the sulfide in that a typical "volcano" plot is obtained when the activity is plotted vs. $\Delta H_f°$. (If the M-S bond is either too strong or too weak, activity drops. In the former case, the surface doesn't release S for the next cycle, and in the latter the S is not abstracted from the organic sulfide.) These authors point out that the average of the Co-S and Mo-S bond strengths falls near the maximum of the volcano curve, i.e. the Co modifies the Mo-S bond strength. This effect has been justified theoretically [11].

Other proposals for the role of the promoter ion(s) include (a) better reduction of the Mo (or W if WS_2) [11,12] thereby, creating more sulfur vacancies (active sites) in the MoS_2 structure, (b) the dispersion of the MoS_2 crystallites is increased by the promotor [13], and (c) the Co ions promote the dissociation of H_2 which "spills over" onto the MoS_2, facilitating either S-removal as H_2S or hydrogenation of the organic sulfide [14]. Mechanism (c) is related to (a), but is more focused on a kinetic effect as opposed to a structural effect. Indeed, much of the recent research has been directed toward structural elucidation, but important kinetic effects (5 - 15 kcal/mol) may not be reflected by detectable structural changes.

Since the actual physical nature of the CoMoS catalyst is not known with certainty, it is little wonder that there are as many proposed mechanisms for the HDS reaction as there are proposals for the active site. Thiophene is often used as a probe molecule to test the HDS activity of catalysts. Even the mode of thiophene adsorption on the catalyst surface is in dispute. Some proposals have the thiophene S-bonded to sulfur vacancies on the MoS_2, other propose an η^2-$(C_2$-$C_3)$ bound thiophene, and still others an η^5- (π-bound) mode of adsorption. These various proposals are summarized in references 1 and 5. Various reaction sequences have also been proposed. The S-bound model predicts that the primary product of thiophene HDS is butadiene, which is then hydrogenated to butenes and butanes. Either butadiene or butenes could be the primary product of HDS if the thiophene were η^2- or η^5- bound. Experimentally, both butadiene and 1-butene have been claimed to be the primary product of HDS of thiophene, even over the "same" catalyst. Thus, although much has been learned of the nature of HDS catalysis in recent years, many questions remain unanswered, particularly concerning the role of the promoter(s) in the reaction mechanism.

B. Contributions of Organometallic Chemistry. Organometallic chemistry aids the study of heterogeneous catalysis in several ways [15]. At the most fundamental level, organometallic structures provide models for possible surface species. Secondly, the spectroscopic properties of organometallic compounds often resemble those of corresponding surface species, e.g. ν_{CH} of μ_3 - ethylidyne, M-NO or M-CO vibrational frequencies, etc. Thirdly, the reactivity of ligands in organometallic complexes can, and often does, mimic that of adsorbate molecules on catalyst surfaces. Finally, organometallics may serve as precursors to supported catalysts that show different activities or selectivities from those prepared by the conventional impregnation procedures. Even if a catalyst prepared from organometallic precursors exhibits the same reactivity as one prepared by the conventional impregnation route, the surface of the former is often "cleaner", i.e. free of extraneous phases, so that interpretation of spectroscopic data is often facilitated. This advantage of organometallic precursors seems to be little appreciated.

Until fairly recently there were no good organometallic models for reactivity patterns on sulfided surfaces, or for binding modes of thiophene compounds, for example. The latter situation has changed dramatically with the detailed and thorough work of Rakowski DuBois, Angelici, Rauchfuss, Jones, Adams, and others [16,17]. Rakowski DuBois has made an extensive study of the reactivity of dimeric molybdenum sulfides, e.g. $Cp_2Mo_2S_4$ and $Cp_2Mo_2(SR)_2(S)_2$ (R=H, alkyl, etc.) [18]. Of particular relevance here are her observations that the complex $Cp_2Mo_2(\mu_2-S_2)(\mu_2-S)_2$ splits H_2 to form $Cp_2Mo_2(\mu_2-SH)_2 (\mu_2-S)_2$, i.e. the sulfide ligands may play an active role in homolytic scission of H_2, that $Cp_2Mo_2(\mu_2-S_2CH_2)(\mu_2-SMe)(\mu_2-S)^+$ heterolytically splits H_2 and that these species are hydrogenation catalysts for a variety of substrates. The complex, $Cp_2Mo(S_2CH_2)(SH)(S)^+$ also catalyzes H/D exchange of the α-hydrogens on thiophene, a reaction also observed over CoMoS HDS catalysts [19]. The significance of this body of work is that it demonstrates substantial *sulfur*-centered reactivity. The role of the metal is nonetheless important, since the Mo can change oxidation states to support changes in the formal oxidation states of the sulfur ligands, and vacancies in the metal coordination sphere facilitate desulfurization reactions [20].

Rauchfuss has made important contributions to understanding the reactivity of polysulfide ligands, e.g. extrusion of S_x [21], interconversion of binding modes of disulfide ligands (μ–η^1,η^1, μ–η^2,η^2) [22], their reactivity with alkynes [23], and the coordination chemistry of thiophene [24]. In connection with the latter, these workers have postulated the importance of the bimetallic, multihapto coordination of thiophenes to the desulfurization process and have also demonstrated a remarkable reactivity of the coordinated thiophenes with the hydroxide ion or the hydrogen ion [25]. These latter demonstrations are especially important since both base (MS⁻, MO⁻, OH⁻ etc.) or Bronsted acids (SH) are present on the surface of MoS_2 catalysts and may serve as agents for thiophene ring opening.

Angelici and co-workers have conducted thorough investigations of the coordination chemistry of thiophenes and have discovered several transformations relevant to its desulfurization reactions over heterogeneous catalysts [16]. In particular, their demonstration that the deuterium exchange pattern on π-bound thiophene is similar to that observed on heterogeneous catalysts [26], of H⁻/H⁺ addition to C=C bonds [27] and C-S bond cleavage in π–bound thiophene [28], the greatly increased basicity of the sulfur atom in η^4-bound thiophene, and interconversions between η^4(C) and η^2(C,S) bonding modes [29] are especially noteworthy. The measurements of relative binding strengths of methyl substituted thiophenes in π- and S- bound coordination modes has also been extremely useful in the interpretation of adsorption and relative HDS rates of substituted thiophenes over supported HDS catalysts [30,31]. Benzothiophene coordination chemistry and reactivity have also received attention [23,32-34].

Most of the recent investigations of reactions of organic sulfur compounds with transition metal complexes have involved mononuclear derivatives, but several reports of C-S bond activation at dinuclear and polynuclear centers have appeared. Rauchfuss et al. showed that a thiaferracyclohexadiene is an intermediate in the desulfurization of thiophene with $Fe_3(CO)_{12}$ [23]. Boorman et al. studied the cleavage of a C-S bond of sulfides coordinated to two metal centers [35], and Rakowski DuBois recently reported that a dinuclear molybdenum complex can desulfurize thiirane [20]. Jones et al. have observed cleavage of the C-S bond of thiophene in dinuclear Co and Rh complexes [36]. An especially interesting result was obtained with the reaction of Cp*$_2$Ir$_2$H$_6$ with thiophene. The sulfur atom was abstracted and formed a μ–S bridge while the C_4H_4 fragment was reduced to butadiene and was also incorporated as a bridging ligand [37]. The sequence of reactions that were proposed to account for the formation of the final product are very similar to those we have proposed for the heterogeneously catalyzed HDS of thiophene [38]. Bianchini et al. have shown that an Ir complex inserts into the C-S bond of benzothiophene or thiophene, and the ring-opened complexes react with acids, e.g. PhSH, to give H_2S, ethylphenyl thiol, or butenyl thiols [39].

Adams et al. [40] have studied the reactions of organic sulfides with osmium clusters. In an extensive series of papers (see ref. 40, 41 for references to the original work), they have isolated new clusters resulting from S-H, C-S, and C-H bond insertions, as well as reductive elimination of alkanes or arenes, dehydrogenation of the organic moiety and oligomerization of the organic sulfide [41]. All these reaction types have been observed over heterogeneous HDS catalysts or clean metal surfaces.

At this point a fair question would be: Have these organometallic studies helped the interpretation of data obtained on the real catalyst systems? I believe the answer to this question is most certainly in the affirmative. The review articles in references 1 and 5 both refer extensively to organometallic model systems. Recent studies on clean and sulfided metal surfaces under ultrahigh vacuum conditions are beginning to uncover aspects of the reactivity of surface bound thiolates that have direct parallels with reaction pathways seen on organometallic centers. Thus, elimination of H-atoms in the β-position w.r.t. the thiol sulfur atom has been observed on Au [42], as has the extrusion of butadiene from 2,5-dihydrothiophene over Mo(110) [43] (similar reactions have been seen on organometallic reactions). Stiefel has also drawn attention to the correspondence of the structural features found in metal sulfido clusters and the MoS_x species believed to exist at the edges of the basal planes of MoS_2 [44]. Hence, there exist organometallic models for the structural features as well as the kinetic or reaction pathways postulated to be important in the catalysis of the HDS reaction.

Model studies on well ordered surfaces [5], the EXAFS, MES, and other spectroscopic studies by Topsøe et al. on sulfided Co-Mo-S catalysts, and the organometallic studies described above have led to a greatly increased understanding of the nature of the catalyst and possible coordination modes of organic sulfides with species likely to be present on metal sulfide catalysts. However, there are many questions yet unanswered. The role of the promoter metals, e.g. Co or Ni, in the catalytic cycle is still not established. Is it electronic (modifying the Mo-S bond strength), physical (promoting *more* active sites), or kinetic (e.g. center for dissociative adsorption of H_2) in nature? Is the cobalt, as some have suggested, the real active site with the MoS_2 acting merely as a support or modifier to the Co-active site(s)? How are the C-S bonds activated on the catalyst surface? What are the kinetically important steps?

The reminder of this article will describe the results that have been obtained in our laboratories over the last several years pertaining to the reactions of discrete Mo/Co/S clusters with a variety of organic sulfur compounds. These studies are unique in that they have focused on the combination of elements that are most commonly used in industrial HDS catalysts, and the structures of the clusters themselves have certain features in common with the edge "clusters" proposed to be present in industrial catalysts [44]. Heterogeneous HDS and CO hydrogenation catalysts may be prepared by adsorbing these sulfido bimetallic clusters on Al_2O_3 or other supports, followed by typical activation steps (reduction and sulfiding). These cluster-derived catalysts have activities very similar to those of conventionally prepared, commercial catalysts, but EXAFS studies show that under these conditions the clusters decompose and appear to agglomerate into structures that resemble closely those found in conventionally prepared samples [38,45,46]. Hence, the solution reactivity of the clusters will be the focus of the following discussion.

II. Mo/Co/S Models for HDS Catalysts

A. Nature of the Cluster Complexes.

A series of cluster complexes of Mo, S, and one of the late transition metals (TM), Fe, Co, or Ni, have been prepared in our laboratories [47], but the two clusters whose reactivity has been most investigated are the two Mo/Co/S clusters, **1** and **2**. Cluster **2** is "electron precise", i.e. it obeys the 18 electron rule as commonly found for organometallic complexes and small clusters.

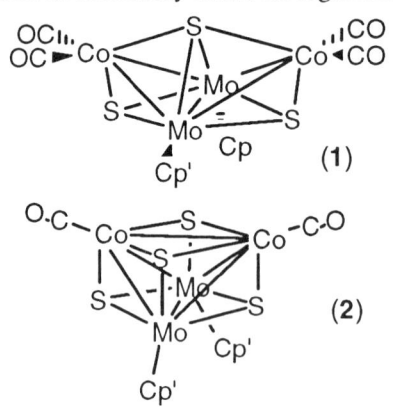

When a cluster obeys the 18 electron rule, each metal must be apportioned its 18 electrons, so a 4-metal cluster would require 72 electrons in the absence of metal-metal bonding. Since each metal that shares a 2-electron (two-center) bond gets to count those two electrons in its total, the two electrons in a metal-metal bond get counted twice, once for each metal. Hence, the number of electrons required to reach the 18-electron rule total is 18n-2m, where n = number of metal atoms and m = number of metal-metal bonds. A tetrahedral cluster of 4 metal atoms with 6 metal-metal bonds, e.g. **2**, thus requires 60 VSE (valence shell electrons) to meet the 18-electron rule. Cluster **1** does not obey the 18-electron rule; it also has 60 VSE but only 5 metal-metal bonds and thus requires 62 VSE to satisfy the 18 electron rule (in these electron counts, the μ_3- or μ_4 bridging S-atoms are counted as 4-electron donors - the inner 3s electrons are two low in energy to participate in cluster bonding). As a consequence, cluster **1** is electron deficient in the same sense as boron clusters are electron deficient. Another way of viewing the electronic structure of cluster **1** is to view the molecule as possessing a latent vacancy in the metal coordination sphere, i.e. the cluster can accept an additional ligand with its two electrons. This concept helps to relate the observed desulfurization activity of cluster **1** to the requirement of vacancies on the surface of heterogeneous catalysts.

B. Homogeneous reactivity of Mo\Co\S Clusters.

We have shown that the *unsaturated* 60 VSE cluster **1** reacts with a variety of organic sulfur compounds, exemplified with thiols in eq. 2.

The organometallic product is the *saturated* 60 VSE cluster **2**, typically formed in quantitative yield [48]. Several features are noteworthy in these reactions. First, aryl-sulfur bonds are cleaved as easily as alkyl sulfur bonds in their respective thiols (eq. 2). Kinetic studies have shown that the rate determining step (r.d.s.) in the desulfurization reactions is the associative attachment of the thiol to cluster **1**, and that the $\Delta H^{\ddagger}$ for this process is typically around 20 kcal/mol [49]. Regardless of the exact mechanism of the desulfurization reaction, the observed $\Delta H^{\ddagger}$ must be the upper limit to the C-S bond dissociation energy (BDE) of the cluster-bound thiol and *represents a decrease of over 75% in the C-S BDE when compared to the free thiol*.

$$\mathbf{1} + \text{RSH} \longrightarrow \mathbf{2} + \text{RH}$$
$$R = \text{Aryl, }^{t}\text{Bu, }sec\text{-pentyl}$$
(eq. 2)

The rates of reaction correlate fairly well with the nucleophilicity of the thiol which explains the relative insensitivity of the rate to the C-S bond strength. This observation has a parallel in heterogeneous catalysis: the overall rate is often composed of an adsorption equilibrium term related to how well the substrate binds to the catalyst surface, and a term representing the rate of reaction of the adsorbed species. From steric effects, etc., the former term may become rate limiting. The kinetic behavior of the reaction of **1** with thiols is reflected also in the reactions of **1** with other nucleophiles, e.g. phosphines and isonitriles [50]. In this case, the kinetic behavior is characteristic of a pre-equilibrium, and, in the reaction of **1** with PMe$_3$, the cluster shown in eq. 3 was isolated and structurally characterized.

(eq. 3)

The attacking nucleophile has added to cluster **1** by displacing one Co-S bond and forming the electronically saturated, 62 VSE cluster with all μ_3-bridging sulfide ligands. Clusters of this type are presumably intermediates in all the associative CO displacement reactions on cluster **1**. Note that a similar process may be operative when active sites of a heterogeneous, metal sulfide catalyst are titrated with a reactive species, e.g. O$_2$ or NO, the two most common reagents for determining the number of active sites on HDS catalysts [46]. The reagent itself may displace a surface sulfide to "create" a site that exists as a "latent" vacancy in the absence of the reagent.

No products expected from carbocation chemistry are observed in the course of the thiol desulfurization reactions. Thus, no propene was observed during the desulfurization of t-butyl mercaptan, a product characteristic of the formation of the t-butyl cation. Consistent with the rate determining step being the attachment of the thiol to the cluster, we have been unable to detect any intermediates in the thiol desulfurization reactions. We therefore studied the desulfurization of thiolate *anions* in an attempt to change the nature of the r.d.s. so that we could isolate and characterize intermediates that would allow us to study the mechanism of the C-S bond cleavage step. These reactions proved to be especially revealing. At low temperature (< -20 °C), a deep red intermediate forms that can be studied by variable-temperature NMR [49]. This initially formed complex has the structure **3** in Scheme I, analogous to the adduct shown in eq. 3 (L=ArS$^-$). Starting near -20 °C, the red complex isomerizes to a green complex of lower symmetry believed to be the complex **4** (Scheme I) with one μ_2- bridging thiolate ligand, and one μ_2- bridging sulfide ligand. Around 0 °C, this latter complex becomes fluxional and the μ_2-bridging thiolate and sulfide ligands begin to "walk" across the surface of the cluster. Note that this fluxional process has a one to one similarity to adsorbates migrating from site to site on the surfaces of heterogeneous catalysts. At about 35 °C, the rate of C-S bond cleavage becomes appreciable and the spectrum begins to lose resolution due to the formation of a paramagnetic complex, shown to be identical to the radical anion of cluster **2**, **2**$^-$, that can be made independently by Na-amalgam reduction of **2**. When the reaction was conducted in a deuterated solvent, the organic product was the mono-deuterated aromatic, ArD formed by abstraction of a deuterium atom from the solvent by the aryl radical, Ar•.

Armed with this information, we then established that the desulfurization of thiols was also occurring be a C-S bond homolysis by determining the fate of a "radical clock" reagent, *cyclo*-propylmethyl thiol. The observed product of the reaction of *cyclo*-propylmethyl thiol with cluster **1** was 1-butene, formed by the rapid rearrangement of the intermediate, *cyclo*-propylmethyl radical, in turn produced by the C-S bond homolysis of the thiol. This same rearrangement was also observed when the *cyclo*-propylmethyl thiolate anion was desulfurized by reaction with cluster **1**, but in this case the 1-butene product was monodeuterated when the reaction was conducted in CD$_3$CN solvent. The lack of deuterium incorporation when the thiol is desulfurized under comparable conditions suggests that the organic radical rapidly abstracts the H-atom from the H-containing cluster before the radical can diffuse into the bulk solution. Thus, the sequence of reaction shown in Scheme II depicts our best picture of the mechanism of desulfurization of thiols by the Mo/Co/S cluster, **1**.

Modeling HDS Catalysis with Sulfido Bimetallic Clusters

Scheme I

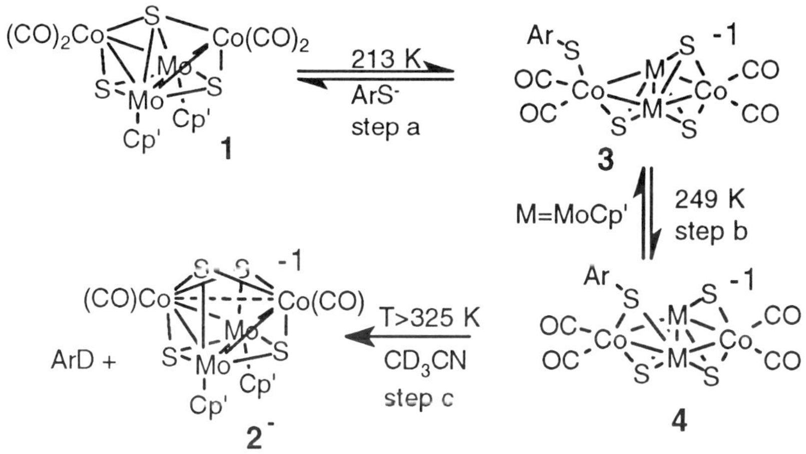

Scheme II

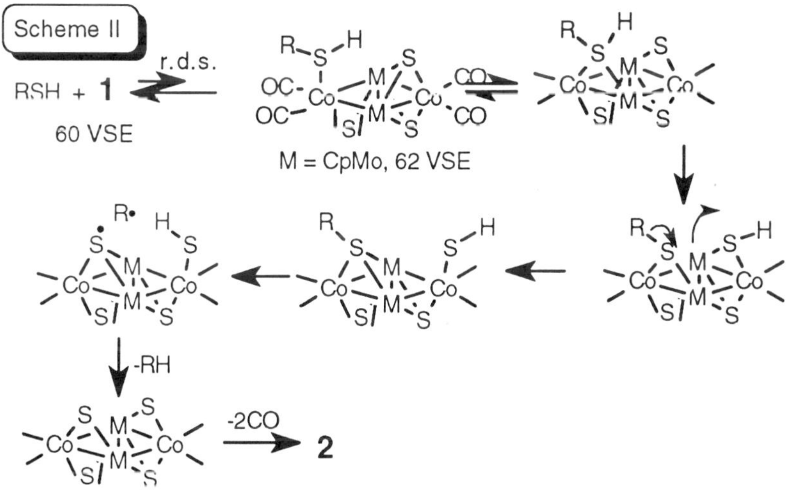

Two key features of the mechanism in Scheme II are the lowering of the C-S bond energy, induced by μ3-coordination of the thiolato sulfur as described above, and the mobility of the thiolate species allowed by the M-S bond breaking and bond making. The observed decrease in bond energy, defined as the difference in the energies of the final and initial states, simply reflects the greater stability of μ3-coordinated sulfur in the organometallic cluster, **2⁻**, as compared to the sulfhydryl radical (compare eq. 4 and 5). The ability of the cluster to accommodate a variable number of electrons by making/breaking M-S bonds as well as its ability to support metals in various oxidation states undoubtedly plays a large part in the observed reactivity [48].

Desulfurization of thiiranes was shown to be stereoregular: *cis*-2,3-dimethylthiirane gave *cis*-2-butene, while the *trans*-isomer gave *trans*-2-butene. This result is consistent with a concerted, topotactic abstraction of the S-atom from the thiirane. The rate profiles in the desulfurization of isothiocyanates suggested that the reaction also proceeded through a single-point (η^1-) coordination of the isothiocyanate followed by the release of free isocyanide.

The four-membered ring thietanes also react readily with cluster **1** to give propenes, as shown in eq. 6. This reaction is stereo*selective*: no matter which isomer of the diphenylthietane is used as the starting material, only cis-1,3-diphenylpropene is obtained as the product. We believe that the mechanism involves the free radical ring opening of the thietane, followed by a stereoselective β-hydrogen elimination. Note that β-elimination essentially converts the thietane into a thiol, so that the desulfurization mechanism depicted in Scheme II is applicable after β-elimination. When the cluster is used in a catalytic amount, the starting thietane is isomerized, presumably by a reversible C-S bond homolysis that opens the ring, allows rotation, and reformation of the ring. Under hydrogen, some of the thietane is catalytically hydrogenated to the saturated diphenylpropane thiol.

The desulfurization of thiophene has several unique features (eq.7). First, due to the decreased basicity of thiophene as a ligand, more forcing conditions are necessary: 150 °C in neat thiophene. Second, the formation of ring "cracking" products, C_3 and C_2 alkenes and alkanes, is even more extensive than is observed during the HDS of thiophene over heterogeneous catalysts at 350 °C! This suggests that metallacycle intermediates are involved since these species are known to exhibit C-C bond cleavage reactions into C_2, and/or C_1,C_3 products. It has been demonstrated that cluster **1** forms metallacycles when reacted with alkynes [51]. A similar metallacycle formation is observed on metal surfaces [52].

In addition to the desulfurizations of organic sulfides by cluster 1, we have recently discovered that cluster 2 also mediates sulfur atom abstraction from the more reactive sulfides, e.g. thiiranes and thietanes [53]. When the Cp' ligand on cluster 2 is C_5H_4Me, the organometallic product of the desulfurization reactions is an insoluble black powder that consistently analyzes as the "adduct" $Cp'_2Mo_2Co_2S_4(RS)$, where RS = cyclohexene sulfide or propylene sulfide. Heating a suspension of the powder in toluene causes loss of the alkene and formation of the pentasulfide, $Cp'_2Mo_2Co_2S_5$. The very soluble derivative with Cp* = C_5Me_4Et leads to a soluble pentasulfide, $Cp*_2Mo_2Co_2S_5$, upon reaction of 2* with thiiranes or thietanes. When this pentasulfide is reacted with CO or CO/H_2, up to a 40% isolated yield of the original cubane, $Cp*_2Mo_2Co_2S_4(CO)_2$, is produced, but a portion of the black solid is not converted back to the cubane.

Our working hypothesis is that $Cp_2Mo_2Co_2S_5$ retains the basic structure of the cubane cluster, but has the carbonyl groups replaced by a bridging sulfido ligand (see structure 5). The replacement of two 2-electron donors (CO) by two 1-electrons donors (S) represents an oxidation of the cluster. This oxidation has been modeled by reactions of 2 with halogens, organic disulfides, and thiols to give the respective dihalide cube clusters, 6, and the dithiolato cluster, 7, respectively. These 58 VSE clusters do not have enough framework electrons to form 6 metal-metal bonds. The X-ray crystal structures show that the chloride and iodide derivatives have Co-Co distances of 2.95 Å, compared to 2.65 Å in the parent carbonyl, 1. The Co-Co distance in the thiolato cluster is 2.74 Å, nearly identical to the Co-Co distance in the radical anion of cluster 2 that has been assigned a bond order of 0.5 on the basis of ESR data, the bond length, and the theoretical results of Harris [54] The deficiency of electron density in the oxidized clusters is also manifested in their magnetic properties: rather than the diamagnetic behavior of the carbonyl clusters, the halo and thiolato clusters exhibit complex paramagnetism [53].

(5)

(6, E = Cl, Br, I)
(7, E = PhS-)

The dithiolato cluster, 7, was shown to react with CO to regenerate the parent carbonyl cluster, 2 and diphenyldisulfide. Therefore, there was the potential to effect the catalytic cycle shown in Scheme III. When 2 was allowed to react with a 200-fold excess of PhSH under 1000 psi of CO, a 170% yield (based on the metal cluster) each of PhSSPh and PhS(CO)Ph was isolated, along with a 57% yield of **7**. The isolation of phenyl thiobenzoate showed that catalytic desulfurization of PhSH occurred along with the catalytic transformation of the benzene thiol to phenyl disulfide. These are the first observations of *catalytic* desulfurizations with organometallic complexes.

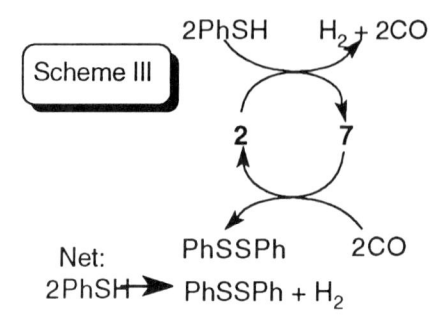

Scheme III

2PhSH → PhSSPh + H$_2$

Net: 2PhSH → PhSSPh + H$_2$

C. Relevance of Homogeneous Cluster Reactivity to Heterogeneous HDS Catalysis.

Our discovery that µ$_3$-coordination of thiols or thiolates to cluster **1** causes a dramatic decrease in the C-S BDE (80 → 20 kcal/mol) suggests that C-S bond breaking is unlikely to be the rate determining step on heterogeneous HDS catalysts at 350 °C if similar coordination sites are present on the heterogeneous catalysts. This conclusion is buttressed by a recent theoretical paper and by some experimental results that suggest the release of H$_2$S from the surface is rate determining [55,56].

In light of these results, it would appear that the key to a more active desulfurization catalyst lies in the creation of more µ$_3$-sulfido ligand vacancies or defects on the catalyst surface. Such sites are almost by definition the most stable binding sites for sulfur and are the last to open up as sulfur is removed from the surface. An analogy exists with cluster **1**, which is electronically unsaturated and contains a *latent* vacant site in the form of the µ$_4$-sulfido ligand. Nucleophiles displace one Co-S bond to create a site vacancy (see formation of **3** in Scheme I). Further "vacancies" are formed by motions of the S atoms around the faces of the cluster; such motions are responsible for the VT-NMR behavior observed for thiolato adducts of **1**. It is possible that cobalt plays a similar role in creating latent site vacancies on the surface of sulfided HDS catalysts. It is also instructive to recall that an early, now largely discarded, hypothesis for the catalytic action of Mo/Co/S HDS catalysts had MoS$_2$ merely acting as a support for the active Co-sulfide phase [57]. Since most of the "action" on our Mo/Co/S clusters actually occurs at the Co, this older hypothesis may need to be re-examined.

The mechanism shown in Scheme II also has the same dichotomy that is associated with the basic chemistry of metal sulfide catalysts: is the chemistry metal-centered or sulfur-centered? In sulfur-centered chemistry, the metal activates the sulfur as a leaving group, either by C-S bond homolysis or by a β-hydride elimination to form a surface sulfhydryl group and an alkene (eq. 8). In the metal-centered mode, oxidative addition reactions with C-S, C-H, and S-H bonds lead to M-H and M-R bonds that can then undergo reductive elimination to the alkane or β-hydride elimination to the alkene (eq. 9). Both types of reactivity have now been observed in model organometallic systems and on surfaces under UHV conditions [58,59].

$$\text{(eq. 8)}$$

$$\text{(eq. 9)}$$

These reaction pathways have some further implications for heterogeneous catalysis. The first lesson is that the "quality" of the S-vacancies are as important as their number. The role of multihapto S-coordination in activating the C-S bond now seems well established. Hence, any structural organization of the surface that promotes 3-fold vacancies rather than 2-fold or terminal vacancies will result in a more active catalyst. The ease with which the C-S bond is cleaved once the sulfur atom is coordinated to a μ_3-site is remarkable. This type of coordination is most likely to occur with thiolato or sulfido groups and the difficulty of desulfurizing thiophenes and benzothiophenes undoubtedly arises from their poor basicity and relatively low affinity to bind to metals. Therefore, catalysts that are more effective at hydrogenating double bonds in thiophenes and benzothiophenes to give more saturated sulfides will also be more active in HDS. This emphasizes the importance of the hydrogenation activity of HDS catalysts and points out a possible role of the promoter element: namely, to provide a higher hydrogenation activity. The latter may be achieved either by acting as a catalyst for the H_2-reduction of Mo(IV) (and thereby providing, indirectly, H^+ for removal of S^{-2} as H_2S) or by acting as a separate hydrogenation site for the unsaturated organics. These proposals have been made many times previously, but the results obtained from model organometallic reactions now provide a factual basis for their importance.

References

1. Chianelli, R. R.; Daage, M.; Ledoux, M. J. *Adv. Catal* **1994**, *40*, 177-232.

2. Hoffman, H. L. *Hydrocarbon Processing* **1991**, *70*, 37. (b)

3. Massoth, F. E. *JAdv. Catal.* **1978**, *27*, 265.

4. Topsøe, H.; Clausen, B. S. *Catal. Rev. - Sci. Eng.* **1984**, *26*, 395.

5. Prins, R.; de Beer, V. H. J.; Somorjai, G. A. *Catal. Rev. - Sci. Eng.* **1989**, *31*, 1.

6. Analysis by the author of original data reported by: Wivel, C.; Candia, R.; Clausen, B. S.; Mørup, S.; Tøpsoe, H. *J. Catal.* **1981**, *68*, 453..

7. Roxlo, C. B.; Daage, M.; Ruppert, A. F.; Chianelli, R. R. *J. Catal.* **1986**, *100*, 176.

8. Chianelli, R. R.; Ruppert, A. F.; Behal, S. K.; Kear, B. H.; Wold, A.; Kershaw, R. *J. Catal.* **1985**, *92*, 56.

9. Ledoux, M. J.; Segura, Y.; Pannisod, P. *ACS Div. Petrol. Chem., Prepr. Papers* **1990**, *35*, 217.

10. Harris, S.; Chianelli, R. R. *J. Catal.* **1984**, *86*, 400, and ibid. **1986**, *98*, 17.

11. Vorhoeve, R. J. H.; Stuiner, J. C. M. *J. Catal.* **1971**, *23*, 228, 243.

12. Konings, A. J. A.; Valster, A.; de Beer, V. H. J.; Prins, R. *J. Catal.* **1982**, *76*, 446.

13. Vrinat, M. L.; de Mourgues, L. *Appl. Catal.* **1983**, *5*, 43.

14. Delmon, B. *Bull. Chim. Soc. Belg.* **1979**, *88*, 979.

15. Muetterties, E. L.; Stein, J. *Chem. Rev.* **1979**, *79*, 479.

16. Review: Angelici, R. J. *Accts. Chem. Res.* **1988**, *21*, 387.

17. Rauchfuss, T. B., *Prog. Inorg. Chem.* **1991**, *39*, 260.

18. Review: Rakowski Du Bois, M. *Chem. Rev.* **1989**, *89*, 1.

19. Lopez, L.; Godziela, G.; Rakowski Du Bois, M. *Organometallics* **1991**, *10*, 2660.

20. Gabay. J.; Dietz, S.; Bernatis, P.; Rakowski DuBois, M. *Organometallics* **1993**, *12*, 3630.

21. Giolando, D. M.; Rauchfuss, T. B.; Reingold, A. L.; Wilson, S. R. *Organometallics* **1987**, *6*, 667.

22. Bolinger, C. M.; Rauchfuss, T. B.; Reingold, A. L. *Organometallics* **1982**, *1*, 1551; idem. *J. Am. Chem. Soc.* **1983**, *105*, 6321.

23. Ogilvy, A. E.; Draganjac, M.; Rauchfuss, T. M.; Wilson, S. R. *Organometallics*, **1988**, *7*, 1171.

24. Luo, S.; Ogilvy, A. E.; Rauchfuss, T. M.; Reingold, A. L.; Wilson, S. R. *Organometallics* **1991**, *10*, 1002.

25. (a) Luo, S.; Rauchfuss, T. B.; Gan, Z. *J. Am. Chem. Soc..* **1993**, *115*, 4943. (b) Krautscheid, H.; Feng, Q.; Rauchfuss, T. B. *Organometallics* **1993**, *12*, 3273. (c) Skaugset, A. E.; Rauchfuss, T. B.; Wilson, S. R. *J. Am. Chem. Soc..* **1992**, *114*, 8521.

26. Spies, G. H.; Angelici, R. J. *J. Am. Chem. Soc.* **1988**, *107*, 5569.

27. Lesch, D. A.; Richardson, Jr., J. W.; Jacobson, R. A.; Angelici, R. J. *J. Am. Chem. Soc.* **1984**, *106*, 2901.

28. Hachgenei, J. W.; Angelici, R. J. *J. Organometal. Chem.* **1988**, *355*, 359.

29. Chen, J.; Angelici, R. J. *Organometallics* **1990**, *9*, 879; and ibid, p. 879.

30. (a) Hachgenei, J.; Angelici, R. J. *Organometallics* **1989**, *8*, 14. (b) Benson, J. W.; Angelici, R. J. *Organometallics* **1993**, *12*, 680.

31. (a) Benson, J. W.; Angelici, R. J. *Organometallics* **1992**, *11*, 922. (b) Choi, M.-G.; Angelici, R. J. *Inorg. Chem.* **1991**, *30*, 1417.

32. Huckett, S.C.; Angelici, R. J.; Ekman, M. E.; Schrader, G. L. *J. Catal.* **1988**, *113*, 36.

33. Kilanowski, D. R.; Tecueven, H.; de Beer, V. H. J.; Gates, B. C.; Schuit, G. C. A.; Kwart, H. *J. Catal.* **1978**, *55*, 129.

34. Jones, W. D.; Dong, L. *J. Am. Chem. Soc.* **1991**, *113*, 559.

35. Boorman, P. M.; Gao, X.; Fait, J. F.; Parvez, M. *Inorg. Chem.* **1991**, *30*, 3886.

36. (a) Jones, W. D.; Chin, R. M. *Organometallics* **1992**, *11*, 2698. (b) Rosini, C. P.; Jones, W. D. *J. Am. Chem. Soc.* **1992**, *114*, 10767. (c) Dong, L.; Duckett, S. B.; Ohman, K. F.; Jones, W. D. *J. Am. Chem. Soc.* **1992**, *114*, 151. (d) Jones, W. D.; Chin, R. M. *J. Am. Chem. Soc.* **1992**, *114*, 9851.

37. Jones, W. D.; Chin, R. M. *J. Am. Chem. Soc.* **1994**, *116*, 198.

38. Curtis, M. D.; Penner-Hahn, J. E.; Schwank, J.; Baralt, O.; McCabe, D. J.; Thompson, L.; Waldo, G. *Polyhedron* **1988**, *7*, 2411.

39. (a) Bianchini, C.; Meli, A.; Peruzzini, M.; Vizza, F.; Frediani, P.; Herrera, V.; Sanchez-Delgado, R. A. *J. Am. Chem. Soc.* **1993**, *115*, 2731. (b) Bianchini, C.; Meli, A.; Peruzzini, M.; Vizza, F.; frediani, P. H.; Sanchez-Delgado, R. A. *J. Am. Chem. Soc.* **1993**, *115*, 7505. (c) Bianchini, C.; Meli, A.; Perizzini, M.; Vizza, F.; Herrera, V.; Sanchez-Delgado, R. A. *Organometallics* **1994**, *13*, 721. (d) Bianchini, C.; Meli, A.; Peruzzini, M.; Vizza, F.; Moneti, S.; Herrera, V.; Sanchez-Delgado, R. A. *J. Am. Chem. Soc.* **1994**, *116*, 4370.

40. Reviews: (a) Adams, R. D. *Polyhedron* **1985**, *4*, 2003. (b) Adams, R. D.; Horvath, I. T. *Prog. Inorg. Chem.* **1985**, *33*, 127.

41. Adams, R. D.; Pompeo, M. P.; Wu, W.; Yamamoto, J. H. *J. Am. Chem Soc.* **1993**, *115*, 8207, and references therein.

42. Jaffey, D. M.; Madix, R. J. *J. Am. Chem. Soc.* **1994**, *116*, 3012.

43. Liu, A. C.; Friend, C. M. *J. Am. Chem. Soc.* **1991**, *113*, 820.

44. Stiefel, E. I.; Halbert, T. R.; Coyle, C. L.; Wei, L.; Pan, W. H.; Ho, T. C.; Chianelli, R. R.; Daage, M. *Polyhedron* **1989**, *8*, 1625-9.

45. (a) Curtis, M. D. *Appl. Organomet. Chem.* **1992**, *6*, 429. (b) Curtis, M. D.; Schwank, J.; Penner-Hahn, J. E.; Thompson, L.; Baralt, O.; Waldo, G. *Mater. Res. Soc. Symp. Proc.* **1988** *111*, 331. (c) Curtis, M. D.;

Penner-Hahn, J. E.; Schwank, J.; Baralt, O.; D. J. McCabe; L. Thompson; and G. Waldo, *Polyhedron* **1988** *7*, 2411.

46. Carvill, B. J.; Thompson, L. T. *Appl. Catal.* **1991**, *75*, 249.

47. (a) Curtis, M. D.; Williams, P. D.; Butler, W. M. *Inorg. Chem.* **1988**, *27*, 2853. (b) Li, P.; Curtis, M. D. *J. Am. Chem. Soc.* **1989**, *111*, 8279.

48. Riaz, U.; Curnow, O. J.; Curtis, M. D. *J. Am. Chem Soc.* **1994**, *116*, 4357.

49. Druker, S. H.; Curtis, M. D. *J. Am. Chem Soc.* **1995**, *117*, 6366.

50. (a) Curnow, O. J.; Kampf, J. W.; Curtis, M. D.; Shen, J.-K.; Basolo, F. *J. Am. Chem. Soc.*, **1994**, *116*, 224. (b) Curtis, M. D. ; Curnow, O. J. *Organometallics* **1994**, *13*, 2489.

51. Riaz, U.; Curtis, M. D. *Organometallics*, **1990**, *9*, 2647.

52. Gentle, T. M.; Tsai, C. T.; Walley, K. P.; Gellman, A. J. *Catal. Lett.* **1989**, *2*, 19.

53. Mansour, M. A.; Curtis, M. D. submitted for publication.

54. Harris, S. *Polyhedron* **1989**, *8*, 2843-2882.

55. Neurock, M.; van Santen, R. A. *J. Am. Chem Soc.* **1994**, *116*, 4427.

56. (a) Moser, W. R.; Rossetti, G. A.; Gleaves, J T.; Ebner, J. R. *J. Catal.* **1991**, *127*(1), 190-200. (b) Moravek, V.; Kraus, M. *Coll. Czech. Chem. Commun.* **1985**, *50*, 2159.

57. (a) DeBeer, V. H. J.; Duchet, J. C.; Prins, R. *J. Catal.* **1981**, *72*, 369. (b) DuChet, J. C.; Van Oers, E. M.; DeBeer, V. II. J.; Prins, R. *J. Catal.* **1983**, *80*, 386. (c) Vissers, J. P. R.; DeBeer, V. H. J.; Prins, R. *J. Chem. Soc. Faraday Trans. I* **1987**, *83*, 2145.

58. (a) Serafin, J. G.; Friend, C. M. *J. Am. Chem. Soc.* **1989**, *111*, 8967. (b) Wiegand, B. C.; Uvdal, P. E.; Serafin, J. G.; Friend, C. M. *J. Am. Chem. Soc.* **1991**, *113*, 6686. (c) Wiegand, B. C.; Uvdal, P.; Friend, C. M. *J. Phys. Chem* **1992**, *96*, 4527.

59. Jaffey, D. M.; Madix, R. J. *J. Am. Chem. Soc.* **1994**, *116*, 3012.

Evidence for H_2S as Active Species in the Mechanism of Thiopene Hydrodesulfurization

J. Leglise, J. van Gestel, and J.-C. Duchet

Laboratoire Catalyse et Spectrochimie, URA.CNRS.414
ISMRA, Université de CAEN
14050 CAEN Cedex, FRANCE. Fax (+33) 31 45 28 77.

ABSTRACT

Examination of the influence of H_2S on the reaction of thiophene under atmospheric pressure between 300 and 400°C shows a complex behaviour : the usual inhibiting effect switches to a promoting effect well evidenced when the reaction is performed at high temperature. Moreover, the apparent activation energy gradually decreases with increasing temperature. These features were modelled by a kinetic expression assuming dearomatization of the thiophene ring as the key step. The reaction is governed by both hydrogen and hydrogen sulfide which provide the proton-like species reacting with adsorbed thiophene. The relative contribution of these two components to the overall rate depends on the experimental conditions.

INTRODUCTION

Hydrodesulfurization of thiophene carried out at atmospheric pressure is widely used as a model reaction for testing sulfide catalysts. However, substantial differences in the operating conditions are encountered among the numerous laboratories involved in this field. Besides the use of various reaction temperatures and thiophene pressures, most laboratories operate at different conversion levels. H_2S produced by the conversion of thiophene is thus present at different concentrations. Since H_2S inhibits the reaction, as always reported, the comparison between catalysts is not rigorous. A better knowledge of the influence of H_2S on the thiophene reaction is therefore needed for standardization of the test.

Numerous systematic studies (mostly in the seventies and eighties) dealing with the kinetics of the thiophene reaction have appeared in the literature (1-4). Although in some cases the investigated range of H_2S concentration was large (up to 15%), modelling was restricted to a few measurements. Moreover the combined effect of H_2S and temperature was scarcely considered, or often restricted to temperatures below 300°C. In all cases H_2S was found to inhibit the thiophene reaction.

Recently, we extended the temperature range up to 400°C (5). Besides the expected inhibition, we evidenced an intriguing promoting effect by H_2S at high temperature. We suggested that H_2S was involved in the mechanism, and that the conversion was dictated by the relative concentration of hydrogen and hydrogen sulfide. The aim of this paper is to introduce kinetics in order to provide a better insight of the phenomenon. A more complete form will be published elsewhere.

EXPERIMENTAL

1. Base catalyst.

The oxidic base catalyst used in this study was a regenerated CoMo/Al_2O_3 (KF 742 1.3Q, AKZO). Elemental analysis (CNRS Vernaison) gave : Co, 2.8 wt%; Mo, 9.3 %; C, 0.5%; S, 0.6%. It was free from Ni and V. Results obtained with a fresh catalyst were strictly comparable.

2. Sulfidation and thiophene reaction.

20 to 100 mg of the oxidic catalyst (sieved 0.2 - 0.5 mm) was in-situ sulfided under a flow (2.2 lh^{-1}) of H_2-H_2S (15 vol% H_2S) in a atmospheric

H₂S as Active Species in Thiopene Hydrodesulfurization

flow micro reactor. The temperature was raised at 3°Cmin⁻¹ upto 400°C and maintained for 2.5 h. Then the total flow was increased to about 6 lh⁻¹ and the H_2S concentration was adjusted in the range 0 - 7 vol% before introducing thiophene. The partial pressure of thiophene was kept constant at 8 kPa with a thermostated saturator.

The conversion of thiophene was measured in differential conditions at temperatures between 300 and 400°C. The major products, H_2S, butenes and butane, were separated on an OV1 column and analysed with a thermistor detector. Traces of thiolane were evidenced after trapping the effluent. The actual H_2S level was taken as the average between the levels at the reactor inlet and outlet.

The rate of thiophene disappearance was expressed as $r = x.F°/W$, where x is the conversion, $F°$ the molar flow of thiophene at the reactor inlet, and W the catalyst weight.

1. Effect of temperature.

Figure 1 represents the Arrhenius diagram for two levels of added H_2S. Clearly, the expected linearity is not observed : in both cases, the apparent activation energy decreases gradually with increasing temperature. Such a behaviour might be caused by mass or heat transfer limitations. However, we checked that external and internal mass transfer were not limiting. This was studied i) by changing the catalyst mass from 20 to 100 mg, at a fixed contact time (total flow 3.6 - 16 lh⁻¹), and ii) by varying the size of the catalyst particles from 0.1 to 0.5 mm. In the same way, heat transfer limitation could be ruled out, as checked by dilution of the catalyst with silicon carbide. Therefore, the continuous change in activation energy is controlled by kinetics.

2. Effect of thiophene and hydrogen partial pressure.

The reaction order with respect to thiophene was determined in the range of partial pressure 2 - 8 kPa with 2% added H_2S. The reaction order was 0.2 at 300°C and increased to 0.7 at 400°C. Similarly, the reaction order for hydrogen (28 - 92 kPa) equals 0.8 at 300°C and 1.1 at 400°C.

3. Effect of H_2S partial pressure.

The influence of H_2S partial pressure on the reaction rate has been studied between 300 and 400°C in the range 0 - 8 kPa. The reaction order with respect to H_2S is -0.3 at 300°C; it becomes almost zero at 400°C. Then, the well known inhibiting effect of H_2S is not observed at high temperature. The

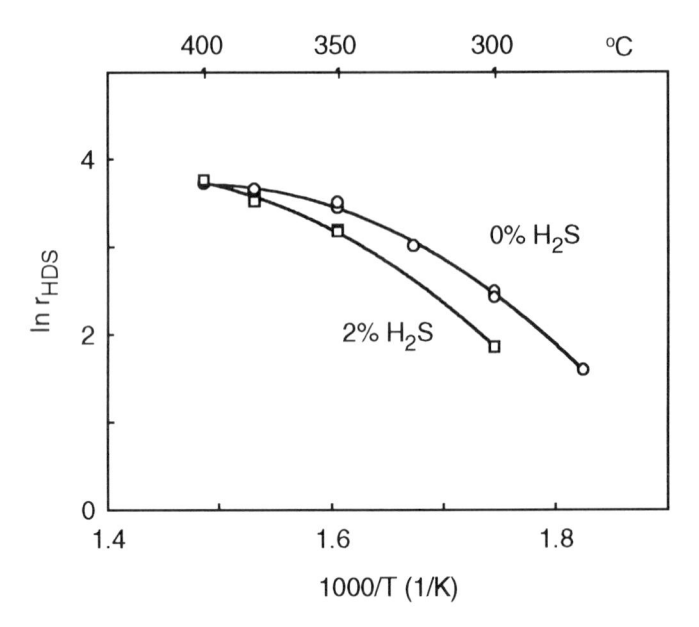

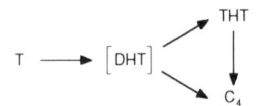

Figure 1. Arrhenius plot of the rate of disappearance of thiophene without and with 2% added H_2S.

plot of the reaction rate versus H_2S partial pressure in Figure 2 shows a slight maximum at 400°C around 2 kPa H_2S, followed by a decrease. These effects are reversible; they correspond to promotion and inhibition by H_2S as already reported (5). The maximum is indicative of a direct participation of H_2S in the reaction mechanism. The interpretation of such an intriguing phenomenon can be investigated by means of a kinetic modelling of the reaction mechanism.

Hydrodesulfurization of thiophene (*T*) produces C_4 hydrocarbons and H_2S. When running under hydrogen pressure, substantial amounts of thiolane (*THT*) are commonly observed, pointing to an intermediate hydrogenation step. Under these conditions, the reaction obeys a triangular scheme. The evidence for thiolane at atmospheric pressure in our experiments as well as in others (6-8), suggests that the triangular scheme operates also at atmospheric pressure :

$$T \longrightarrow [DHT] \begin{array}{c} \nearrow THT \\ \searrow \downarrow \\ C_4 \end{array}$$

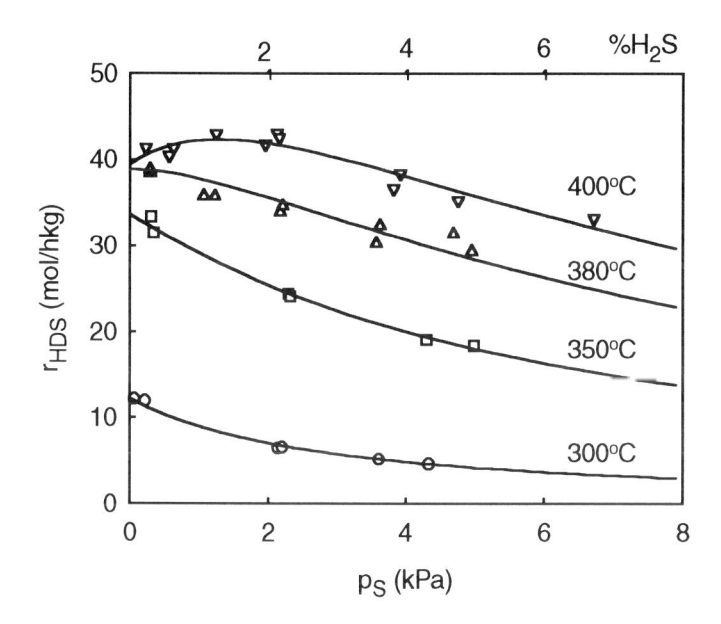

H$_2$S as Active Species in Thiopene Hydrodesulfurization

Figure 2. Effect of the H$_2$S partial pressure on the rate of disappearance of thiophene at 400, 380, 350 and 300°C.

Following Joffre et al (9), a preliminary dearomatization step of the thiophene ring is considered, since hydrogenation of thiophene to thiolane could not proceed directly. Dihydrothiophenes (*DHT*) may represent the semi hydrogenated compounds (10, 11). They are not detected, owing to their high reactivity.

According to Moreau et al (12), dearomatization occurs by an electrophilic (e.g. protonic) attack of the aromatic ring. In the case of thiophene, the attack on the α-carbon is favoured (13). Based on this picture, the disappearance of thiophene may be controlled by the concentration of active protonic species on the catalyst surface, reacting with adsorbed thiophene.

4. Derivation of the kinetic expression; elementary steps.

It is usually postulated that sulfide catalysts bear two distinct types of sites: one for hydrogenolysis and one for hydrogenation (1). Several forms of rate equations, most of them based on the Langmuir Hinshelwood formalism, were derived on the hypothesis that the rate determining step corresponds to the hydrogenolysis of the carbon-sulfur bond, which is assumed to occur directly from the thiophene molecule. Conversely, we consider the dearomatization step as rate limiting so that the hydrogenating function of the catalyst will be primarily involved in the reaction.

We assume that thiophene and H_2S compete on the same site, usually described as a surface vacancy on the mixed sulfide (14). The adsorption of hydrogen is less clear. Literature gives examples for hydrogen adsorption on the thiophene site (15) or on a separate site (3, 10, 16, 17). We consider first that hydrogen adsorbs separately from thiophene and H_2S. According to this hypothesis, the sequence of elementary steps may then be written as follows, using α and β exponents for adsorbed species on the corresponding sites, and the subscripts T, S, and H for thiophene, H_2S, and hydrogen respectively:

(1) $\quad T + \alpha \rightleftarrows T^\alpha \qquad\qquad b_T = T^\alpha/\alpha.p_T$

(2) $\quad H_2S + \alpha \rightleftarrows H_2S^\alpha \qquad\qquad b_S = H_2S^\alpha/\alpha.p_S$

(3) $\quad H_2 + \beta \rightleftarrows (H\text{-}H)^\beta \qquad\qquad b_H = H_2^\beta/\beta.p_H$

(4) $\quad T^\alpha + (H\text{-}H)^\beta \rightarrow \ldots \rightarrow DHT^\alpha + \beta$

(5) $\quad DHT^\alpha + (H\text{-}H)^\beta \rightarrow H_2S^\alpha + C_4 + \beta$

(6) $\quad DHT^\alpha + (H\text{-}H)^\beta \rightarrow \ldots \rightarrow THT^\alpha + \beta$

H_2S is involved in step (2) as a competitor with thiophene on the α sites. The first stage of step (4) is dearomatization of thiophene; it is rate limiting compared to the subsequent formation of dihydrothiophene. We could not distinguish between dissociative and molecular hydrogen adsorption; in the latter case, dissociation may take place during the reaction steps. The low conversion allows us to neglect the C_4 hydrocarbons adsorption in this sequence.

According to this sequence, the rate equation for thiophene disappearance takes the following form:

$$r_1 = k_1 \frac{b_T p_T}{1 + b_T p_T + b_S p_S} \cdot \frac{b_H p_H}{1 + b_H p_H} \qquad \text{Eq. I}$$

The kinetic parameters are readily accessible, and account for the observed reaction orders at low temperature. However, the equation cannot predict the singularity of the experimental data, i.e. the promoting effect of H_2S observed at high temperature.

In that respect, adsorption of hydrogen and sulfur compounds on the same type of site as a second hypothesis, yielding the rate equation II would not be more satisfactory :

$$r_{1'} = k_{1'} \frac{b_T p_T b_H p_H}{(1 + b_T p_T + b_S p_S + b_H p_H)^2} \qquad \text{Eq.II}$$

Going further, we consider a particular role of H_2S in the mechanism. Recalling that dearomatization requires an electrophilic species, it is noteworthy that protonic species can be issued either from hydrogen or from hydrogen sulfide. Indeed, there is general agreement that H_2S dissociates into H^+ and HS^- species on sulfide catalysts (18). Then, we may write a second sequence of elementary steps, where H_2S, adsorbed on the same type of site as thiophene, provides the required electrophilic species.

(4') $T^\alpha + (H\text{-}SH)^\alpha \rightarrow \ldots \rightarrow DHT^\alpha + S^\alpha$

(6') $DHT^\alpha + (H\text{-}SH)^\alpha \rightarrow \ldots \rightarrow THT^\alpha + S^\alpha$

Here again, dissociation of H_2S will occur during the reactions. In the above sequence, one sulfur atom remains on the α site, which should not be confused with sulfidation of the catalyst. This sulfur is liberated by reaction with hydrogen :

(7) $\quad S^\alpha + H_2 \rightleftharpoons \alpha + H_2S \qquad\qquad b_{2S} = \dfrac{S^\alpha \cdot p_H}{\alpha \cdot p_S}$

The dearomatization rate r_2 of thiophene by H_2S is then expressed as :

$$r_2 = k_2 \frac{b_T p_T b_S p_S}{(1 + b_1 p_1 + b_S p_S + b_{2S} p_3 / p_H)^2} \qquad \text{Eq.III}$$

This form reflects a promoting effect by H_2S.

Hence, the global rate of disappearance of thiophene is given by the sum of the two contributions $r_1 + r_2$ from H_2 and H_2S respectively. The reaction order for hydrogen may exceed unity, as observed at 400°C.

The kinetic parameters of this complex rate expression were obtained by a non-linear regression technique, using the complete set of data simultaneously. They are gathered in Table 1, together with their 95% confidence intervals.

TABLE 1 : Optimized parameters with their 95% confidence interval.

$$r = k_1 \frac{b_T p_T}{1 + b_T p_T + b_S p_S + b_{2S} p_S / p_H} \cdot \frac{b_H p_H}{1 + b_H p_H} + k_2 \cdot \frac{b_T p_T b_S p_S}{(1 + b_T p_T + b_S p_S + b_{2S} p_S / p_H)^2}$$

Rate constants (k : mol.h^{-1}.kg^{-1}) and adsorption coefficients (b : kPa^{-1})		Activation energies (E : kJ.mol^{-1}) and heats of adsorption (Q : kJ.mol^{-1})	
k_1	70000 ± 7000	E_1	251 ± 71
k_2	580 ± 30	E_2	213 ± 17
b_T	(8.8 ± 0.1).10^{-2}	Q_T	-82 ± 2
b_S	(9.5 ± 0.3).10^{-2}	Q_S	-116 ± 3
b_H	(1.9 ± 0.1).10^{-5}	Q_H	-201 ± 67
b_{2S} *)	21 ± 0.2	Q_{2S}	0 ± 40

*) dimensionless

In kinetics, the intrinsic values of constants are linked to the chosen model, making comparison with literature data difficult. However, the magnitude of adsorption coefficients provides information on the strength of the gaseous molecule interacting with the surface. In this respect, thiophene and H$_2$S appear equivalent; they adsorb at least 1000 times more strongly than hydrogen. This is in line with others (8).

5. Simulation of the experimental data.

The curves drawn in Figure 2 for the influence of H$_2$S partial pressure on the reaction rate show the quality of the fit for all the temperatures examined. Similar agreement is found for the effect of hydrogen or thiophene partial pressures.

In Figure 3, we simulated the contribution of r_1 and r_2 to the overall rate at 400°C (Figure 3A) and 300°C (Figure 3B) as a function of the H$_2$S pressure. It can be seen that the mechanism with hydrogen (r_1) operates almost exclusively at low temperature resulting in a continuous inhibition by H$_2$S. This is a consequence of the adsorption coefficients: at low temperature, the strong adsorption of H$_2$S and thiophene makes the second term (r_2) almost ineffective. The contribution of r_1 is still important at 400°C, but the

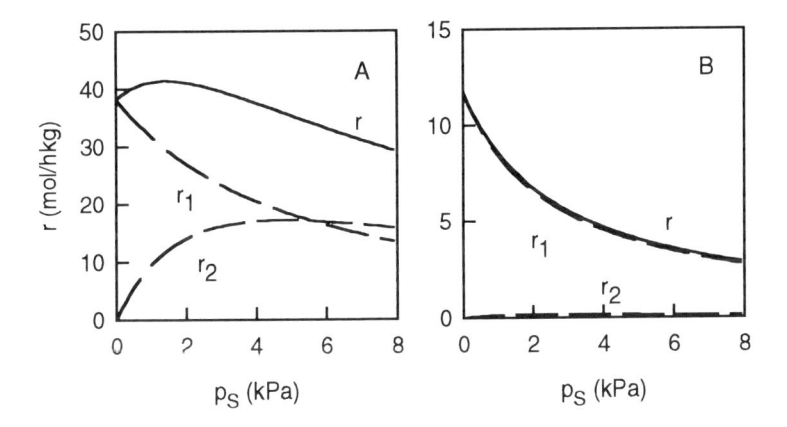

Figure 3. Simulated curves of the effect of H_2S, showing the contribution of the two terms r_1 (H_2) and r_2 (H_2S) to the overall rate at 400°C (A) and 300°C (B).

promotion by H_2S (r_2) increases rapidly at low H_2S pressure; the net effect results in a slight maximum. Beyond 2 kPa, the inhibition in r_1 prevails.

Finally, the kinetic modelling we propose well describes the change in apparent activation energy (Figure 1). Results obtained without added H_2S are easy to interpret. In that case there is no contribution of H_2S to the reaction of thiophene, and r_1 relative to hydrogen is the only rate expression to consider. At high temperature, thiophene and hydrogen are weakly adsorbed; the rate expression r_1 ultimately reduces to $k_1 b_T p_T b_H p_H$ above 400°C. In this temperature region, the Arrhenius plot gives a negative apparent activation energy $E_{app} = E_1 + Q_T + Q_H$. By the same token, at sufficiently low temperature (<250°C), the rate r_1 reduces to k_1 and the true activation energy E_1 will be accessible. Within these two limiting cases, intermediate values of the apparent activation energy are measured, yielding a curve in the Arrhenius diagram. The same features appear upon adding H_2S. The calculated curves in Figure 1 fit the experimental data. Some authors reported also a curvature in the Arrhenius plot (14, 16, 17). Chemical kinetics provides a plausible explanation, bypassing the interpretation based on diffusion phenomena, as often invoked.

6. Comparison with other models. Mechanism.

The rate expressions reported in the literature only focus on the inhibiting action of H_2S on the thiophene conversion. Hydrogen is the sole desulfurization agent and the agreement with the experimental data hardly distinguishes between single and dual sites for hydrogen, thiophene and H_2S. The sophisticated model with interconversion of hydrogenation and hydrogenolysis sites (2) does not improve the modelling of the experimental data under low pressure.

Our results showing both an inhibiting and a promoting action of H_2S provide new information on how the reaction operates. Although kinetics is not infallible, the modelling can be used to support the suggested mechanism.

The key point in our interpretation is the dearomatization of the thiophene ring. Some groups already inferred that this step is critical (9, 19) on energetic grounds; many others postulated the reaction to proceed as such (10, 20). Experimental evidence for thiolane is the most convincing.

With respect to the intimate mechanism, we followed the organic approach performed by the group of Moreau (12), who demonstrated the specific action of proton-like species in dearomatization over sulfide catalysts. On the contrary, several authors advocate the major role of hydridic species in the thiophene reaction (21). The promoting effect of H_2S reported here, would favour the former, since H_2S provides more easily H^+ than H^-. Moreover, the optimized values for the true activation energy associated with the rate constant k_2 in the contribution of H_2S to the reaction rate differs only by a few kJ/mol from that relative to the reaction of thiophene with hydrogen (k_1). This is strongly indicative for the same operating active species.

Finally, our kinetic study of thiophene conversion shows that the reaction is much more complex than often considered in this fast screening activity test. One simple rate equation is not sufficient to describe the results in a wide range of temperature and H_2S pressure. For practical purposes, care should be taken in the use of this reaction as a catalytic test at atmospheric pressure. H_2S produced by the reaction greatly influences the rate at low temperature. It is recommended to operate at very initial conversion and at a temperature of 380°C which proved to be less sensitive to low levels of H_2S.

A preliminary kinetic analysis of the reaction of thiophene under pressure shows also that the electrophilic dearomatization may be the key step. Several terms in the rate expression are needed according to the contributions of H_2 and H_2S in the mechanism. Such a generalisation may reasonably be applied to other thiophenic compounds involved in hydrotreating, i.e. benzothiophene and dibenzothiophene.

REFERENCES

1. M. L. Vrinat, Appl. Catal., 6, 137 (1983) and references therein.
2. I. A. van Parijs and G. F. Froment, Ind. Eng. Chem. Prod. Res. Dev., 25, 431 (1986).
3. S. K. Ihm, S. J. Moon and H. J. Choi, Ind. Eng. Chem. Res., 29, 1147 (1990).
4. M. J. Ledoux, C. P. Huu, Y. Segura and F. Luck, J. Catal., 121, 70 (1990).
5. J. Leglise, J. van Gestel and J. C. Duchet, J. C. S. Chem. Comm., 5, 611 (1994).
6. J. Kraus and M. Zdrazil, React. Kinet. Catal. Lett., 6, 475 (1977).
7. J. A. De Los Reyes and M. Vrinat, Appl. Catal. A : General, 103, 79 (1993).
8. H. C. Lee and J. B. Butt, J. Catal., 49, 320 (1977).
9. J. Joffre, P. Geneste, J. B. Mensah and C. Moreau, Bull. Soc. Chim. Belg., 100, 865 (1991).
10. F. E. Massoth, J. Catal., 47, 316 (1977).
11. V. Moravek and M. Kraus, Collect. Czech. Chem. Comm., 50, 2159 (1985).
12. C. Moreau, J. Joffre, C. Saenz, J. C. Afonso and J. L. Portefaix, in Actas Simp. Iberoam. Catal. (S. I. Catal., ed.), Segovia, 1992, p.1261.
13. R. Houriet, H. Schwarz, W. Zummack, J. G. Andrade and P. Von Ragné Schleyer, Nouv. J. Chim., 5, 505 (1981).
14. A. N. Startsev, Russian Chemical Reviews, 61, 175 (1992).
15. B. Radomyski, J. Szczygiel and J. Trawczynski, Appl. Catal., 39, 25 (1988).
16. Y. Kawaguchi, I. G. Dalla Lana and F. D. Otto, Can. J. Chem. Eng., 56, 65 (1978).
17. S. Morooka and C. E. Hamrin Jr., Chem. Eng. Sci., 32, 125 (1977).
18. F. E. Massoth and P. Zeuthen, J. Catal., 145, 216 (1994).
19. M. Zdrazil, Collect. Czech. Chem. Comm., 40, 3491 (1975).
20. S. P. Ahuja, M. Derrien and J. F. Le Page, Ind. Eng. Chem. Prod. Res. Dev., 9, 272 (1970).
21. N. N. Sauer, E. J. Markel, G. L. Schrader and R. J. Angelici, J. Catal., 117, 295 (1989).

Role of the Sulfiding Procedure on Activity and Dispersion of Nickel Catalysts

J.P. Janssens [°], A.D. van Langeveld, R.L.C. Bonné [b], C.M. Lok [b], and J.A. Moulijn

Delft University of Technology, Faculty of Chemical Technology and Materials Science, Julianalaan 136, 2628 BL, Delft, The Netherlands

[b] Unilever Research Port Sunlight Laboratory, Quarry Road East, Bebington, Wirral L63 3JW, United Kingdom

[°] to whom all correspondence should be addressed, e-mail: J.P.Janssens@stm.tudelft.nl

INTRODUCTION

Alumina supported nickel catalysts are used in petrochemical and refining processes, for example in pyrolysis gasoline hydrogenation. The conversion of di-olefinic hydrocarbons into mono-olefinic hydrocarbons is necessary to prevent the formation of undesired gums and resins. However, the hydrogenation reaction has to be selective because hydrogenation of mono-olefines result in a reduction of the octane number of the gasoline. In order to obtain the desired selectivity nickel has to be (partially) sulfided. In general, three different procedures can be applied to sulfide and activate hydrotreating catalysts [1]:
(1) Reduction of the oxidic catalyst precursor followed by sulfiding;
(2) Simultaneous reduction and sulfiding of the oxidic catalyst precursor;
(3) Sulfiding of the oxidic catalyst precursor followed by reduction.
Also, the nature of the activation molecules can be various. For the sulfiding step a spiked feed or pure sulfiding agents such as thiophene, H_2S, dimethyldisulfide and CS_2 can be employed. For the reduction step H_2 is generally used.

Present work describes the characterisation of the oxidic nickel catalyst precursors as well as the activated sulfidic nickel species with Temperature Programmed Reduction (TPR), Temperature Programmed Sulfiding (TPS) and X-ray Photoelectron Spectroscopy (XPS). The simultaneous sulfiding-reduction process with H_2S/H_2 will be elucidated and the influence of the sulfiding step on the dispersion of the active phase will be assessed. Thiophene hydrodesulfurisation (HDS) is used as a model reaction to determine the activity of the sulfided catalyst samples.

EXPERIMENTAL

Commercial θ-Al_2O_3 supported nickel catalysts are provided by Crosfield Catalysts. Table 1 summarizes their characteristics.

Table 1. Characteristics of investigated catalysts.

Catalyst	HTC100	HTC200	HTC400	HTC500
nickel content in wt.% Ni	7.8	12.5	15.6	18.6
nickel surface area in 10^3 m^2/kg Ni (from H_2 chemisorption)	161	146	165	167

The reduction and sulfiding behaviour of the oxidic nickel catalyst precursor are studied with TPR and TPS. Experiments are performed in an atmospheric flow reactor. The heating rate is 0.167 K/s and the sample weight varies between 50 and 100 mg.

For TPR a gas mixture consisting of 66 mole% H_2 in Ar is employed. For TPS a gas mixture consisting of 3 mole% H_2S, 25 mole% H_2 and 72 mole% Ar is used. The H_2 consumption of the catalyst samples is monitored with a Thermal Conductivity Detector (TCD) and the H_2S consumption or production is monitored with an UV-spectrophotometer tuned at 205 nm.

XPS spectra of the oxidic nickel catalyst precursors and of the sulfided catalysts are recorded. The sulfided samples are prepared by sulfiding up to 623 K, followed by cooling to room temperature in H_2S-H_2/Ar flow. The samples are transported to the XPS apparatus under air exposure.

Thiophene HDS activity measurements of the activated catalysts are carried out with a gas mixture consisting of 6 mole% thiophene in H_2 in an atmospheric flow reactor at 673 K. The oxidic catalysts are activated *in situ* using a mixture of 10 mole% H_2S in H_2, the applied heating rate is 0.10 K/s from 293 to 673 K. The sample weight used is approximately 200 mg (particle sieve fraction: 63-160 μm) diluted with an equal amount of SiC. The reaction products are analysed with an on-line gas chromatograph during three hours time on stream. The thiophene activity tests are run at constant space velocity.

RESULTS AND DISCUSSION

Figure 1 shows the TPR profiles of the oxidic nickel catalyst precursor. The oxidic precursors show a reduction peak over a wide temperature range from 400 up to 850 K. Scheffer et al. [2] assigned the TPR peaks to various nickel species depending on the temperature range in which they appear:

up to 600 K   : reduction of bulk nickeloxide;
700 - 1000 K   : reduction of disperse nickeloxide species interacting with the
                              support;
1000 - 1200 K : reduction of nickel aluminates.

The broad reduction peak from 400 K up to 850 K indicates the presence of several disperse nickeloxide species. Since no reduction peaks in the temperature region 1000 K up to 1200 K are observed, it can be concluded that no nickel aluminates are present [2]. The use of θ-Al_2O_3 support eliminates the formation of surface spinel species.

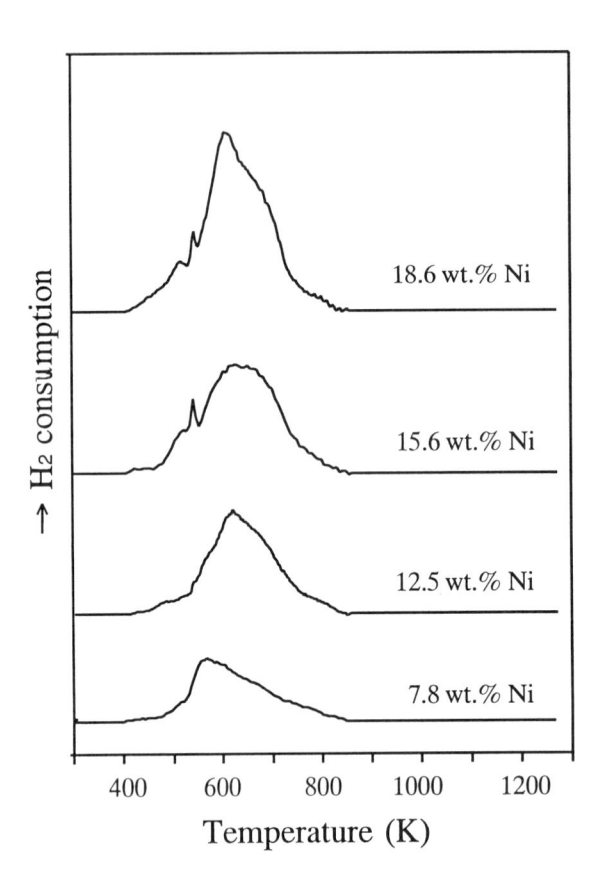

Figure 1. TPR profiles of the oxidic nickel catalyst precursor. (all TPR profiles are normalized to 100 mg of catalyst sample)

The Ni(15.6 wt.%) and Ni(18.6 wt.%) catalysts show an additional amount of reduction in the temperature region 400 K up to 600 K with a sharp reduction peak at approximately 540 K, which is not present in the Ni(7.8 wt.%) and Ni(12.5 wt.%) catalysts. This indicates that the Ni(15.6 wt.%) and Ni(18.6 wt.%) catalysts contain some large nickeloxide clusters which are not present in the Ni(7.8 wt.%) and Ni(12.5 wt.%) catalysts. As can be observed from the relative peak area, the amount of these nickeloxide clusters is very small.
Quantitative TPR revealed that nickel on the oxidic nickel catalyst precursors is not present as NiO, but rather as nickeloxide species with a O/Ni ratio varying between 1.2 and 1.5, dependent on the nickel loading.

Figure 2 shows the TPS profiles of the oxidic nickel catalyst precursor. For all samples a H_2S consumption was measured during isothermal sulfiding at room temperature. The θ-Al_2O_3 support adsorbs small amounts of H_2S, whereas the oxidic nickel precursors show a considerable H_2S uptake, indicating that nickeloxide is already sulfided at this low temperature. When the temperature was raised slightly above room temperature some H_2S desorbs.
In the temperature region from 350 K up to 500 K, all catalyst samples show a large H_2S consumption. The major part of the H_2S consumption takes place in this region. For the Ni(7.8 wt.%) and Ni(12.5 wt.%) catalysts no significant change in H_2 concentration is observed in this region, whereas the Ni(15.6 wt.%) and Ni(18.6 wt.%) catalysts show H_2 consumption simultaneous with H_2S consumption. It is suggested that the H_2 consumption in this temperature region is due to the hydrogenation of sulfur. Apparently, at higher nickel loadings sulfur is easily hydrogenated due to the presence of larger nickel crystallites and, hence, a decreased influence of the support on the metal-sulfur interaction [3]. It is suggested that the observed nett H_2S consumption consists of superposition of a H_2S production due to the hydrogenation of sulfur and a large H_2S consumption due to sulfiding by O-S exchange [4].
In the temperature region from 500 K up to 600 K, all catalyst samples show H_2 consumption and H_2S production in stochiometic amounts. This indicates the hydrogenation of excess sulfur S_x according to $S + H_2 \rightarrow H_2S$. Apparently, excess sulfur, which is chemisorbed on coordinatively unsaturated sites, is

formed at temperatures below 500 K [5].

In the temperature region from 700 K up to 1000 K, all catalyst samples also show H_2 consumption and H_2S production in stochiometric amounts, indicating the reduction of sulfided nickel species. No significant H_2S consumption was observed during isothermal sulfiding at 1273 K, suggesting that difficult sulfidable nickel aluminates are absent.

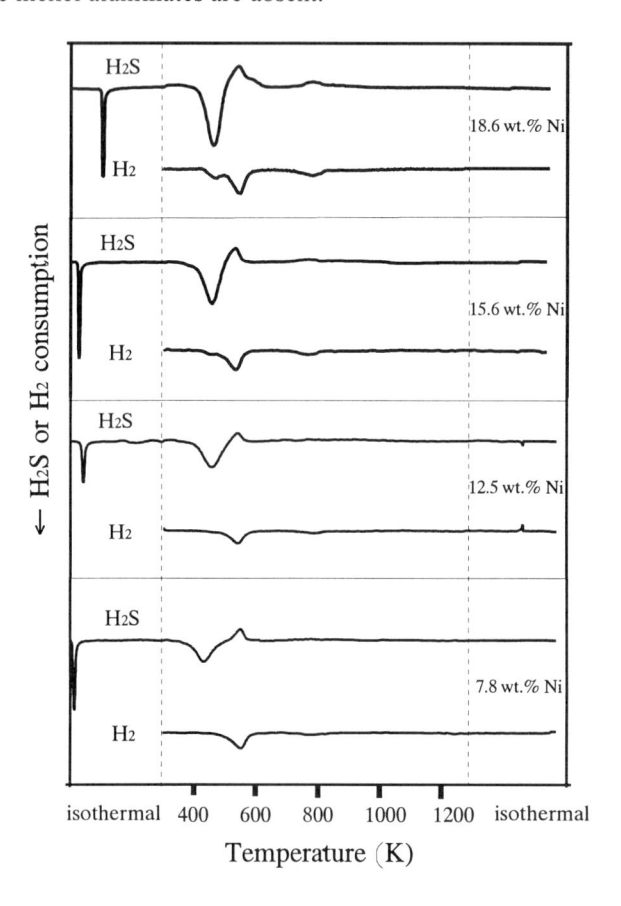

Figure 2. TPS profiles of the oxidic nickel catalyst precursor.
(all TPS profiles are normalized to 100 mg catalyst sample)

From quantitative TPS the amount of sulfur consumed by the oxidic nickel catalyst precursor can be determined. Table 2 lists the S/Ni ratios obtained at 623 and 1273 K, assuming that all sulfur is bonded to nickel as nickelsulfide species.

Table 2. Overall stoichiometry of nickelsulfide from TPS.

Catalyst	HTC100	HTC200	HTC400	HTC500
S/Ni ratio (623 K) in mol/mol	0.77	0.72	0.81	0.70
S/Ni ratio (1273 K) in mol/mol	0.69	0.65	0.74	0.60

From XPS line position analysis on the oxidic nickel catalyst precursor it is inferred that Ni_2O_3 is present [6]. This supports the quantitative TPR results which show that nickeloxide is not present as NiO, but rather as nickeloxide species with a O/Ni ratio varying from 1.2 to 1.5.

XPS analysis of the sulfided nickel catalyst gives no clear identification of the nickelsulfide species present on the catalyst. Due to transport through air the nickelsulfide species are (partly) oxidized into sulphates. Clearly, in situ XPS is necessary to characterize the sulfided nickel catalysts.

Kerkhof & Moulijn [7] proposed a model which gives the possibility to predict a theoretical XPS intensity ratio of the active phase and the support for solid catalysts. From the difference between the theoretical and experimental XPS intensity ratio the crystallite size of the active phase can be calculated.

Figure 3 shows the measured XPS intensities of the oxidic and sulfided nickel catalysts as a function of the nickel loading.

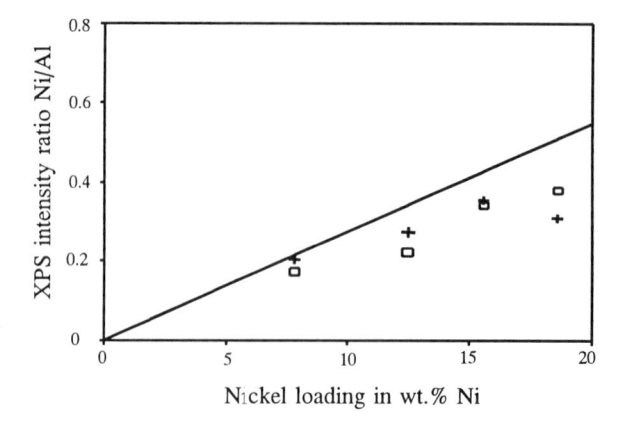

Figure 3. Measured XPS intensities of oxidic catalyst precursor and sulfided precursor.
(+ oxidic nickel catalyst precursor; □ sulfided nickel catalyst; — theoretical)

An almost linear dependence of the XPS intensity ratio with the nickel loading is observed at low nickel loadings. An increasing discrepancy between the measured XPS intensities and the theoretical line at higher nickel loadings can be understood from a reduced dispersion of the active phase or the formation of bulk active phase, from which only part is observed with XPS. As expected, an increase in crystallite size is observed with increasing nickel loading, data evaluation indicates crystallite sizes range from 0.3 to 1.0 nm.

Comparison of the calculated crystallite sizes of the oxidic and sulfided nickel catalysts shows no significant difference between the two states. Clearly, no lumping of the surface nickel species is observed during sulfiding with a strong reagent as H_2S up to 623 K.

Main product of the thiophene hydrodesulfurisation with sulfided nickel catalysts is butene, whereas butane can not be detected. This indicates that the hydrogenation activity of completely sulfided nickel catalysts is very low. The decrease in activity of 'over-sulfided' nickel catalysts was also observed by Bourne at al. [8].

It is suggested that mild surface sulfiding is necessary to obtain selective hydrogenation catalyst. Furthermore, core sulfiding of the nickel particulates leads to dramatic decrease in hydrogenation activity.

All samples show comparable thiophene HDS activity except for the Ni(7.8 wt.%) catalyst. When expressed as pseudo first order reaction rate constant, normalized to the nickel surface area, the lowest nickel content catalyst shows a significantly higher activity, which may suggest a structure dependence for the thiophene HDS reaction. The thiophene HDS activity normalized on the total weight of catalyst increases with increasing nickel loading.

REFERENCES

1. Prada Silvy, R., Fierro, J.L.G., Grange, P., and B. Delmon, in Preparation of Catalysts IV, (B. Delmon, P. Grange, P.A. Jacobs and G. Poncelet, eds.), Elsevier Science Publishers, Amsterdam, 1987, pp. 605-617.
2. Scheffer, B., Molhoek, P., and J.A. Moulijn, Appl. Catal., 46: 11 (1989).
3. Bonné, R.L.C., in Hydrodemetallisation of Ni-TPP and VO-TPP over sulfided molybdenum and vanadium catalysts, Chapter 7, PhD thesis, University of Amsterdam, pp. 108-127 (1992).
4. Mangnus, P.J., Poels, E.K., Langeveld, A.D. van, and J.A. Moulijn, J. Catal., 137: 92 (1992).
5. Mangnus, P.J., Riezebos, A., Langeveld, A.D. van, and J.A. Moulijn, J. Catal., 151: 178 (1995).
6. Wagner, C.D., Riggs, W.M., Davis, L.E., Moulder, J.F., and G.E. Muilenberg, in Handbook of X-ray Photoelectron Spectroscopy, published by Perkin-Elmer Corporation, Minnesota, 1979.
7. Kerkhof, F.P.J.M., and J.A. Moulijn, J. Phys. Chem., 83: 1612 (1979).
8. Bourne, K.H., Holmes, P.P., and R.C. Pitkethly, in Proceedings of the third International Congress on Catalysis Vol. II, (W.M.H. Sachtler, G.C.A. Schuit and P. Zwietering, eds.), North Holland Publ. Co., Amsterdam, 1965, pp. 1400.

Ruthenium Sulfide Based Catalysts

Michèle Breysse, Christophe Geantet, Michel Lacroix, Jean Louis Portfaix, and Michel Vrinat   Institut de Recherches sur la Catalyse, CNRS, 2, Avenue Albert Einstein, F-69626, Villeurbanne Cédex, France

I. ABSTRACT

New economical and ecological constraints call for the development of more active catalysts than the conventional ones for all hydrotreating reactions. In this respect, ruthenium sulfide catalysts have received attention due to their prominent catalytic activities. The present paper gives an overview of the studies carried out in our group concerning the properties of ruthenium sulfide supported on alumina or dispersed in zeolite, its ability to form solid solutions either in the supported or unsupported state and the utilization of ruthenium sulfide as a model compound for fundamental studies dealing with the interaction of reactants and sulfide phases.

II. INTRODUCTION

The large number of research works devoted to the hydrotreating processes carried out in industrial or academic laboratories have led to the present CoMo, NiMo or NiW catalysts supported on alumina often doped by phosphorus or fluorine. New ecological and economic constraints require the development of a new generation of catalysts much more efficient than the present ones. As a matter of fact, new regulations have appeared or will appear in the near future concerning the maximal concentration of sulfur in fuels and gasoils. It is most likely that the amount of nitrogen and aromatics will also be strongly limited. In addition, there is a change in the nature of feedstocks to be processed, either heavier crudes containing more sulfur and nitrogen, or distillates from conversion processes which also contain more heteroatoms than straight run distillates. Using the present catalysts, it would be necessary to build new plants working in more severe conditions to fulfill the new regulations. The significant investment required will probably imply that these new plants will be build only in the most important refineries. In these conditions the development of more efficient catalysts represents a very important challenge for the refiners in the next few years. This problem can be addressed by replacing alumina by other supports, changing the active phase using other transition metal sulfides or compounds such as nitrides. For all these directions of research the exploratory state has not been passed yet and still an important effort concerning the fundamental aspects of the hydrotreating catalysts and reactions appears necessary.

Over the last few years extensive research on other transition metal sulfides in hydrotreating reactions have been carried out. Ruthenium sulfide was found to exhibit prominent catalytic activities. According to Pecoraro and Chianelli *(1)* bulk RuS_2 appeared to be 13 times more active than MoS_2 for dibenzothiophene hydrodesulfurization (HDS). Lacroix et al *(2)* reported an hydrogenation activity 6 times higher than that of MoS_2. In both studies unsupported RuS_2 displayed one of the highest activities in comparison with other transition metal sulfides. These high catalytic properties were also observed for carbon supported ruthenium sulfide *(3,4)*. Several interesting properties could be expected if RuS_2 catalysts were to be deposited on other supports like alumina or zeolites. In addition to these promising properties of RuS_2 in catalysis, it has been shown that this chalcogenide may form ternary compounds with other pyrite type sulfides like CoS_2 *(5)*, CoS_2, NiS_2 or FeS_2 *(6)*. This means that the properties of mixed sulfide solid solutions can be studied by contrast to molybdenum or tungsten sulfide catalysts for which these ternary sulfide phases whose slabs are only decorated by the promoter atoms. Lastly, ruthenium sulfide is a good candidate for more fundamental studies concerning the interaction of reactants and sulfide catalysts due to both its structural properties and outstanding catalytic properties.

The objective of the present paper is to give an overview of the studies carried out in our laboratory concerning i) the characterization of active sites and the nature of adsorbed hydrogen species, ii) the properties of ruthenium sulfide supported on alumina or dispersed in zeolite. iii) the ability to form solid solutions either in the supported or unsupported state.

III. RESULTS AND DISCUSSION

A. Active sites characterization, interaction with hydrogen

Numerous studies were devoted to the numbering of active sites by probe molecule adsorption but the results are rather contradictory. Several reasons account for these discrepancies. One of the most important reasons is related to the lack of physico-chemical characterization of the interaction between the molecules and the catalyst. As a matter of fact, the industrial CoMo or NiMo supported on alumina catalysts are very difficult to characterize due to their structural complexity and the existence of support interaction. Similarly, the unsupported CoMo or NiMo are biphasic and pure MoS_2 catalyst presents relatively poor catalytic properties.

As stated above, ruthenium sulfide presents outstanding properties for hydrogenation and hydrodesulfurization reactions and its activity levels are close to those of molybdenum sulfide promoted by nickel or cobalt. This is the reason why we have chosen ruthenium sulfide for these studies. As a first step, it was decided to use unsupported catalysts in order to avoid any difficulties coming from the sulfide - support interaction.

1. Nature of active sites

It has been recognized that for lamellar sulfide catalysts the activity can be related to the presence of coordinatively unsaturated cations formed by removal of superficial sulfur atoms in the conditions used for hydrotreatment reactions. For example, this was shown by Kasztelan et al. by modifying the number of anionic vacancies by progressive desulfurization of a MoS_2 catalyst *(7)* and simultaneous measurement of the catalytic activity for isoprene hydrogenation. A similar study was carried out on ruthenium sulfide with a high surface area ($70m^2g^{-1}$) *(8, 9)*. The fully sulfided state was taken as the starting point and the influence of progressive desulfurization on the activities for the 1-butene hydrogenation and the H_2-D_2 exchange reaction was evaluated. These reactions were selected because they occur at temperatures lower than the temperatures required for catalyst reduction.

The study of the reduction process of ruthenium disulfide has shown that about 50% of the total sulfur content can be eliminated from the solid without any noticeable modification of the structural and morphological properties. But for an amount of H_2S evolved corresponding to a higher degree of reduction, bulk solid reduction occurs which provokes surface collapse and phase segregation leading to metallic ruthenium. The changes in the *activities* of the catalyst as a function of the amount of H_2S evolved are represented in Fig. 1. These results clearly show that as long as the stability of the catalyst is preserved an increase of the degree of desulfurization brings about a large increase of activity for both reactions.

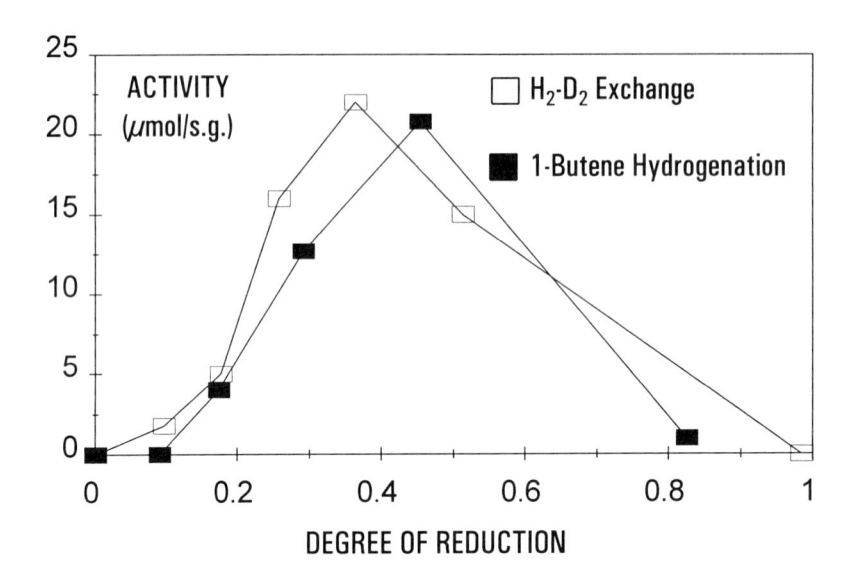

Fig.1. H$_2$-D$_2$ exchange and 1-butene activities as a function of the degree of reduction of the catalysts (From Ref. 8.).

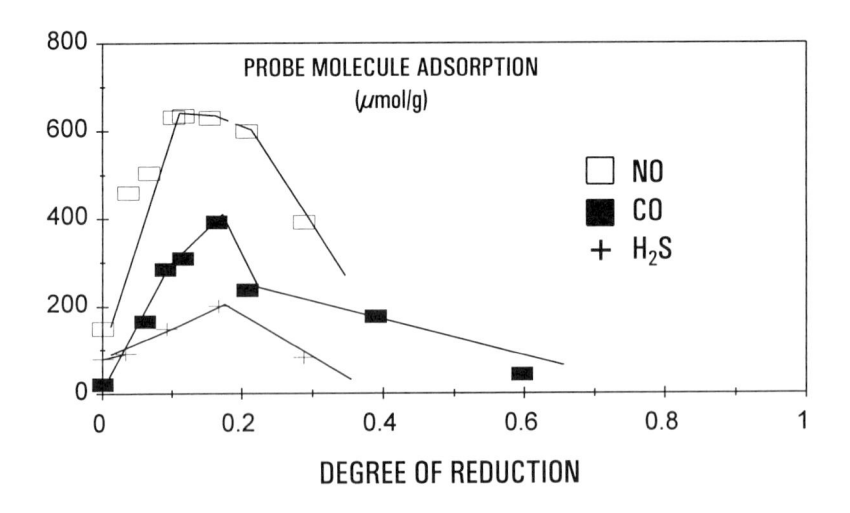

Fig.2. Adsorption of CO, NO and H$_2$S as a function of the degree of reduction of the catalyst (From Ref. 9.).

The major increase of the catalytic properties with desulfurization demonstrates the role of anionic vacancies as for the lamellar sulfides. Using the same samples, we tried to make the numbering of active sites by probe molecule adsorption, i.e. CO, NO and H_2S *(9)*. The adsorption of these molecules increased with the degree of reduction of the solid. However, the maximum of adsorption capacities did not correspond to the maximum of catalytic activity observed for the model hydrogenation reactions, i.e. H_2-D_2 exchange and 1-butene hydrogenation (Fig. 2). This suggested that distinct sites are required for the adsorption of these probe molecules and catalysis. The use of hydrogen for characterizing active sites appeared then much more realistic since this molecule is involved as reactant in all hydrotreating reactions.

2. Interaction with hydrogen

Two different species were detected by thermoflash desorption of hydrogen : one was assigned to hydrogen adsorbed on surface sulfur anions while the other one could be retained on coordinatively unsaturated ruthenium cations. In agreement with these results, ^1H NMR study of the reduced samples has evidenced two signals located at 5.1 and -7.4 ppm, with respect to tetramethylsilane, ascribed to SH groups and to hydride - type adsorbed species respectively *(10)*. The existence of these two different kinds of hydrogen adsorbed was further confirmed by inelastic neutron scattering with bending modes of linear RuH species appearing at 540 and 823 cm^{-1} and two non degenerated SH bending modes of a simple state located at 645 and 720 cm^{-1} *(11)*. The relative concentrations of the hydridic and protonic (SH) species, denoted by H^a and H^b respectively, depend on the degree of reduction of the samples (Fig. 3). The comparison of these adsorption results with those obtained for the H_2-D_2 exchange and the 1-butene hydrogenation evidences a striking similarity between the concentration of hydridic species and the catalytic activity in hydrogenation reactions (Fig. 4). This strongly favours the hypothesis that the hydridic species is the active one for hydrogenation reactions.

This work gives the first experimental support to the often proposed hypothesis on the role of hydrogen adsorbed on coordinatively unsaturated sites in catalysis by sulfides. Moreover, this opens a way to determine turn over numbers as it is done for metallic catalysts.

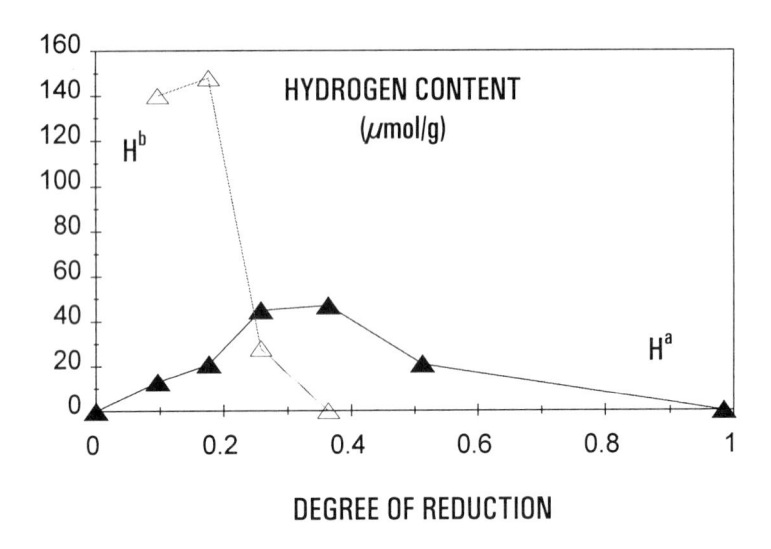

Fig.3 Amount of H^a and H^b adsorbed as a function of the degree of reduction of the catalyst (From Ref. 10.).

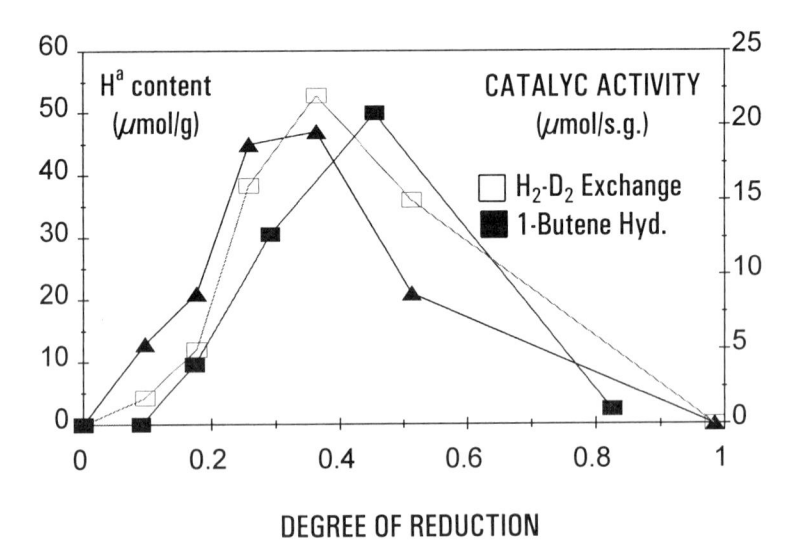

Fig.4. Correlation between catalytic activity and H^a adsorption (From Ref. 10.).

B. Catalytic properties of supported ruthenium sulfide

1. Alumina supported catalysts

Literature on the preparation and activation of supported RuS_2/Al_2O_3 is rather scanty and the relatively low activities reported seem to be related to an incomplete sulfidation of ruthenium *(12 - 14)*. In a preliminary work *(15)*, we demonstrated that sulfidation in the absence of hydrogen and without prereduction or calcination allows the genesis of the pyrite phase and that sulfidation under 15% H_2S in nitrogen leads to a sulfided, well dispersed and very active catalyst for thiophene hydrodesulfurization. Monitoring the size of the RuS_2 particles may be obtained by using the same sulfiding mixture but at various temperatures *(16)*. The variation of the average crystal size as a function of the sulfiding temperatures is given in Table 1. The properties of the same catalyst for the hydrodesulfurization (HDS) of thiophene and the hydrogenation of biphenyl are also given in Table 1.

Table 1. Particle sizes and catalytic activities of a ruthenium sulfide supported on alumina catalyst containing 7 weight % of Ru sulfided at various temperatures (From Ref. 16.).

Sulfiding temperatures	l (nm)	r HYD (1)	r HDS (2)
573	>3.0	10	160
673	3.0	8	580
773	3.2	2	730
873	4.0	1.5	750
973		1.0	610

(1) Biphenyl HYD rate (10^{-8}mol.s^{-1}.g^{-1})
(2) Thiophene HDS rate (10^{-8}mol.s^{-1}.g^{-1})

Striking differences appear since the HDS and hydrogenation activities vary in opposite ways. The TPR patterns of these catalysts present also important differences. On the basis of crystallographic considerations, it was proposed that the sulfidation temperature could affect not only the size but also the nature of the planes exposed at the surface. Low sulfidation temperatures would lead to small particles with cubo-octahedrical and icosahedrical shape, preferentially exhibiting (111) planes which would be more active in hydrogenation. High sulfidation temperature induces large RuS_2 particles, mostly with (210) faces, which would be more active in HDS.

To obtain higher hydrogenation activities, it would be interesting to prepare better dispersed RuS_2 particles. Nevertheless, this was not possible on alumina where the smallest particle size obtained was around 3 nm. Using zeolites, higher dispersion might be achieved if it would be possible to introduce ruthenium sulfide particles inside the silico-aluminate framework.

2. Ruthenium sulfide dispersed in zeolites

Ruthenium sulfide in Y zeolite has good properties for hydrodenitrogenation reactions, which were first reported by Harvey and Matheson *(17)*. An activity similar (or slightly better than) that of conventional $NiMo/Al_2O_3$ was obtained for hydrodenitrogenation of quinoline with metal loadings much lower than those used commonly in commercial hydrodenitrogenation catalysts. Unexpectedly, the high activity of these catalysts for HDN was not reflected in a correspondingly high activity for HDS. From XRD examinations and TEM imaging of the sulfided catalysts, the authors concluded that the ruthenium sulfide was dispersed as aggregates composed of only few atoms.

The very high dispersion which can be achieved using zeolites is also reported in a recent work where the properties of sulfided Ru, Rh, Pd and Mo in KY zeolite were compared *(18)*. In most of the samples, the particle sizes were equal to or smaller than 1 nm (in some cases, the size of the particles were so small that their identification was at the limit of the resolution of the microscope) and homogeneously dispersed at the interior of the framework of the zeolite. The sulfur on metal ratios determined by EDX analysis on the particles of Mo and Ru were close to or greater than the stoichiometric values expected for the corresponding sulfides. By contrast, it appears that Rh and Pd catalysts were not completely sulfided. Results given in Table 2 evidence the superiority of Ru by comparison to the other transition metal sulfides dispersed in zeolites for the conversion of nitrogen containing molecules, particularly for the diethylaniline conversion. This very high activity can be related to the ability of particles of ruthenium sulfide partially desulfurized to activate hydrogen as explained above. As a matter of fact, TPR observations showed that these small particles are highly reducible and the determination of the sulfur to ruthenium ratio after test evidenced a partial desulfurization of the catalyst. This variation of the sulfur to metal ratio was also observed in the TPR of molybdenum catalysts but results obtained in our group showed that ruthenium sulfide is about three orders of magnitude more active than molybdenum sulfide for model hydrogenation reactions (at the desulfurization state corresponding to the maximum of activity) *(19)*. The catalytic properties presented in Table 2 are also related to the hydrogenation function of the catalyst since the first step in the conversion of the model molecule is an hydrogenation reaction. The catalysts containing Rh and Pd would behave as metals which are more or less poisoned with sulfur since the physico-chemical characterizations utilized never evidence a real sulfide structure. It is interesting to note that for all these reactions, the activities of the alumina supported catalysts were lower. Moreover, the high activity of Ru/KY was not observed in HDS. This confirms the previous results reported by Harvey and Matheson and our hypothesis concerning the different nature of hydrogenation and hydrodesulfurization sites.

Table 2. Catalytic activities of M/KY, M = Mo, Ru, Rh and Pd containing in weight % Mo :3.3, Ru : 3.2, Rh : 2.7, Pd : 2.7, for the conversion of nitrogen containing molecules (From Ref. 18.).

Catalyst	r Pyridine	r Diethylaniline	Quinoline conversion
Mo/KY	5	3	6
Ru/KY	28	1690	91
Rh/KY	28	150	52
Pd/KY	6	50	52

rate of pyridine conversion (molec/ metal atom.s)10^4
rate of 2,6 diethylaniline conversion (molec/ metal atom .s)10^4
quinoline conversion in %

Nevertheless, the results presented above concern only one kind of zeolite with low acidic properties. It is obvious that the properties of these clusters of ruthenium sulfide can also be influenced by the zeolite environment. Work is in progress in our laboratory concerning these problems.

III. Ternary sulfide phases

A. Unsupported system

As mentioned above, ruthenium sulfide may form ternary sulfide phases with other pyrite compounds. Members of the system $M_xRu_{1-x}S_2$ (M = Ni, Co, Fe) were prepared by sulfidation of mixtures of hydroxides at 673 K, the homogeneity range of the pyrite structure was examined and the catalytic properties were determined for the hydrodesulfurization of thiophene and the hydrogenation of biphenyl *(20)*. The XRD patterns of $M_xRu_{1-x}S_2$ (M = Ni, Co, Fe) can be interpreted on the basis of a cubic pyrite structure. As reported in Fig.5, the lattice parameters were found to verify the Vegard's law. The linear change of the cell parameter with the composition is a proof of a solid solution formation. The catalytic properties of $Ni_xRu_{1-x}S_2$ are much higher than those of RuS_2 even for high Ni content (Ni = 0.8). Cobalt has also a promoting effect in HDS but the introduction of iron in the pyrite phase leads to a decrease of the activity. The HDS activities of the ternary sulfides, the RuS_2 as well or those obtained for the molybdenum based catalysts (r = M/(M+Mo) = 0.3) are schematized in Fig.6. The ruthenium solid solution present higher activities than the corresponding NiMoS, CoMoS or FeMoS catalysts. Some parallelism may be drawn between the catalytic activities of the two systems (ruthenium and molybdenum) the components of which behave as synergetic pairs although in the first case a bulk solid solution exists and in the second case the promoter atoms are only located at the edges of the lamellar disulfide.

Among these new phases, the members of the system $Ni_xRu_{1-x}S_2$ appeared to be the most promising, since their catalytic properties for both types of reactions were found to be higher than those for both RuS_2 and the mixed NiMo unsupported catalyst. Consequently, it was of particular interest to examine if such ternary phases can be transposed to the supported state.

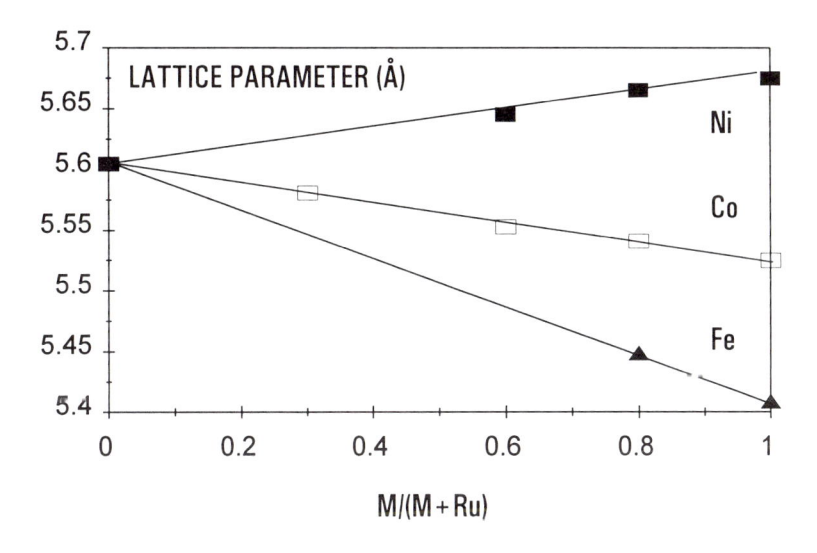

Fig.5. Lattice parameters of ruthenium sulfide compounds containing Ni, Co and Fe as a function of the atomic concentration (From Ref. 20.).

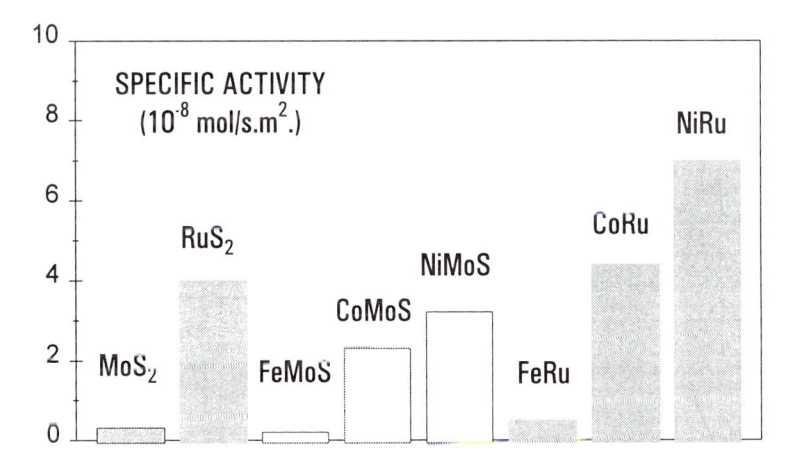

Fig.6. HDS of thiophene. Comparison of the properties of Mo and Ru based catalysts ($M_{0.3}Mo_{0.7}S_y$ and $M_{0.8}Ru_{0.2}S_2$ with M = Fe, Co or Ni), (From Ref. 20.).

B. Alumina supported $Ni_xRu_{1-x}S_2$

Mixed nickel-ruthenium sulfides supported over alumina, with different metal compositions, were prepared by stepwise impregnation and further sulfidation. XRD, XPS, TPR and STEM measurements indicate that ternary nickel-ruthenium-sulfur compounds with a pyrite like structure are synthesized up to an atomic ratio of Ni/(Ni+Ru) near 0.7 *(21)*. The variation of the biphenyl hydrogenation rate versus the atomic composition r is shown in Fig. 7. The hydrogenation rate sharply increases with r, reaches a maximum between r = 0.35 -0.45 and then decreases continuously down to the low value observed for pure nickel sulfide. The maximum of activity is ca. 30 times higher than the sum of the rates displayed by the corresponding amounts of ruthenium and nickel supported sulfides. However, this enhancement of the hydrogenation properties is much higher than for supported catalysts which presented only a two fold increase of the activity by comparison to RuS_2 *(22)*. On that account, the carrier would play an important role by improving the dispersion of the active phase. As already mentioned for ruthenium sulfide supported on alumina, the formation of small particles induces some preferential exposed planes, favoring hydrogenation properties.

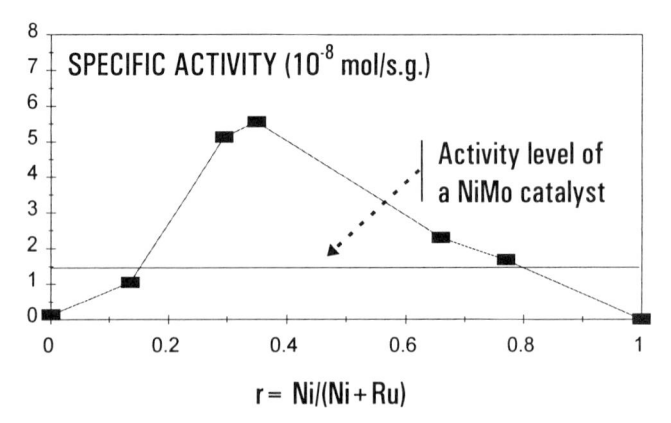

Fig.7. Evolution of the catalytic properties of mixed NiRu sulfide phases supported over alumina in biphenyl hydrogenation with the atomic composition r = Ni/(Ni+Ru). Comparison with the activity of a NiMo supported on alumina reference catalyst (From Ref. 21.).

The synergetic effect is ascribed to the existence of mixed NiRu sites with different electronic properties from those of either Ni or Ru sites themselves as proposed for unsupported catalysts.

It should be noted that this work represents the first example of the transposition of a well defined ternary sulfide phase to a supported state.

IV. CONCLUSION

The objective of the present review was to summarize the comprehensive work carried out in the last few years in our laboratories concerning the ruthenium sulfur systems. This work includes fundamental studies and some more applied ones since the catalytic properties of some of these active phases is much higher than those of the present industrial catalysts. Although due to economic problems, it does not appear yet sensible to utilize ruthenium as the main active phase of future hydrotreating catalysts. This can be done in some applications by combination with other metals or possibly by dispersing small amounts in zeolites. Moreover, the catalytic properties of ruthenium sulfide can be considered as a target which permits to evaluate what kind of activity can be achieved with transition metal sulfide catalysts.

REFERENCES

1. T.A. Pecoraro, and R.R. Chianelli, J.Catal. 120 : 473 (1989).
2. M. Lacroix, N. Boutarfa, C. Guillard, M. Vrinat, and M. Breysse, J.Catal., 102 : 275 (1986).
3. .J. Ledoux, O. Michaux, J. Agostini and P.J. Panissod, J.Catal., 102 : 275 (1986).
4. P.R. Vissers, C.K. Groot., E.M. van Oers, V.J.H de Beer and R. Prins, Bull. Soc. Chim. Belg., 93 : 813 (1984).
5. J.Passaretti, R.R. Chianelli, A.Wold, K. Dwight, J. Covino, J.Solid State Chem., 64 : 365 (1986).
6. A. Bellaloui, L. Mosoni, M. Roubin, M. Vrinat, M. Lacroix, M. Breysse,. C.R.Acad. Sci. Paris, 304 : 1163 (1987).
7. A. Wambeke, L.Jalowiecki, S.Kasztelan, J.Grimblot, J.P.Bonnelle, J.Catal., 109 : 320, (1988)

8. S.Yuan, T.Decamp., M.Lacroix, C.Mirodatos and M. Breysse, M., J.Catal., 132 : 253 (1991).
9. M. Lacroix, C. Mirodatos, M.Breysse, T.Decamp, S. Yuan, New Frontiers in Catalysis, Guczi, L., et al Ed. , Proceedings of the 10 th International Congress on Catalysis, July 1992, Budapest, Hungary, pp. 597.
10. M.Lacroix, S.Yuan, M. Breysse, C. Dorémieux-Morin, J.Fraissard, J.Catal., 138 : 409 (1992).
11. H. Jobic, G. Clugnet, M. Lacroix, S.Yuan, C. Mirodatos and M. Breysse, J. Am. Chem. Soc., 115 : 3654 (1993).
12. K. Wakabayashi, H. Abe and Y.Orito, Kogio Kagaku Asshi, 74 : 1317 (1971).
13. P.C.H. Mitchell, C.E. Scott, J.P. Bonnelle and J.G. Grimblot, J.Catal., 107: 482 (1987).
14. Y.J. Kuo, R.A. Cocco. and B.J. Tatarchuk, J.Catal., 112 : 250 (1988).
15. J.A. De Los Reyes, S Göbölös, M. Vrinat and M.Breysse, Cat.Lett., 5 :17 (1990).
16. J.A.De Los Reyes, M. Vrinat, C.Geantet, M.Breysse, Catalysis Today, 10 : 645 (1991).
17. T.G.Harvey and T.W. Matheson, J.Catal., 101 : 253 (1986).
18. J.L.Zotin, M.Cattenot, J.L.Portefaix, and M Breysse, Proceedings of the Ibero-American Congress on Catalysis, 1994, pp. 1431.
19. Decamp, T., Ph.D.Thesis, Université Claude Bernard, Lyon (1992)
20. M.Vrinat, M.Lacroix, M.Breysse, L.Mosoni, L., and M. Roubin, Catt. Lett. 3 : 405 (1989).
21. J.A. De Los Reyes, M.Vrinat, C.Geantet, M. Breysse and J. Grimblot, J.Catal., 142 :455 (1993).
22. M. Vrinat, M. Lacroix, M. Breysse, A. Bellaloui, L. Mosoni, and M. Roubin, in "Proceedings, 9th International Congress on catalysis, Calgary, 1988", (M.J.Philips and M.Ternan, Eds.,) Vol.1.p.88. Chem. Inst. of Canada, Ottawa, 1988.

Comparison of Sulfided CoMo Al$_2$O$_3$ and NiMo/Al$_2$O$_3$ Catalysts in Hydrodesulfurization of Gas Oil Fractions and Model Compounds

Xiaoliang Ma, Kinya Sakanishi, Takaaki Isoda, and Isao Mochida

Institute of Advanced Material Study, Kyushu University, Kasuga, Fukuoka 816, Japan

Introduction

Deep desulfurization of petroleum products with the least severity is desired to protect the environment (1,2,3,4). CoMo catalyst has conventionally been applied in the hydrodesulfurization of gas oil to remove sulfur in products, since under conventional hydrodesulfurization (HDS) processing of gas oil it shows slightly higher catalytic activity for HDS and less hydrogen consumption than NiMo catalyst. The latter catalyst is more commonly applied in hydrogenation and hydrodenitrogenation because of its higher hydrogenation activity than the former (5,6,7). Consequently, relatively less attention has been paid to the HDS of gas oil than NiMo catalysts (8).

In a previous study (9), the superiority of NiMo over the CoMo at the second stage HDS of gas oil was observed in a multi-stage HDS, although both catalysts showed similar activity at the first stage HDS. It is inferred that the NiMo could be more effective for HDS of refractory sulfur compounds, which are hardly desulfurized at the first stage with the CoMo. In another paper (10), we reported that more than 60 kinds of sulfur compounds exist in a gas oil that exhibit very different HDS reactivities. Some alkyldibenzothiophenes with alkyl groups at the 4 and/or 6 positions showed reactivities as low as two orders of magnitude less than that of benzothiophenes.

The reaction networks for dibenzothiophene have been proposed by Houalla et al. (11) over a CoMo catalyst and by Aubert et al. (12) over a NiMo catalyst. However, little work has addressed the networks and mechanisms of the alkyldibenzothiophenes with alkyls at 4- and/or 6-positions (13), although the removal of them from gas oil is considered as an important challenge in deep HDS (10). Specifically, no work examining the influence of different catalysts on the reaction networks and mechanisms of such compounds has been found in the obtainable literature.

In the present paper, an attempt was made to compare the catalytic activities of CoMo and NiMo for the HDS of fractions containing different sulfur species, and for model compounds dibenzothiophene (DBT) and 4,6-dimethyldibenzothiophene (4,6-DMDBT). The reaction networks of alkyl dibenzothiophene were proposed and the rate constants of hydrogenolysis and hydrogenation respectively over the NiMo and the CoMo were estimated and compared. The electron densities of the sulfur atoms in some sulfur compounds were calculated by Computer Aided Chemistry (CAChe) and further correlated with their relative desulfurization reactivities in order to clarify the HDS networks and mechanisms of refractory sulfur compounds and the influences of the CoMo and NiMo catalysts on them.

Experimental

A gas oil feed (T-F) (sulfur content: 1.02 wt %, boiling range: 227–377°C) was obtained from a Middle East crude. T-F was separated by distillation at intervals of 20°C into 5 fractions with boiling ranges of <280, 280–300, 300–320, 320–340, and >340°C, which were denoted as F-1, F-2, F-3, F-4, and F-5,

Table 1 Composition of Gas Oil and Its Distribution Fractions

	F-1	F-2	F-3	F-4	F-5	T-F
Boiling range (°C)	<280	280–300	300–320	320–340	>340	227–377
Density (15°C)(g/mL)	0.8409	0.8512	0.8545	0.8675	0.8843	0.8589
Fraction distribution (wt %)	29.0	16.6	15.8	13.1	25.4	
Sulfur content (wt %)	0.544	0.819	0.925	1.262	1.582	1.006
Nitrogen content (wt ppm)	11	34	62	126	380	143
Group composition (HPLC) (wt %)						
Saturates	74.7	70.3	66.4	57.5	44.7	63.1
Mono-aromatics	15.8	16.6	18.2	18.2	19.4	18.1
Di-aromatics	9.4	12.6	14.2	21.8	30.2	16.9
Poly-aromatics	0.1	0.5	1.3	2.5	5.6	1.9

Source: Ref. 16.

respectively. The properties and composition of T-F and distillation fractions are summarized in Table 1. DBT (GC purity > 98%) and biphenyl (GC purity > 98%) were obtained from Wako Chemicals Co. 4,6-DMDBT was synthesized according to the reference (14), and purified by recrystallization.

The catalysts used in the present study were commercial CoMo/Al$_2$O$_3$ and NiMo/Al$_2$O$_3$ supplied by Nippon Ketjen Co. Their properties and composition are listed in Table 2. The catalysts were presulfided with a 5% H$_2$S/H$_2$ flow under atmospheric pressure at 360°C for 6 h before use.

Within 25 minutes, T-F and its fractions were separately desulfurized over the NiMo and the CoMo catalysts at 360°C in a 50-ml magnetically stirred (1000 rpm) batch-autoclave with the catalyst-to-oil weight ratio of 0.10. The heating rate was ca. 25°C/min, and the cooling rate was ca. 30°C/min. The hydrogen gas-to-oil ratio was 110 vol/vol. The total reaction pressure was controlled at 2.9 MPa throughout the reaction by adding gaseous hydrogen into the autoclave to compensate its consumption. HDS of model sulfur compounds, DBT and 4,6-DMDBT was performed in the batch-autoclave at 320°C, 2.5 MPa, using decalin as a solvent. The concentration of the model compounds in the solvent was 0.3 wt % and the catalyst-to-solvent weight ratio was 0.10.

The sulfur content of the oils was analyzed by Automatic Sulfur Determination Apparatus (YOSHIDA Science Co.) after the H$_2$S dissolved in the product oil was purged by nitrogen gas. The sulfur compounds in the oils were

Table 2 Chemical Composition and Physical Properties of Catalysts

	CoMo	NiMo
Chemical composition (wt %)[a]		
MoO$_3$	14.9	14.9
CoO	4.4	
NiO		3.1
SiO$_2$	0.95	4.8
SO$_4$	0.50	0.65
Physical properties:		
Surface area (m^2/g)	268	273
Pore volume (mL/g)	0.53	0.52
Shape	four leaves	four leaves
Average diameter	1.4 × 1.2	1.3 × 1.1
Average length	2.8	3.5

[a]Remainder Al$_2$O$_3$.

analyzed by a Yanaco Gas Chromatograph (G6800) equipped with a methyl silicone capillary column (0.25 mm × 50 m) and a flame photometric detector (FPD). The HDS products of the model compounds were analyzed by the same gas chromatograph equipped with a flame ionization detector (FID).

The CAChe was used to calculate the molecular parameters using the Molecular Orbital Package (MOPAC) program. The optimized geometry and the electronic properties of representative sulfur compounds were determined by solving the Schrödinger equation using the PM3 semi-empirical Hamiltonian (Version 6.10) developed by Stewart (15), where the standard parameters were employed.

Results

1. Catalytic activities of NiMo and CoMo for fractional HDS

The HDS conversions of T-F and each fraction over the NiMo and CoMo catalysts, respectively, at 360°C, 2.9 MPa for 25 minutes are illustrated in Figure 1. The CoMo showed slightly higher catalytic activity than that of the NiMo for HDS of both F-1 and F-2. In contrast, the NiMo catalyst showed significantly higher catalytic activity than that of the CoMo for HDS of the heavier fractions (F-3, F-4, and F-5). The HDS conversion of F-5 over the NiMo was 59%, whereas it was only 48% over the CoMo catalyst. HDS conversions of T-F over the NiMo and CoMo were very similar in this study.

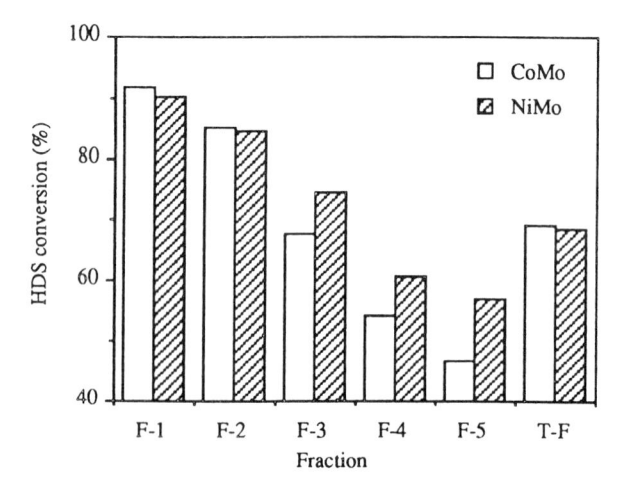

Fig. 1 Comparison of HDS activities of CoMo and NiMo catalysts. Reaction conditions: 360°C, 2.9 MPa, 25 min (from ref. 16).

2. Sulfur compounds existing in each fraction

FPD chromatograms of the total feed and each fraction are shown in Figure 2, which represents more than 60 kinds of sulfur compounds existing in the gas oil. Benzothiophene (BT), DBT, and their alkyl substituted derivatives were dominant. The sulfur compounds existing in each fraction were very different: F-1 primarily contained alkyl BTs with two or three alkyl carbon atoms and some alkyl BTs with four alkyl carbon atoms without any DBTs. F-2 primarily

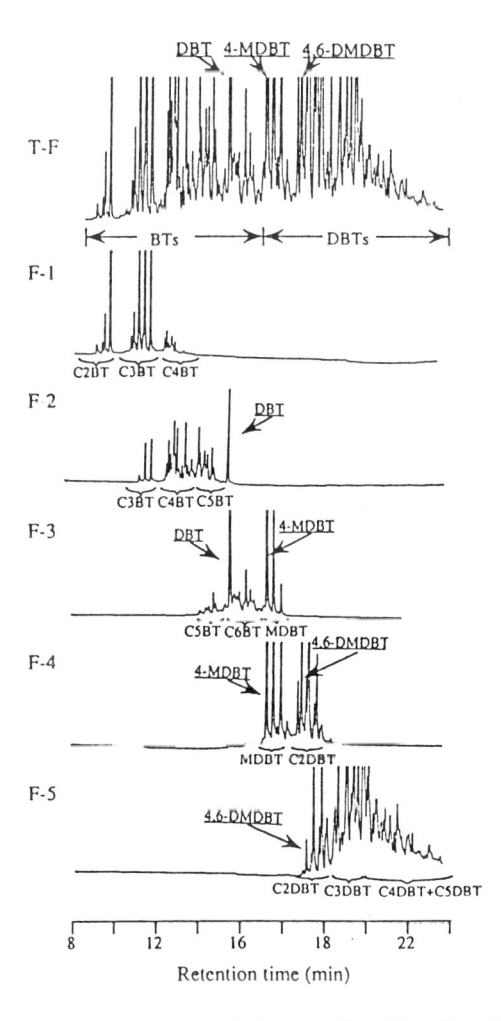

Fig. 2 Sulfur compounds in gas oil and its distribution.

Table 3 HDS Product Distribution of Model Compounds

Catal.	DBT (mol %)			4,6-DMDBT (mol %)			
	DBT	BP	CHB	4,6-DMDBT	DMBP	DMCHB	DMBCH
NiMo	1.7	48.1	50.2	28.2	8.7	50.8	12.3
CoMo	2.4	74.9	22.7	44.4	16.3	34.2	5.1

contained BTs with three, four, and five alkyl carbon atoms and some DBTs. In F-3, the dominant sulfur compounds were DBT and its methyl substituted DBTs (MDBTs), the rest being alkyl BTs with five and six alkyl carbon atoms. In F-4, MDBTs and alkyl DBTs with 2 alkyl carbon atoms (C2DBTs) were the dominant sulfur compounds. Alkyl DBTs with two or more alkyl carbon atoms were dominant in F-5.

3. HDS of model sulfur compounds

The HDS product distribution of model sulfur compounds, DBT, and 4,6-DMDBT at 320°C in 60 minutes over the CoMo and NiMo catalysts, respec-

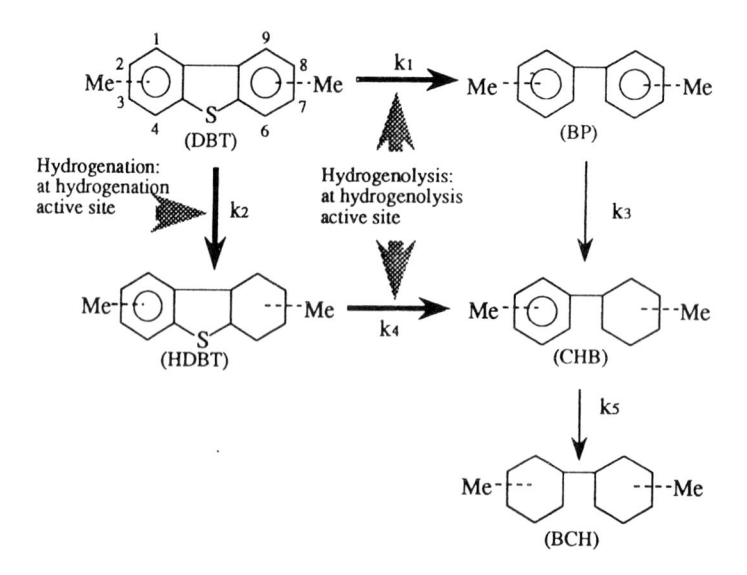

Fig. 3 HDS networks of alkyl DBTs.

tively, are listed in Table 3. On the basis of the product distribution, HDS networks of DBTs or alkyl DBTs over both catalysts were assumed as shown in Figure 3, similar to those by Houalla et al. (11) and Aubert et al. (12).

Under commercial reaction conditions, the kinetics of hydrogenation and hydrogenolysis of aromatic and sulfur compounds can be described by the pseudo-first-order model (17). In this study, the pseudo-first-order model was assumed to be suitable for each step in the networks. The following rate equations can then be given:

$$\frac{dC_{DBT}}{dt} = -(k_1 + k_2) C_{DBT} \tag{1}$$

$$\frac{dC_{BP}}{dt} = k_1 \cdot C_{DBT} - k_3 \cdot C_{BP} \tag{2}$$

$$\frac{dC_{HDBT}}{dt} = k_2 \cdot C_{DBT} - k_4 \cdot C_{HDBT} \tag{3}$$

$$\frac{dC_{CHB}}{dt} = k_3 \cdot C_{BP} + k_4 \cdot C_{HDBT} - k_5 \cdot C_{CHB} \tag{4}$$

$$\frac{dC_{BCH}}{dt} = k_5 \cdot C_{CHB} \tag{5}$$

where C_{DBT}, C_{BP}, C_{HDBT}, C_{CHB}, and C_{BCH} are the molar concentrations of the reactants biphenyl (or alkyl biphenyl), hexahydrodibenzothiophene (or alkyl hexahydrodibenzothiophene), cyclohexylbenzene (or alkyl cyclohexylbenzene), and bicyclohexyl (or alkyl bicyclohexyl) in the products within the reaction time of t (min). k_i (i=1,2,3,4,5) is a rate constant corresponding to the reaction step as shown in Figure 3.

Hydrogenation of biphenyl over both catalysts under comparable conditions was also performed. The results showed that the conversion of biphenyl to cyclohexylbenzene and bicyclohexyl were only 3.6 and 0.9 mol % over the NiMo and CoMo catalysts, respectively [similar to the results reported by Singhal et al. (18)]. Minimal conversion of biphenyl allows for neglecting the hydrogenation of biphenyl (or alkyl biphenyl) in the networks in order to simplify the kinetic model. Another assumption was $k_2 \ll k_4$, based on the experimental observation that the detected concentration of hexahydrodibenzothiophene was very small (<1 mol %) during all reaction processes due to its much higher hydrogenolysis reactivity. This implies that the hydrogenation step in the series of reactions (hydrogenation followed by hydrogenolysis) was a rate controlling step. Such an assumption was also supported by the results reported by Houalla et al. (11) in which k_2 was about four orders of magnitude less than k_4 in HDS of DBT over a CoMo catalyst.

According to the above assumptions, the model can give the following integrated rate of expression:

$$(k_1 + k_2) \cdot t = \ln\left(\frac{C_{DBT_0}}{C_{DBT_t}}\right) \quad (6)$$

$$\frac{k_1}{k_2} = \frac{C_{BP}}{C_{CHB} + C_{BCH}} \quad (7)$$

where the subscripts 0 and t represent the reaction times of zero and t minutes, respectively. k_1 and k_2 represent the rate constants of direct desulfurization (hydrogenolysis) and indirect desulfurization (hydrogenation followed by hydrogenolysis) respectively, which are related to the reactivities of the reactant and catalytic activities of the catalysts for separate hydrogenolysis and hydrogenation.

The k_1 and k_2 values calculated by equations [6] and [7] in accordance with the experimental data for HDS of DBT and 4,6-DMDBT are summarized in Table 4. 4,6-DMDBT showed much lower HDS activity than DBT over both of the catalysts [$(k_1 + k_2)_{4,6\text{-DMDBT}}/(k_1 + k_2)_{DBT}$ being 0.31 and 0.22 over NiMo and CoMo, respectively], concurring with the foregoing results with the fractional HDS. Such a reduction in the reactivity was predominantly attributed to the reduction in the hydrogenolysis reactivity (k_1), k_1 over the NiMo being reduced from 0.033 for DBT to 0.0027 for 4,6-DMDBT. The reduction extent was almost identical for the NiMo and CoMo catalysts, $k_{1,4,6\text{-DMDBT}}/k_{1,DBT}$ being 0.077 and 0.083, respectively. The hydrogenation reactivity was slightly reduced by comparison with hydrogenolysis. The extent was also similar for the two catalysts, $k_{2,4,6\text{-DMDBT}}/k_{2,DBT}$ being 0.53 and 0.66, respectively. Thus it indicates that reduction in the hydrogenolysis and hydrogenation reactivities are independent of the catalysts.

The total activity ($k_1 + k_2$) for HDS of DBT over two catalysts was almost identical. The hydrogenolysis activity over the CoMo ($k_1 = 0.048$) was significantly higher than that of the NiMo ($k_1 = 0.033$), whereas hydrogenation ac-

Table 4 Rate Constants of DBT and 4,6-DMDBT for Hydrogenolysis and Hydrogenation

Catal.	DBT (min^{-1})				4,6-DMDBT (min^{-1})			
	k_1	k_2	k_1+k_2	$k_2/(k_1+k_2)$	k_1	k_2	k_1+k_2	$k_2/(k_1+k_2)$
NiMo	0.033	0.035	0.068	0.57	0.0026	0.019	0.021	0.88
CoMo	0.048	0.015	0.062	0.23	0.0040	0.010	0.014	0.71

tivity over the former ($k_2 = 0.0150$) was significantly lower than that over the latter ($k_2 = 0.035$). However, for 4,6-DMDBT, total catalytic activity over the NiMo was about 2 times higher than that over the CoMo since the HDS of 4,6-DMDBT through the path of hydrogenation followed by hydrogenolysis occupied 88% of total HDS due to a strong inhibition of hydrogenolysis by the methyls at the 4 and 6-positions. Other interesting results are that the value of $k_{1,NiMo}/k_{1,CoMo}$ for DBT (0.69) is almost the same as that for 4,6-DMDBT (0.65), and the value of $k_{2,NiMo}/k_{2,CoMo}$ for DBT (2.39) is also close to that for 4,6-DMDBT (2.0), thereby indicating that the relative activity of the two catalysts for both hydrogenolysis and hydrogenation is independent of the reactants.

4. Correlation between electron densities of sulfur of representative sulfur compounds and their relative desulfurization reactivities

The calculated electron densities of the sulfur atoms were correlated with their desulfurization reactivities reported in our previous paper (10) and other three references (19-21). In order to compare the reactivities of various sulfur compounds obtained under different reaction conditions by different authors, their hydrogenolysis reactivity (RHR) relative to DBT (the ratio of rate constants of sulfur species to that of DBT under the same reaction conditions) was employed in the present study. The electron densities on the sulfur atoms are compared with their RHR as shown in Figure 4. A fair correlation is observed, except the 4-methyl and 4,6-dimethyl DBTs increasing with higher electron density. The higher the electron density, the higher its RHR. For example, the RHR of 2,3-dihydrobenzothiophene is 9.67, much higher than that of benzothiophene (1.14). The electron density of the former is also higher than that of the latter, being 5.913 and 5.739, respectively. Similar results are also observed between $5_b,6,11,11_a$-tetrahydrobenzo[b]naphtho[2,3-d]thiophene and benzo[b]naphthol[2,3-d]thiophene, their RHRs being 8.69 and 2.13, respectively.

The electron densities of the sulfur atom of both 4-methyl and 4,6-dimethyl DBTs are almost the same as that of other DBTs, but their RHR are much lower than the others, about 3 times and 1 order of magnitude less, respectively. Strong steric hindrance of methyl groups rather than electron effect is indicated.

Discussion

This study reveals that the lighter fractions have higher reactivities than those of heavier fractions over both CoMo and NiMo. Such results can be explained as follows: 1) The gas oil contains more than 60 kinds of sulfur compounds which appear to have very different reactivities. The BTs, more reactive sulfur

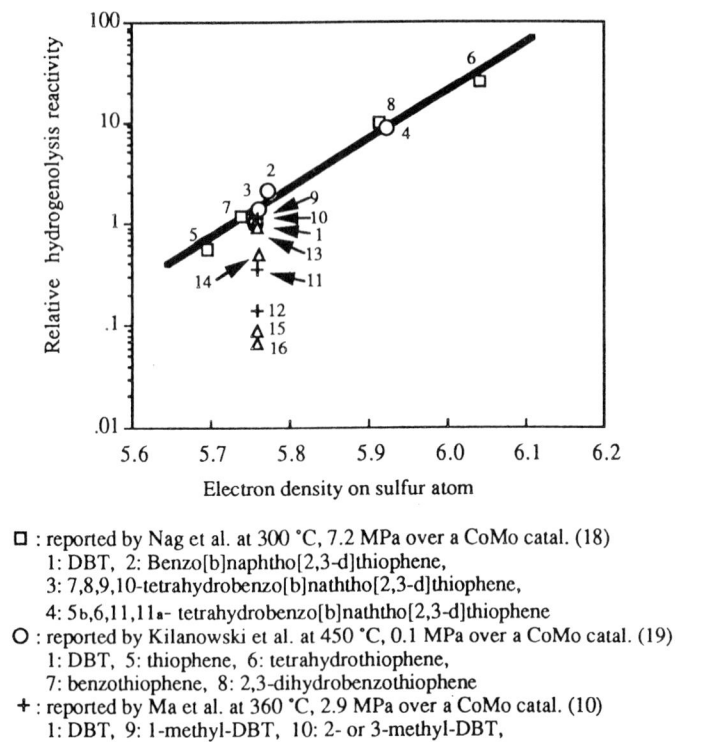

Fig. 4 Correlation between relative hydrogenolysis reactivity (RHR) and the electron density on sulfur atom (from Ref. 22).

□ : reported by Nag et al. at 300 °C, 7.2 MPa over a CoMo catal. (18)
  1: DBT, 2: Benzo[b]naphtho[2,3-d]thiophene,
  3: 7,8,9,10-tetrahydrobenzo[b]naththo[2,3-d]thiophene,
  4: 5b,6,11,11a- tetrahydrobenzo[b]naththo[2,3-d]thiophene
○ : reported by Kilanowski et al. at 450 °C, 0.1 MPa over a CoMo catal. (19)
  1: DBT, 5: thiophene, 6: tetrahydrothiophene,
  7: benzothiophene, 8: 2,3-dihydrobenzothiophene
+ : reported by Ma et al. at 360 °C, 2.9 MPa over a CoMo catal. (10)
  1: DBT, 9: 1-methyl-DBT, 10: 2- or 3-methyl-DBT,
  11: 4-methyl-DBT, 12: 4,6-dimethyl-DBT
Δ : reported by Houalla et al. at 300 °C, 10.3 MPa over a CoMo catal. (20)
  1: DBT, 13: 2,8-dimethyl-DBT, 14: 3,7-dimethyl-DBT,
  15: 4-methyl-DBT, 16: 4,6-dimethyl-DBT

compounds than DBTs (10,19), exist dominantly in the F-1 and F-2 fractions. In contrast, the alkyl DBTs, especially those with one or two of the alkyl groups at the 4- and 6-positions [which are recognized as the most unreactive among them (10,20,21)], are dominant in the heavier fractions (F-3, F-4, and F-5). 2) The heavier fractions contain more aromatic hydrocarbons, particularly the poly-aromatics, which compete for the hydrogenation active site of the catalyst surface with the refractory sulfur compounds. Consequently, it is obvious that HDS of alkyl DBTs, particularly those with alkyls at 4- and/or 6-positions in the heavier fractions, are strongly inhibited by the coexistent poly-aromatics (23–26) because such alkyl DBTs are almost completely desulfurized by the hydrogenation followed by hydrogenolysis. 3) The higher content of nitrogen species in the heavier fractions than in the lighter fractions also results in lower reactivities of the heavier fractions because basic organic nitrogen compounds strongly inhibit HDS (27–29).

In this study, the authors are interested in the superiority of NiMo over the CoMo for HDS of heavier fractions. (The former was slightly less active than the latter for HDS of lighter fractions.) As described above, the heavier fractions dominantly contain alkyl DBTs, among which the alkyl DBTs with the alkyl groups at the 4- and/or 6-positions have been proved to be the most difficult to desulfurize. Higher catalytic activity of the NiMo for the heavier fractions implies that the former is more effective for HDS of refractory alkyl dibenzothiophenes, which is further demonstrated by the following HDS of model compounds:

The remarkable correlation of the RHR for the aromatic sulfur compounds (without steric hindrance of the alkyl groups) with their electron density of sulfur atom illustrates that hydrogenolysis is strongly dependent upon the electron characteristics of the sulfur atom. The electron density of the sulfur atom is involved with an adsorbed intermediate by chemical reaction of the sulfur atom with the anion vacancy on the catalyst surface.

The difference of desulfurization reactivities between aromatic sulfur compounds and their hydrogenated species can be explained by the fact that hydrogenation of benzene ring adjacent to the thiophene ring, such as 2,3-dihydrobenzothiophene and $5_b,6,11,11_a$-tetrahydrobenzo[b]naphtho[2,3-d]thiophene, enhances the electron density of the sulfur atom, thereby promoting interaction between the sulfur atom and the anion vacancy.

Much lower reactivities of 4-MDBT and 4,6-DMDBT can be ascribed to steric hindrance of their methyl groups rather than electronic effect, since the electron densities of their sulfur atoms are hardly influenced by methyl substituents at the positions. On the basis of such results, the reaction networks of alkyl-substituted DBTs are proposed as shown in Figure 3. In the networks, which reaction path dominates over the total HDS is dependent on two factors: 1) whether the sulfur compound has a steric hindrance of alkyls to the hydrogenolysis, and 2) hydrogenolysis-to-hydrogenation activity ratio of the catalyst, i.e., the relationship of k_1 and k_2, is dependent not only on the properties of the catalyst but also on the structure of the reactant. When no alkyl substituent is present at the 4- and/or 6 positions, HDS occurs dominantly by hydrogenolysis due to $k_1 \gg k_2$, in agreement with the reaction networks of DBT proposed by Houalla et al. (11). On the contrary, it occurs dominantly when hydrogenation is followed by hydrogenolysis if there are one or two alkyl substituents at 4- and/or 6-positions acting as a shield to restrict access of the sulfur atom to the anion vacancy on the catalyst surface.

High hydrogenation activity of catalysts makes HDS take place dominantly by the hydrogenation path. This is favorable for the HDS of alkyl DBTs with the steric hindrance of alkyls, although higher hydrogen consumption is required. The superiority of the NiMo over the CoMo for HDS of refractory sulfur compounds existing mainly in the heavier fractions is attributed to its higher hydrogenation activity, as revealed in HDS of the model compounds, which accelerate the HDS by the path of hydrogenation followed by hydrogenolysis. The partial hydrogenation of the benzene ring adjacent to the thiophene ring not only increases the electron density on sulfur atom by suppressing the conjugation of π-electrons on the benzene ring with the lone-pair electrons on sulfur atom, but also diminishes the alkyl steric hindrance by molecular puckering, which is further discussed in our next paper (30). Thus the catalyst with the higher hydrogenation activity is favorable for the deep HDS of refractory sulfur compounds with alkyl steric hindrance. Consequently, for deep HDS of gas oil, the catalyst with higher hydrogenation activity should be designed, especially for selective hydrogenation of the refractory sulfur compounds.

References

1. G. Parkinson. Chem. Eng. 96:42 (1989).
2. Federal Register. Regulation of Fuels and Fuel Additives: Fuel Quality Regulations for Highway Diesel Fuel Sold in 1993 and Later Calendar Years. Government Printing Office: Washington, DC, Aug. 21 1990; Vol. 55, No. 162, pp. 34120.
3. G. M. Wallace. Am. Chem. Soc. Div. Fuel Chem. Prepr. 35:1080 (1990).
4. G. F. Froment, G. A. Depauw, and V. Vanrysselberghe. Ind. Eng. Chem. Res. 33:2975 (1994).
5. J. P. Frank and J. F. Le Page. Proc. 7th Intern. Congr. Catal. Elsevier, Amsterdam, 1981, pp. 792.
6. A. Stanislaus and B. H. Cooper. Catal. Rev. Sci. Eng. 36:75 (1994).
7. R. Prins, V. H. J. De Beer, and G. A. Somorjai. Catal. Rev. Sci. Eng. 31:1 (1989).
8. M. J. Girgis and B. C. Gates. Ind. Eng. Chem. Res. 30:2021 (1991).
9. X. Ma, K. Sakanishi, and I. Mochida. Fuel 73:1667 (1994).
10. X. Ma, K. Sakanishi, and I. Mochida. Ind. End. Chem. Res. 33:218 (1994).
11. M. Houalla, N. K. Nag, A. V. Sapre, D. H. Broderick, and B. C. Gates. AIChE J. 24:1015 (1978).
12. C. Aubert, R. Durand, P. Geneste, and C. Moreau. J. Catal. 97:169 (19886).
13. T. Isoda, X. Ma, and I. Mochida. Sekiyu Gakkaishi. 37:368 (1994).
14. R. Gerdil and E. A. C. Lucken. J. Am. Chem. Soc. 87:213 (1965).
15. J. J. P. Stewart. J. Comp. Chem. 10:209 (1989).
16. X. Ma, K. Sakanishi, T. Isoda, and I. Mochida. Ind. Eng. Chem. Res. 34:748 (1995).
17. M. J. Girgis and B. C. Gates. Ind. Eng. Chem. Res. 30:2021 (1991).

18. G. H. Singhal, R. L. Espino, J. E. Soble, and G. A. Huff, Jr. J. Catal. 67:457 (1981).
19. N. K. Nag, A. V. Sapre, D. H. Broderick, and B. C. Gates. J. Catal. 57:509 (1979).
20. D. R. Kilanowski, H. Teeuwen, V. H. J. de Beer, B. C. Gates, G. C. A. Schuit, and H. Kwart. J. Catal. 55:129 (1978).
21. M. Houalla, D. H. Broderick, A. V. Sapre, N. K. Nag, V. H. J. de Beer, B. C. Gates, and H. Kwart. J. Catal. 61:523 (1980).
22. X. Ma, K. Sakanishi, T. Isoda, and I. Mochida. Energy & Fuels 9:33 (1995).
23. H. S. Lo. Kinetic Modeling of Hydrotreating. Ph.D. Dissertation, University of Delaware, Newark, 1981.
24. V. La Vopa and C. N. Satterfield. J. Catal. 110:375 (1988).
25. M. Nagai and T. Kabe. J. Catal. 81:440 (1983).
26. T. Isoda, X. Ma, and I. Mochida. Sekiyu Gakkaishi. 37:508 (1994).
27. M. V. Bhinde. Ph.D. Dissertation, University of Delaware, Newark, 1979.
28. C. N. Satterfield, M. Modell, and J. F. Mayer. AIChE J. 21:1100 (1975).
29. M. Nagai, T. Sato, and A. Aiba. J. Catal. 97:52 (1986).

Competitive Coversion of Nitrogen and Sulfur Compounds in Naphtha with Transition Metal Sulfide Catalysts

Shuh-Jeng Liaw, Ajoy Raje, Rongguang Lin and
Burtron H. Davis

Center for Applied Energy Research
University of Kentucky
3572 Iron Works Pike
Lexington, KY 40511

ABSTRACT

The hydrotreatment of naphtha derived from the Illinois No. 6 coal has been carried out over supported transition metal sulfide catalysts (Row 2 and 3). Ruthenium sulfide is the most active catalyst for both HDS and HDN and is much more active than supported molybdenum sulfide catalyst. A supported Ruthenium (5.1 wt.%) sulfide catalyst is more active than commercial Ni-Mo and Co-Mo catalysts for HDN and as active for HDS.

INTRODUCTION

The liquefaction of coal, found in large amounts in the U.S.A. and other locations, is an attractive alternative to crude petroleum as a source of liquid fuel. Also, in recent years the need to more effectively utilize heavier resids has led to hydrotreating processes for heavier feedstocks. However, coal-derived naphtha, compared with the typical crude petroleum-derived naphtha, contains much higher amounts of nitrogen and sulfur heteroatoms (1). A number of investigations show that more severe conditions or more active catalysts are necessary to hydrotreat coal-derived naphtha than is the case for petroleum-derived naphtha (1-6).

Chianelli et al. (7) showed that certain Group VIII unsupported transition metal sulfides of the second and third row of the Periodic Table exhibited an activity for the hydrodesulfurization (HDS) of dibenzothiophene that was at least 10-times higher than that of molybdenum sulfide. Following this study, Vissers et al. (8) and Ledoux et al. (9) reported similar volcano curves for the HDS of thiophene over carbon supported transition metal sulfides. Subsequently, Eijsbouts et al. (10) and Ledoux and Djellouli (11) demonstrated that similar volcano curves are observed for the hydrodenitrogenation (HDN) of pyridine and quinoline over carbon supported transition metal sulfides.

The studies mentioned above used only model compounds for the HDN or HDS reactions that are catalyzed by transition metal sulfides. Since a coal-naphtha contains hundreds of nitrogen and sulfur compounds, inhibition and competitive adsorption may affect the rate of nitrogen and sulfur removal. It is therefore of interest to learn whether the sulfur and nitrogen compounds that are present in a coal-derived naphtha exhibit the same type of volcano curves for the transition metal sulfide catalysts as was obtained for the model compounds.

In this study, we have carried out hydrotreatment of an Illinois No. 6 coal-derived naphtha using supported (γ-Al_2O_3) transition metal sulfides of the second and third rows of the Periodic Table. The activities of the catalysts prepared for this study are compared under the same experimental conditions to those of commercial Ni-Mo/Alumina and Co-Mo/Alumina catalysts.

EXPERIMENTAL

Catalyst Preparation
The catalysts were prepared by impregnation of a γ-alumina (surface area 210 m^2/g, pore volume 0.66 cm^3/g) with appropriate solutions of the second and third row transition metal salts listed in Table 1, dried in air at 110°C, then calcined in air at 400°C. The catalysts were then sulfided in a gas flow consisting of 10% H_2S in H_2 at 600 psig and heated at a rate of 1°C/min to 375°C and maintained at this temperature overnight. The chloride added during catalyst preparation may alter the behavior of the support. After catalyst pretreatment using H_2/H_2S, the chloride level in each catalyst was determined and for all catalysts, was lower than could be measured reliably with the ion exchange technique employed (e.g., 0.1 wt.%).

Naphtha
The naphtha derived from the liquefaction of a bituminous Illinois No. 6 was obtained from the Wilsonville, Alabama Advanced Integrated Two Stage Liquefaction Plant. The sample was collected during Run 261. The elemental analyses of the naphtha are shown in Table 2.

Table 1
Composition of Catalyst Prepared

Catalyst	Metal Salt	Metal, wt.%	Symbol
Zr	$ZrCl_4$	0.69	Zr(0.69)
Mo	$(NH_4)_6Mo_7O_{24} \cdot 4H_2O$	0.73	Mo(0.73)
Ru	$(NH_4)_2RuCl_5 \cdot H_2O$	0.77	Ru(0.77)
Rh	$RhCl_3$	0.78	Rh(0.78)
Pd	$(NH_4)_2PdCl_6$	0.81	Pd(0.81)
W	$(NH_4)_{10}W_{12}O_{41} \cdot 5H_2O$	1.40	W(1.40)
Re	$ReCl_5$	1.42	Re(1.42)
Os	$(NH_4)_2OsCl_6$	1.45	Os(1.45)
Ir	$H_2IrCl_6 \cdot H_2O$	1.46	Ir(1.46)
Pt	$H_2PtCl_6 \cdot xH_2O$	1.49	Pt(1.49)

Table 2
Elemental Composition of the Naphthas

Element	Illinois No. 6
C	85.57%
H	13.24%
N	1470 ppm
S	820 ppm
O	1.26%

Reactor System

A bench scale fixed-bed reactor, operated in concurrent downflow mode, was used for these studies. A Brooks Mass flow controller, Model 5850 E, was used to deliver a constant gas flow. A Milton Roy mini Pump Solvent Delivery system was used to pump the feedstocks at high pressure. The reactor was 1/4 inch O.D. (wall thickness = 0.035 inch). The reactor was placed in a 1" x 12" Lindberg 1200°C tube furnace held in a vertical position. A hand-loaded back-pressure regulator, rated to 4000 psig, was used to regulate the reaction pressure.

Hydrotreatment

The three grams of catalyst were sulfided *in-situ*. The gas flow was switched to helium for 1 hour after sulfiding and prior to the start of liquid and H_2 flow. The heteroatom removal, HDS and HDN, was performed over the 10 catalysts as the temperature was varied over the range of 300 to 400°C while holding constant the total pressure (660 psig), weight hourly space velocity (WHSV = 1g of naphtha/hr/g of catalyst), and hydrogen to naphtha g-mole ratio (2.6). Hydrotreatment of the Illinois No. 6 naphtha was carried out over the catalyst with each condition maintained constant during a 24 hour period; three samples were taken at steady state conditions during the last 6 hours of the 24 hours period. Prior to analysis, the sample was washed three times with distilled water to remove dissolved H_2S and NH_3.

Analytical

The feedstock and products were analyzed for total sulfur and nitrogen contents and for the individual sulfur and nitrogen compounds. Total nitrogen was obtained using a Xertex DN-10 total nitrogen analyzer equipped with a chemiluminescence detector. Total sulfur content was determined using a C-300 microcoulometer. Carbon and hydrogen analyses were performed using a Leco CHN analyzer. Individual sulfur compounds were determined using a Sievers Model 350B Sulfur Chemiluminescence Detector (SCD) coupled with a HP5890 Series II gas chromatograph containing a SPB-1 Column (30m x 0.32 mm). Individual nitrogen compounds were determined using a Thermionic Specific Detector (TSD) coupled with a Varian 3700 gas chromatograph using a KOH treated Carbowax column.

Catalysts

Supported catalysts were prepared using 10 transition metals. The catalysts were prepared to contain the same moles of transition metal atoms per unit weight of support. The content of metal atoms supported on the alumina was chosen to be equivalent to that of the 1.49 wt% of Pt on alumina catalyst; this corresponds to 0.22 metal atoms per nm^2 support surface area, assuming an uniform atomic dispersion. The actual metal loadings are given in Table 1.

RESULTS AND DISCUSSION

Hydrodesulfurization

The trend of the HDS activity (Figure 1) shows that at 300°C, ruthenium sulfide is the most active catalyst. For the second and third row

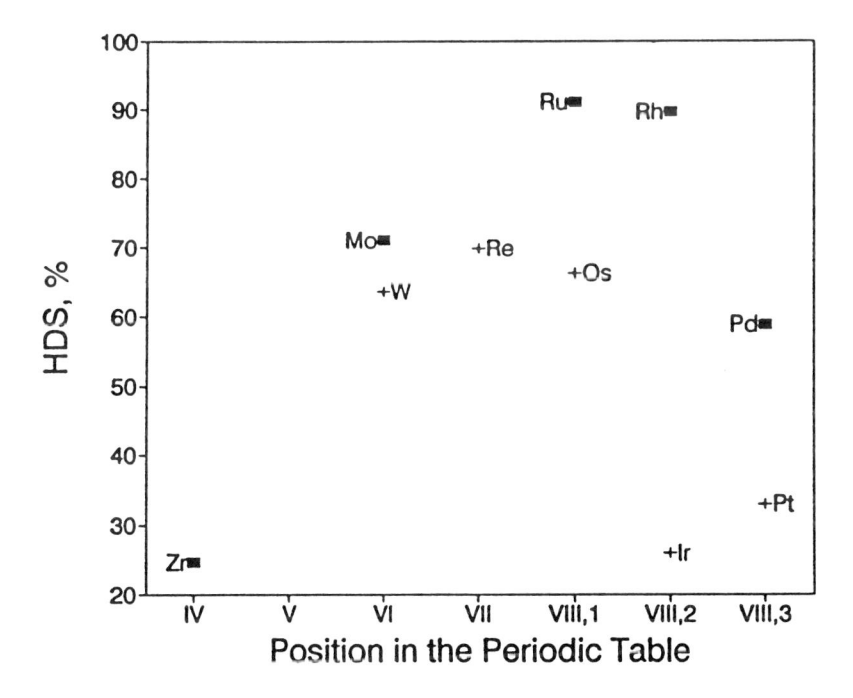

Figure 1. Trends within the Periodic Table of Row 2 (■) and Row 3 (+) metal sulfide catalysts for HDS of Illinois No. 6 at 300°C.

transition metal sulfides, a volcano type curve describes the activity for HDS. The transition metal sulfides of the second row are more active for HDS than those of third row; however, generalizations when the conversions are above about 50% are risky. These results are comparable to those reported by Chianelli *et al.* (7) for the HDS of dibenzothiophene and by Vissers *et al.* (8) and Ledoux *et al.* (9) for carbon supported transition metal sulfide catalysts for the for the HDS of thiophene. Although the naphtha contains hundreds of nitrogen and sulfur compounds, the trend for the activity of HDS of the naphtha is similar to that reported for a single compound, thiophene or benzothiophene. Similar curves for the HDS activity are obtained for second row transition metal sulfide catalysts at 350°C except that the conversion is higher than shown in figure 1. Ru, Rh, Pd, Re and Os sulfides are the most active for HDS at 350°C. At temperature 400°C, the extent of HDS is so great that all catalysts exhibit 95% or higher sulfur removal except for the Zr, Ir, and Pt sulfide catalysts.

Hydrodenitrogenation

The variation of the HDN activity with the position of the metal in the Periodic Table is illustrated in Figures 2-4 for temperatures of 300, 350 and 400°C, respectively. At 300°C, ruthenium sulfide is the most active catalyst (Figure 2) as was observed for the HDS reaction. For the second row

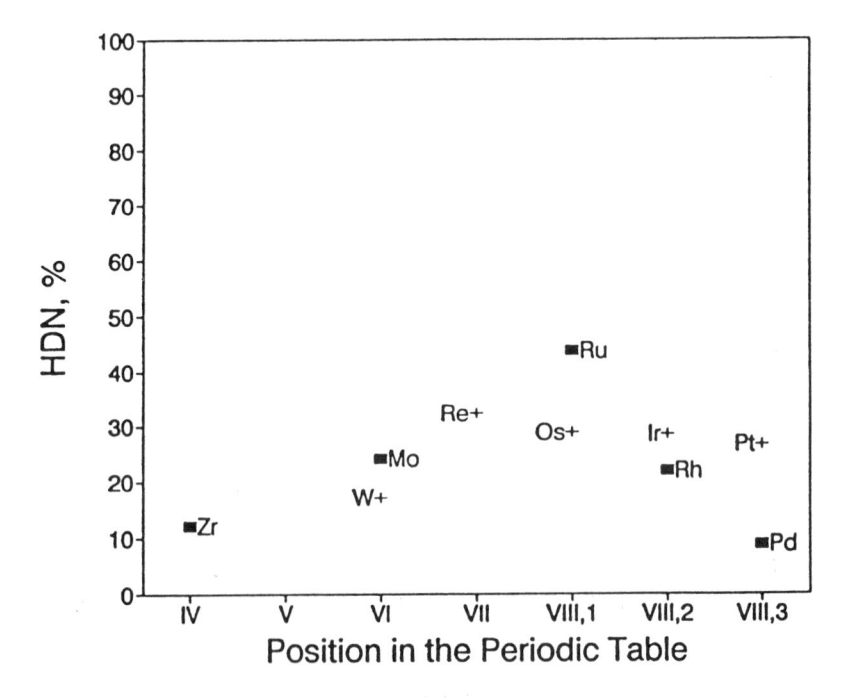

Figure 2. Trends within the Periodic Table of Row 2 (■) and Row 3 (+) metal sulfide catalysts for HDN of Illinois No. 6 at 300°C.

Competitive Coversion of Compounds in Naphtha

transition metal sulfides, a volcano curve for the HDN activity is observed. However, for the third row transition metal sulfide catalysts, the catalysts exhibit a similar activity for HDN, except for the lower activity for the tungsten sulfide catalyst. The data in Figure 3 for HDN at 350°C show a volcano curve applies for the second and third row transition metal sulfide

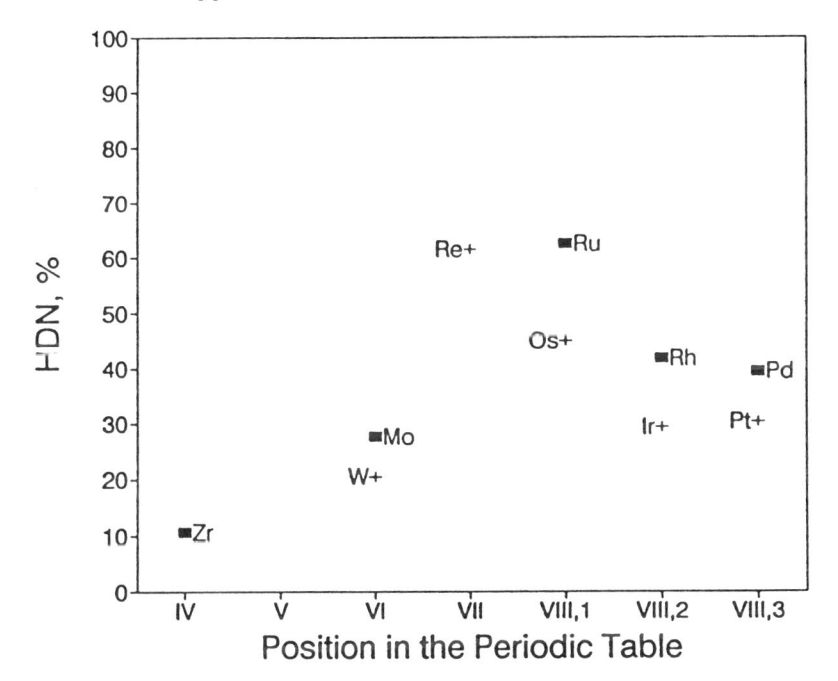

Figure 3. Trends within the Periodic Table of Row 2 (■) and Row 3 (+) metal sulfide catalysts for HDN of Illinois No. 6 at 350°C.

catalysts. The catalyst with maximum activity for the second row is ruthenium sulfide, which removed approximately 70 % of the nitrogen compounds. Rhenium sulfide is the most active for the third row catalysts, and has an activity that is similar to that of ruthenium sulfide. At 400°C, the periodic trend for HDN is similar to that at 350°C (Figure 4) except that osmium and rhodium sulfides are about as active as ruthenium and rhenium sulfide catalysts for HDN. The differences at 400°C may be due to the high conversion levels at this temperature.

A comparison of the HDS to HDN data show that it is much easier to remove sulfur than nitrogen from the naphtha (Figure 5). A similar trend was observed while hydrotreating Illinois No. 6 and Black Thunder naphtha

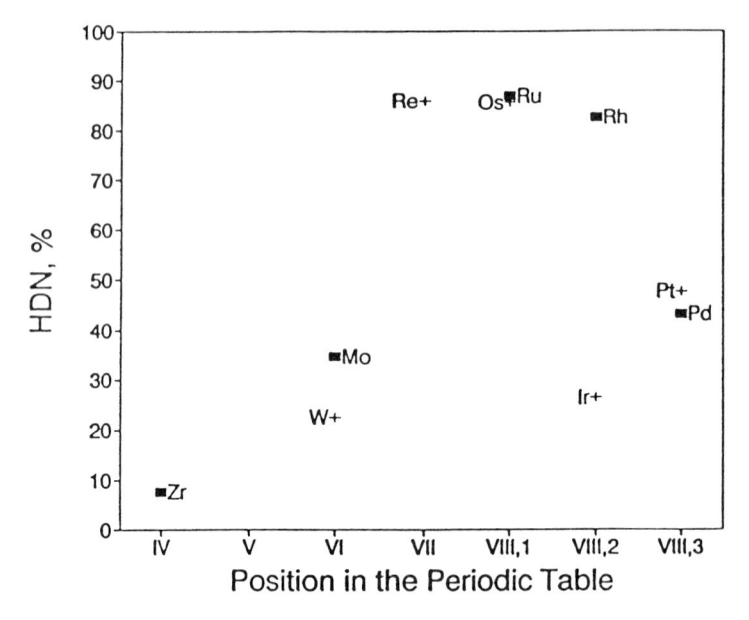

Figure 4. Trends within the Periodic Table of Row 2 (■) and Row 3 (+) metal sulfide catalysts for HDN of Illinois No. 6 at 400°C.

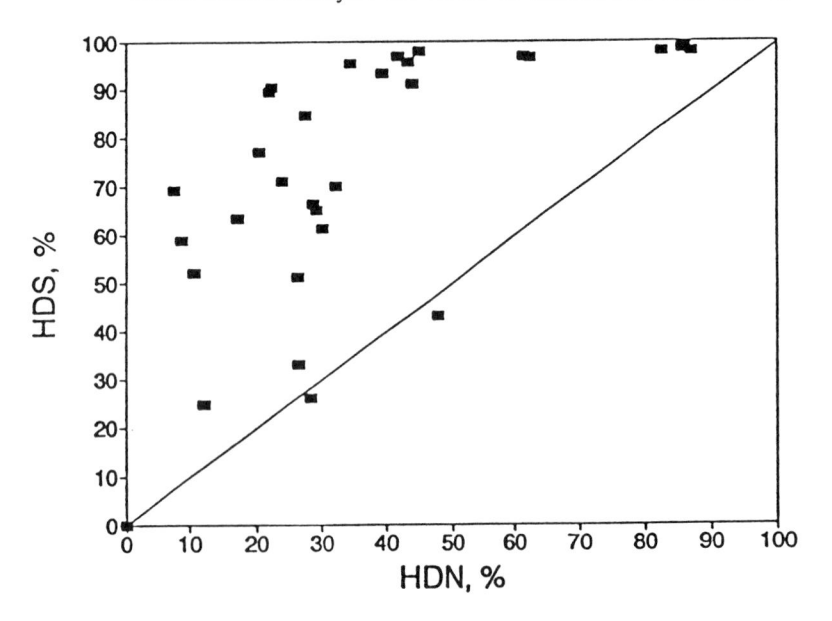

Figure 5. Comparison of the extents of HDS and HDN.

samples using three commercial catalysts (12). Since the ruthenium sulfide is much more active than the molybdenum sulfide in the HDS and HDN of naphtha, it is of interest to learn whether the supported ruthenium sulfide catalyst is more active than the molybdenum-based commercial catalysts.

Comparison of Ruthenium Sulfide to Commercial Catalysts

Two commercial catalysts, a Co-Mo on alumina (American Cyanamid HDS-1442A, 1/16" x 1/4" pellets) and a Ni-Mo on alumina (Akzo KF-840 1.3 Q) have been utilized together with two alumina supported ruthenium sulfide catalysts (0.77 and 5.1 wt. % of Ru). The Ru (0.77) catalyst is the one utilized for the comparisons with other transaction metal catalysts. Analytical data for the commercial catalysts are presented in Table 3. The catalysts (prepared in-house and commercial) were ground and the fraction passing through a 200 mesh sieve was used; therefore, the granular size of all catalysts were less than 74 μm. The flow region using the catalysts prepared in-house and the commercial materials should therefore be essentially the same. Since the Ru and commercial catalysts were ground to produce a fine powder of similar size, diffusion resistance is not a significant

Table 3 Analytical Data for Commercial Catalysts	
Co-Mo-Alumina	Ni-Mo-Alumina
Co 2.7%	Ni 3.1%
Mo 11.1%	Mo 13.3%
S = 288 m^2/g	S = 138 m^2/g
S = BET Surface Area	

problem. Two different reactors, 1/4" with 0.035" wall thickness and 1/2" with 0.045" wall thickness, were used in the initial runs to identify the reactor flow regime. Mass transfer resistance was observed for the 1/2" reactor (a laminar flow); however, no significant mass transfer resistance was found using the 1/4" reactor. Therefore, a 1/4" O.D. with 0.035" wall thickness reactor was chosen for this study.

The Ru (0.77) catalyst, compared to the commercial catalysts on a total weight basis, at the three temperatures for the removal of sulfur is only slightly less active than either commercial catalyst; at 400°C the performance of the Ru (0.77) catalyst approaches that of the Ni-Mo catalyst. For HDS, the Co-Mo catalyst that was utilized was superior to the Ni-Mo catalyst; the differences between these two catalysts are too small to permit this observation to be extended to a generalization concerning the relative performance of Co-Mo and Ni-Mo catalysts. The Ru (5.1) catalyst is as good as the Co-Mo catalyst and, on a total catalyst weight basis, is superior to the Ni-Mo catalyst at the three temperatures used in this study. However, the extent of HDS is above 90% for all of the temperatures utilized, and this should be kept in mind.

For HDN, the Ru (0.77) and Ru (5.1) catalysts are not as active as either the Co-Mo or Ni-Mo catalysts at 300°C; the Co-Mo catalyst is less active for HDN than the Ni-Mo catalyst (Figure 6). At 350°C, the nitrogen content of the naphtha produced for the Ru (5.1) catalyst is about the same as that of the two commercial catalysts. At 400°C, the Ru (5.1) catalyst

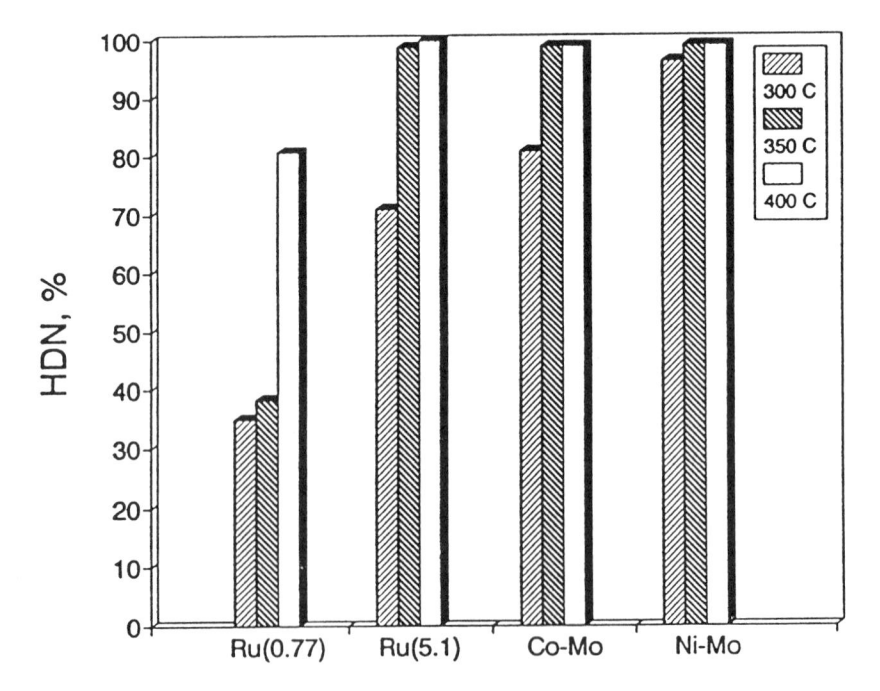

Figure 6. Comparison of ruthenium sulfide catalysts and commercial catalysts for HDN on the basis of total catalyst weight.

Competitive Coversion of Compounds in Naphtha

reduces the nitrogen in the naphtha products to a lower level than either of the commercial catalysts do; however, even at 400°C, the Ru (0.77) catalyst is not as active as the commercial catalysts when the removal is based on the total weight of the catalyst.

Another approach is to make the comparison of activity on the basis of the weight of active metal. When this is done the two ruthenium catalysts are more active than either the Co-Mo or Ni-Mo catalysts (Figures 7 and 8). The Ru (5.1) catalyst is less active per gram of Ru than the Ru (0.77) catalyst is, and this is probably due to the higher dispersion of the Ru in the catalyst with the lower Ru loading.

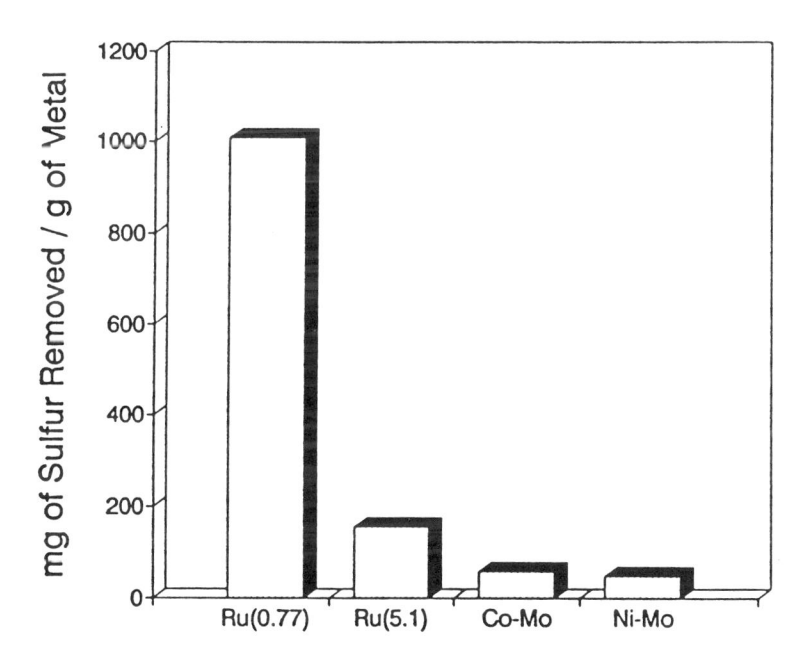

Figure 7. Comparison of ruthenium sulfide and commercial catalysts for HDS on the basis of equal weight of active metal.

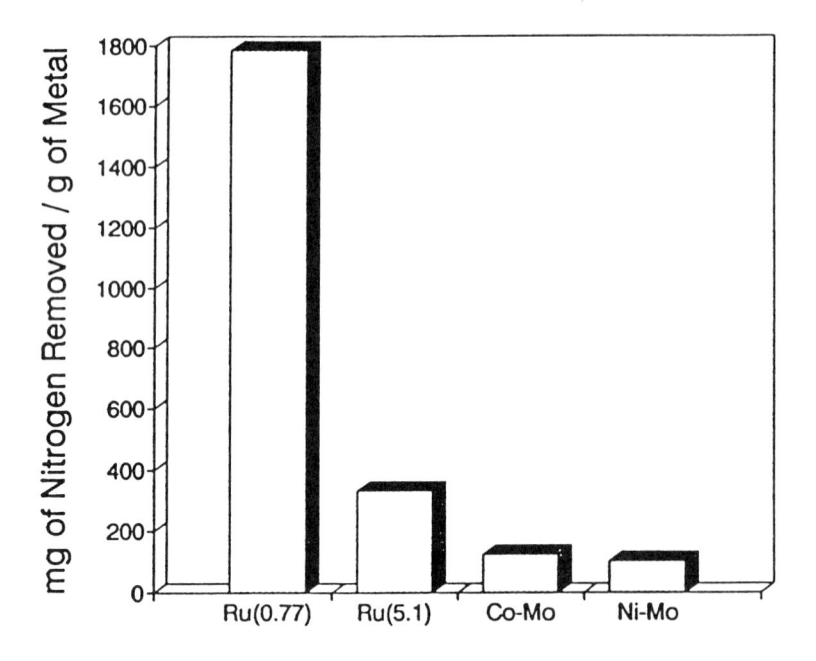

Figure 8. Comparison of ruthenium sulfide and commercial catalysts for HDN on the basis of equal weight of active metal.

CONCLUSION

The activity of second and third row transition metal sulfides for the HDS and HDN of Illinois No. 6 naphtha show that ruthenium sulfide is the most active catalyst. The comparison between a supported ruthenium sulfide and commercial Co-Mo and Ni-Mo catalysts shows that ruthenium sulfide is more active for HDN at higher temperatures and is as active for HDS. The competitive adsorptions involved in the conversion of N and S heterocompounds in a coal-derived naphtha produce volcano type curves that are similar to those obtained from model compound studies. Furthermore, a common metal sulfide (Ru) is the most active for both HDS and HDN.

ACKNOWLEDGMENT

This work was supported by the DOE contract #DE-AC22-90PC91058 and the Commonwealth of Kentucky.

REFERENCES

1. de Rosset, A. J., Tan, G. and Hilfman, L., "*Upgradinging Primary Coal Liquids by Hydrotreatment,*" in *Refining of Synthetic Crudes,*" (M. L. Gorbaty and B. M. Harner, Eds.), Advances in Chemistry Series, **179**, pp 108-119, (1979).

2. Frumkin, H., Sullivan, R. F., and Strangeland, B. E., in *Upgrading Coal Liquids*", (R. F. Sullivan, ed.), ASC Symp. series 156, American Chemistry Society, Washington, DC, 1981, p. 75.

3. Smith, V. E., Cha, C. Y., Merriam, N. W., Faky, J. and Guffy, F., *Proceedings of the Third Annual Oil Shale, Tar Sand and Mild Gasification Contractors Review Meeting*, **166**, 1988.

4. Gaeser, U. R., Holighaus, R., Dohms, K. D., and Langhoff, J., *Am. Chem. Soc. Prepr., Div. Fuel Chem.*, **33**, 339 (1988).

5. Parker, R. J., Mohammed, P. and Wilson, *J. Am. Chem. Soc. Prepr., Div. Fuel Chem.*, **33**, 135 (1988).

6 Su, L., Keogh, R. A., Huang, C., Spicer, R. L., Sparks, D. E., Lambert, S., Thomas, G. A. and Davis, B. H., *Am. Chem. Soc. Prepr., Div. Fuel Chem.*, **36**, 1909 (1991).

7. Chianelli, R. R., Prestridge, E. B., Pecoraro, T. A. and DeNeufville, *Science*, **203**, 1105, (1979).

8. Vissers, J. P. R., Groot, C. K., Van Oers, E. M., De Beer, V. H. J. and Prins, R., *Bull. Soc. Chim. Belg.*, **93**, 813, (1984).

9. Ledoux, M. J., Michaux, O. and Agostini, G., *J. Catal.*, **102**, 275, (1986).

10. Eijsbouts, S., De Beer, V. H. J., and Prins, R., *J. Catal.*, **127**, 619, (1991).

11. Ledoux, M. J. and Djellouli, B., *J. Catal.*, **115**, 580, 1989.

12. Liaw, S. J., Keogh, R. A., Thomas, G. A., and Davis, B. H., *Energy & Fuels*, in press.

The Effect of Additives and Impregnation Stabilizers on Hydrodesulfurization Activity

E. P. Dai and L. D. Neff

TEXACO R&D-Port Arthur, P. O. Box 1608, Port Arthur, Texas 77641-1608

ABSTRACT

Our results showed that the incorporation of Li, Mg, and B additives into alumina supports has a deleterious effect on the benzothiophene hydrodesulfurization activity of NiMo catalysts. In the present work, the effects of additives and impregnation stabilizers were investigated with an aim to identify the best combination of additive with stabilizer for optimizing the activity. The effectiveness of impregnation stabilizers depends on the additives and their loadings. With a proper choice of stabilizer, the activities of NiMo catalysts on alumina supports containing various amounts of silica were significantly improved. Phosphorus as an impregnation stabilizer showed a more pronounced promoting effect on the activities of NiMo catalysts than CoMo catalysts. The combined use of phosphorus and citric acid exhibited the best enhancement in the activity of CoMo catalysts.

INTRODUCTION

The increasing demand for progressively lower sulfur contents in transportation and other fuels has resulted in motivation for the improvement of hydrotreating catalysts currently in use. These catalysts incorporate almost exclusively some type of variation of the $NiO-MoO_3/Al_2O_3$ or $CoO-MoO_3/Al_2O_3$ base constituents. Modification of hydrodesulfurization (HDS) catalyst activity has been accomplished through the use of a variety of additive materials (1-5). These additives may either directly result in a change in the chemistry of the catalyst or affect the manner in which the active elements are distributed on the catalyst surface. The latter class of additives will be here termed impregnation stabilizers as their principal function is to stabilize the active metals in the impregnating solution.

Phosphorus has probably been the most commonly used stabilizer element in HDS catalysts and many studies (6-10) have been carried out to both assess its effect on HDS activity and characterize its chemical and physical state. Chadwick et al (7) studied the influence of phosphorus on the HDS activity of Ni-Mo/gamma-alumina catalysts and concluded that the activity for thiophene HDS increased with phosphorus content reaching a maximum at about 1 wt.% P. Their XPS studies indicated that the phosphorus is in monolayer form and that it influences the repartition of Mo in the catalysts. Lewis et al (8) concluded that phosphorus enhances the thiophene HDS activity of $Ni-Mo/Al_2O_3$ catalysts particularly when the P is impregnated before the metals. FTIR spectroscopic studies of Mo/Al_2O_3 and $P-Mo/Al_2O_3$. The catalysts used in this study were prepared by Criterion Catalyst Co., as part of a joint catalyst development effort with Texaco. Samples of NiMo and CoMo on alumina and modifications of these were prepared by Criterion using proprietary procedures.

A set of samples of NiMo and CoMo (prepared with and without phosphorus stabilization) were examined in the oxide and sulfide forms.

The Effect of Additives and Impregnation Stabilizers

Sulfiding the catalysts was accomplished in a modified Berty reactor. These catalysts were examined using high resolution transmission electron microscopy (HRTEM) at Arizona State University in addition to the following techniques for all the other catalysts.

Nitrogen porosimetry analyses were obtained using a Micromeritics ASAP 2400, an automated nitrogen porosimeter. Porosity calculations involved a BJH (14) analysis of the desorption isotherms at 77K. The samples (0.2g) were outgassed in a vacuum at 0.01 Torr (1 Torr = 133.3 Nm-2) at 623K for 16 h. before analysis. Pore mode calculations were obtained by fitting the desorption pore volume distribution function with a gaussian and determining the curve maximum.

Mercury porosimetry analyses were obtained using a Micromeritics Autopore 9220, an automated mercury porosimeter. The samples (approximately 0.8g) were heated in an oven at 727K for 2 h. and then outgassed on the instrument at 0.01 Torr. An arbitrary contact angle of 130 degrees and a surface tension of 484 dynes/cm were used to calculate the pore volume distribution data from the mercury intrusion curves. Pore mode calculations were obtained by fitting the pore volume distribution function (for the intrusion curve) with a gaussian and determining the curve maximum.

X-ray photoelectron spectroscopy (XPS) was used to determine the molybdenum gradient for the catalysts. This was done using a Vacuum Generators (V. G. Scientific Ltd.) ESCALAB Mark II, instrument equipped with a 1253.6 volt magnesium X-ray source. The molybdenum gradient is arbitrary defined as the ratio of the "exterior" molybdenum/aluminum atomic ratio to the "interior" molybdenum/aluminum atomic ratio, [Mo gradient = (Mo/Al)ext / (Mo/Al)int]. The Mo/Al atomic ratio of a given catalyst pellet exterior, (Mo/Al)ext, was determined using whole extrudate pellets, whereas the "interior" Mo/Al atomic ratio, (Mo/Al)int, was obtained after crushing the extrudates into a powder. Atomic percentage values were calculated from the peak areas of the Mo 3p(3/2) and Al 2p(3/2) signals using sensitivity factors supplied by V. G. Scientific Ltd. The value of 74.7 electron volts for aluminum was used as a reference binding energy.

The activities for benzothiophene hydrodesulfurization were determined using an HDS microactivity test (HDS-MAT) developed at our laboratory. The catalysts were evaluated as a powder (30 - 60 mesh) to eliminate possible diffusion effects. The catalysts (0.5g) were presulfided in situ at 672K using a 10% H_2S/H_2 mixture for 2 h. The presulfided catalysts were then exposed to a benzothiophene/heptane (0.86M benzothiophene, 3.7 wt% S) feed at 561K and at a liquid feed rate of 4 cc/h in flowing hydrogen (50 cc/min, atmospheric pressure) for four hours. The product was analyzed by GC for ethylbenzene formation and the HDS activity expressed as the percentage of benzothiophene converted to ethylbenzene During this four-hour test period, no catalyst deactivation was ever experienced The average activity is calculated on both a catalyst weight and catalyst volume basis to account for any density differences between catalysts. At the abovestated temperature and space velocity, a commercially available HDS-1443B NiMo catalyst, used as the standard, was found to have a %HDS value of 67% on a weight basis and have a standard deviation of 1%. The kinetics of the reactor used in the HDS-MAT are first-order, plug flow. The activation energies for typical fresh commercial NiMo and CoMo catalysts are 24±2 kcal/mole.

RESULTS AND DISCUSSION

Tables 1 and 2 list the composition, physical properties, and HDS microactivity of the modified NiMo/alumina catalysts. The catalysts in Table 1 are pairs which were made with and without phosphoric acid as stabilizer.

Table 1. Effect of H_3PO_4 on Preparation of NiMo/Alumina Catalysts

Sample No.	NiO (wt%)	MoO$_3$ (wt%)	Stabilizer[a]	Mo Gradient	Surface Area (m^2/g)	Macropore[b] Volume (mL/g)	Pore Mode (nm)	HDS (%)
1-A1	2.9	14.8	NONE	3	179	0.02	11	78
1-A2	2.9	15.2	P	1	180	0.02	11	86
1-B1	3.1	14.8	NONE	3	203	0.05	11	85
1-B2	3.2	14.8	P	1	185	0.05	11	98
1-C1	3.4	14.5	NONE	2	189	0.12	11	83
1-C2	3.1	15.0	P	1	185	0.12	11	92
1-D1	3.3	15.2	NONE	1	269	0.12	6	84
1-D2	3.2	15.6	P	1	242	0.12	7	91

[a] P = phosphoric acid. [b] Pore diameter >25 nm.

Table 2 Effect of Additives and Stabilizers on NiMo/Alumina Catalysts

Sample No.	NiO (wt%)	MoO (wt%)	Additive	Stabilizer[a]	Mo Gradient	Surface Area (m²/g)	Pore Mode (nm)	HDS (%)
2-A	3.2	14.7	NONE	NONE	3	189	11	88
3-A	3.3	15.2	1% Li$_2$O	C	1	187	12	78
3-B	3.3	15.0	1% Li$_2$O	H	4	174	12	80
3-C	3.0	15.1	1% Li$_2$O	P	1	173	11	85
4-A	3.2	14.9	10% MgO	C	3	168	11	66
4-B	3.2	15.2	10% MgO	H	5	173	9	57
4-C	3.1	14.8	10% MgO	P	2	162	11	70
5-A	3.3	15.7	6% B$_2$O$_3$	C	12	175	11	84
5-B	3.4	15.3	6% B$_2$O$_3$	H	15	146	11	61
6-A	3.3	14.1	4% SiO$_2$	H	6	194	11	93
6-B	3.3	14.3	8% SiO$_2$	H	5	171	14	87
6-C	3.2	14.8	8% SiO$_2$	P	2	179	9	83
6-D	3.1	14.2	16% SiO$_2$	C	4	193	11	92
6-E	3.2	14.4	16% SiO$_2$	H	8	198	9	64
6-F	3.1	15.0	16% SiO$_2$	P	1	173	8	50

[a] C = citric acid; H = hydrogen peroxide; P = phosphoric acid.

The Effect of Additives and Impregnation Stabilizers

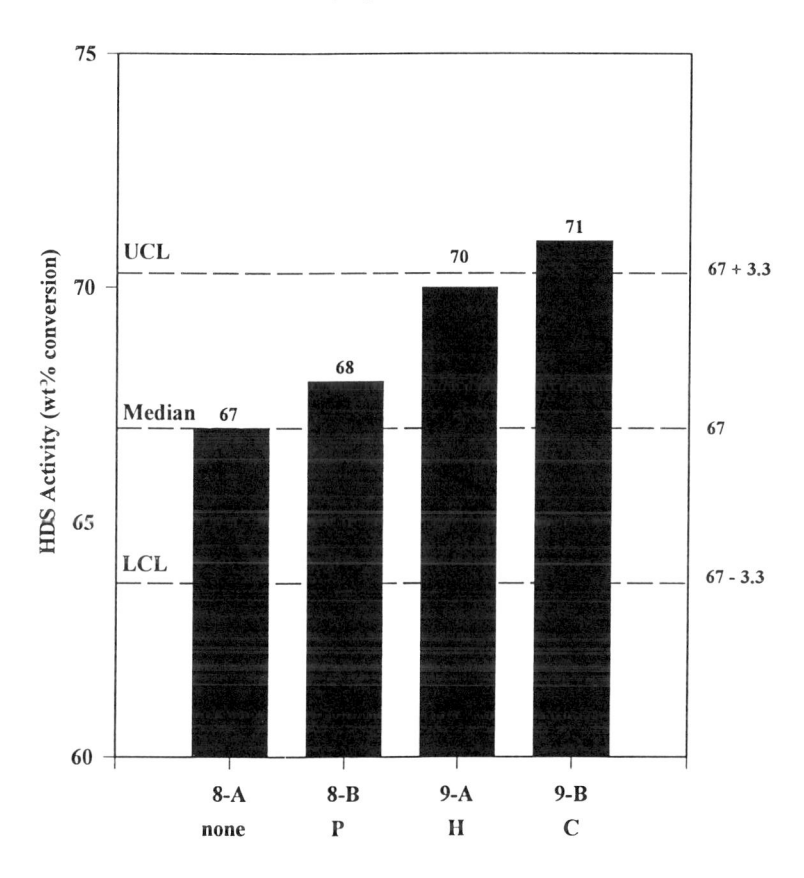

Figure 1. Statistical control chart of HDS-MAT test illustrating effect of impregnation stabilizers on CoMo catalysts. UCL: upper control line; LCL: lower control line, P: phosphoric acid, H: hydrogen peroxide, C: citric acid.

The phosphorus level in the P-containing catalysts was 2 wt % as P_2O_5. The groups in Table 2 were made with the same alumina and in comparable ways, except for the additives and stabilizers used. Fig. 1 illustrates the accuracy of benzothiophene HDS-MAT test by plotting the HDS activities of four catalysts (8-A, 8-B, 9-A, and 9-B) along with the upper and lower control lines (67±3.3%, 2σ=3.3%, σ is the standard deviation). It is seen that 8-A, 8-B, and 9-A were similar with 9-B showing some improvement over 8-A in HDS activity.

Effect of Phosphorus Stabilizer

In this study phosphoric acid was used as the solution stabilizer during the impregnation of the aluminas. Each pair of catalysts, 1-A through 1-D, differed in the amount of macropore (>25 nm) volume or pore mode. Catalysts 1-A through 1-C had pore modes of 11 nm and macropore volumes of 0.02, 0.05, and 0.12 mL/g, respectively. Catalysts 1-D had the same macropore volume as catalysts 1-C, but with a smaller pore mode. In every case the use of phosphoric acid in the catalyst preparation resulted in ca. 10% higher benzothiophene HDS activity. There was no significant difference in the percentage increase in the HDS activity with respect to the macropore volume of the catalysts. Atanasova et al. (10) reported that use of phosphoric acid in the preparation of NiMo/alumina catalysts improved the activity for the HDS of thiophene at concentrations up to 8% P_2O_5 in the catalyst with the maximum activity at about 2% P_2O_5.

Additives

The objective of this work was to determine how surface acidity factors affect the performance of hydroprocessing catalysts. It was assumed that addition of alkali and alkaline earth elements into alumina would decrease the surface acidity. In contrast, the incorporation of B and Si would increase the surface acidity. It is generally accepted that the surface acidity does not affect HDS activity, but does affect the HDN activity. Li and Mg were found to increase the conversion of heavy hydrocarbons (15). It has been reported that the addition of alkali metals such as K to MoO_3 lowers the extent of reduction and sulfidation (16), thereby lowering the hydrogenation function of MoS_2 catalysts. In this work an attempt was made to determine the relative importance of surface acidity and Mo distribution on HDS activity.

Lithium and magnesium as representatives of the alkali metal and alkaline earth metal groups were added to the alumina support before

The Effect of Additives and Impregnation Stabilizers 219

impregnation of the Ni and Mo to determine the effect of these groups on HDS activity. Comparison of the catalysts in groups 3 and 4, in Table 2, with catalyst 2-A shows that using Li or Mg as an additive is detrimental to the HDS activity irrespective of which stabilizer is used in the preparation of the catalysts. Duchet et al. (17) reported that a NiMo/magnesium aluminate catalyst was less active than the NiMo/alumina catalyst for the HDS of thiophene.

With NiMo catalysts we found that 6% boria had a loss of HDS activity whether made using hydrogen peroxide or citric acid as the stabilizer. This is seen by comparing catalysts 5-A, 5-B, and 2-A in Table 2. Both of the boron containing catalysts had an unusually high Mo gradient of ca. 14. Occelli and Debies (11) found by microprobe analysis a similar preferential deposition of Mo on the exterior of a NiMo/boria/HY zeolite/alumina catalyst. Their catalyst contained a 1:1 ratio of HY zeolite:alumina and was prepared with phosphomolybdic acid.

Effect of Other Impregnation Stabilizers

Impregnation stabilizers were studied to try to improve the HDS activity of NiMo catalysts. The effect of stabilizers was found to depend on the additive present. The effectiveness of stabilizers for improving the HDS activity of NiMo catalysts on Li-containing alumina can be seen by comparing samples 3-A and 3-B and 3-C, in Table 2I. The HDS activity decreased in the order phosphoric acid > hydrogen peroxide ≅ citric acid. With catalysts on magnesium-containing alumina supports (4-A, 4-B, and 4-C), the phosphoric acid was the best among those three stabilizers with hydrogen peroxide giving the lowest HDS activity.

The effectiveness of citric acid and hydrogen peroxide stabilizers for improving HDS activity of NiMo catalysts on the boron-containing alumina supports is shown by comparing samples 5-A and 5-B. The citric acid stabilizer resulted in a catalyst with considerably higher activity than the catalyst made with hydrogen peroxide.

Comparison of catalysts 6-D, 6-E, and 6-F with a 16% silica content or catalysts 6-B and 6-C with 8% silica shows the effectiveness of the three stabilizers for improving the HDS activity of NiMo catalysts on silica-containing alumina supports. As shown in Fig. 2(a), the catalyst HDS activities follow the order of citric acid (6-D) >> hydrogen peroxide (6-E) > phosphoric acid (6-F).

The effect of the phosphoric acid and hydrogen peroxide stabilizers depends on the silica concentration. For phosphoric acid, the HDS activity decreases from 83% (6-C) to 50% (6-F) in going from 8% to 16% silica. Likewise, for the hydrogen peroxide stabilizer, the HDS activity decreases as the silica content of the support increases. This is shown by comparing samples 6-A, 6-B, and 6-E. Muralidhar et al. (18, 19) reported that HDS activity decreased with increasing SiO_2 content from 10 to 75%. Only one silica concentration was used with the citric acid stabilizer, so a possible effect of silica concentration with this stabilizer was not detected.

Effect of Phosphorus Stabilizer on CoMo Catalysts

The influence of phosphorus was investigated using three types of alumina supports with distinct pore size distributions. Samples 7-A and 7-B in Table 3 were made with a monomodal alumina support with pore mode of 11 nm and very low macroporosity. Addition of phosphorus at the nominal 2.0 wt% P_2O_5 level seems to improve the HDS activity. Samples 7-C and 7-D were made with a monomodal alumina support with pore mode of 11 nm and low macroporosity. Addition of phosphorus at the nominal 2.0 wt% P_2O_5 level appears to have no effect on the HDS activity.

Samples 8-A and 8-B made with a monomodal alumina support with pore mode of 8 nm and medium macroporosity were used to illustrate the effects

TABLE 3. EFFECT OF PHOSPHORUS ON HDS ACTIVITIES OF CoMo CATALYSTS ON ALUMINA SUPPORTS

Sample No.	Co (wt%)	Mo (wt%)	Additive	Stabilizer	Surface Area (m²/g)	Mo Gradient	Macropore Volume (cc/g)	Pore Mode (nm)	HDS (%)
7-A	3.2	15.2	NONE	NONE	194	1	0.01	11	55
7-B	3.2	15.2	NONE	P	182	1	0.01	11	63
7-C	3.3	15.1	NONE	NONE	186	2	0.04	10	62
7-D	3.2	15.2	NONE	P	178	1	0.04	11	63
8-A	3.3	15.1	NONE	NONE	262	1	0.11	8	67
8-B	3.2	15.2	NONE	P	273	1	0.12	8	68
8-C	3.3	14.6	NONE	P	242	17	0.13	7	54

TABLE 4. EFFECT OF IMPREGNATION STABILIZERS ON HDS ACTIVITIES CoMo CATALYSTS ON ALUMINA SUPPORTS

Sample No.	Co (wt%)	Mo (wt%)	Additive	Stabilizer	Surface Area, (m²/g)	Mo Gradient	Macropore Volume (cc/g)	Pore Mode (nm)	HDS %
8-B	3.2	15.2	NONE	P	273	1	0.12	8	68
9-A	3.4	15.2	NONE	H	288	2	0.12	8	70
9-B	3.4	15.3	NONE	C	274	1	0.08	8	71
10-A	3.2	15.2	NONE	P	260	1	0.14	7	67
10-B	3.2	15.2	NONE	P + C	260	1	0.11	8	77

of surface area and pore mode on HDS activity. In comparison with the sample 7-C, the HDS activities of samples 8-A and 8-B were higher, however, they were lower on a per unit surface area basis. The dispersion of Mo and impregnation profile of active metals may be affected by the change in pore mode. The addition of phosphorus at the nominal 2.0 wt% P_2O_5 level does not seem to have any effect on the HDS activity. The effect of Mo gradient is demonstrated by comparing samples 8-B and 8-C. Sample 8-C having a high Mo gradient of 17 showed a lower HDS activity relative to 8-B.

Effect of Other Impregnation Stabilizers

The influence of impregnation stabilizers was studied with an aim to further improve the HDS activity of CoMo catalysts. A comparison of samples 8-B, 9-A, and 9-B in Table 4 and Fig. 1 with 8-A for their HDS activities showed that the use of individual hydrogen peroxide and citric acid as the impregnation stabilizer appears to have a small, but beneficial effect on HDS activity. It is seen that the combined use of citric acid and phosphoric acid as the impregnation stabilizer clearly gives a significant improvement in HDS activity over the use of phosphoric acid alone when comparing samples 10-A and 10-B in Fig. 2(b). The reason for improvement in HDS activity is not yet known. Moulijn et al. (20) reported that the poisoning of the low pressure thiophene HDS activity of carbon supported CoMo catalysts by phosphorus can be explained by either the formation of "Co-PO_4" or CoP and Co_2P species. The presence of citrate anion could form complexes with cobalt and thereby reducing the formation of of "Co-PO_4" or CoP and Co_2P species. Since there is no difference in the value of Mo gradient and the XPS atomic ratio of (Mo/Al)int between the catalysts 10-A and 10-B, it is less likely that citric acid improves the Mo dispersion in CoMoP catalysts.

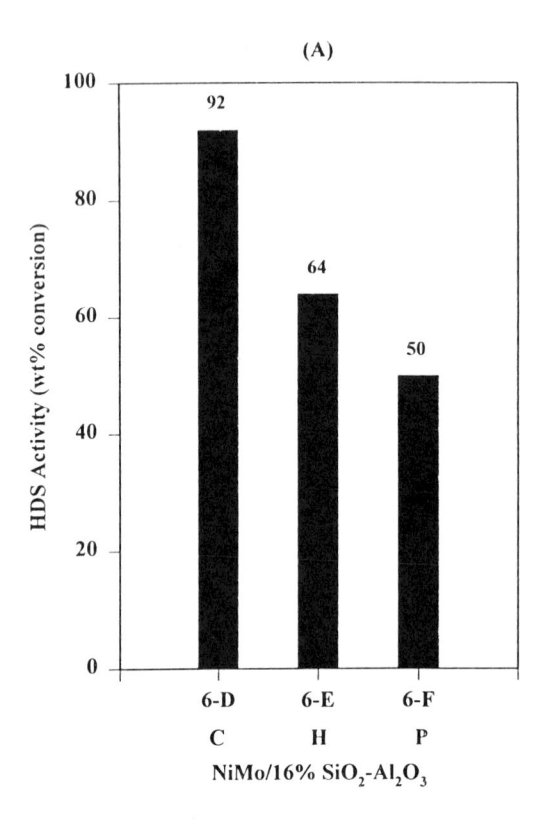

Figure 2. Effect of impregnation stabilizers on a) NiMo/16% SiO$_2$-Al$_2$O$_3$ catalysts and b) CoMo/alumina catalysts.

(B)

HDS Activity (wt% conversion)

10-A: P — 67
10-B: P + C — 77

CoMo/Alumina

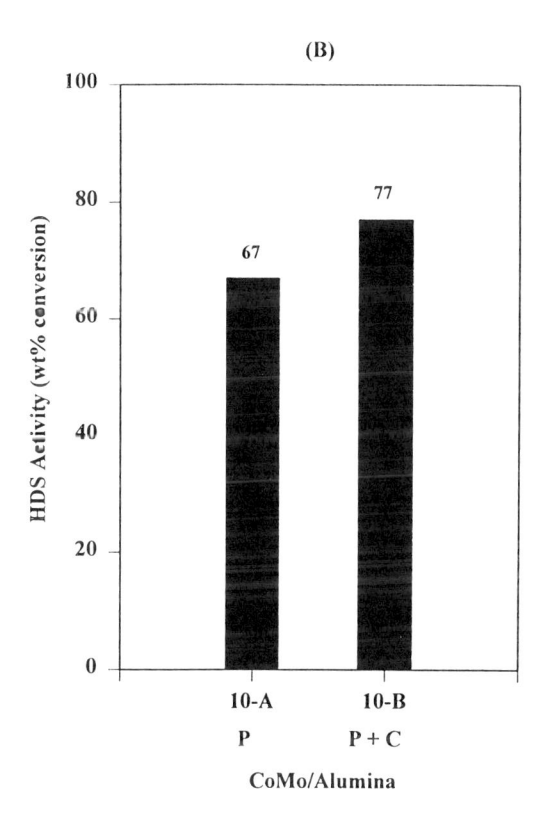

Characterization of Sulfided Catalysts

All of the catalysts in their oxide state are analyzed for their elemental compositions of the exterior surfaces of the catalyst extrudates and the crushed powders by XPS. The $Mo3p_{3/2}$ XPS spectra for CoMP catalysts, 8-B and 8-C in Table 3, were shown in Fig. 3 to illustrate the difference in Mo gradients, between these two samples. The sample 8-B had a uniform distribution of Mo

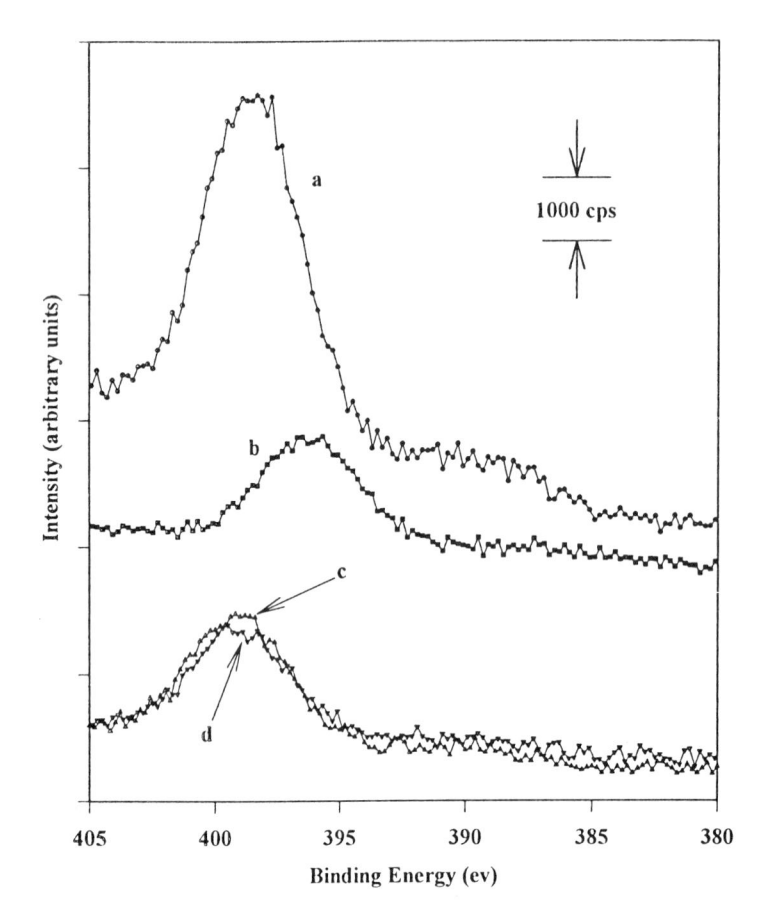

Figure 3. Mo 3p$_{3/2}$ XPS Spectra for CoMoP Catalysts. a) Extrudate surface of catalyst, 8-C, with concentration gradient; b) Powder surface of catalyst 8-C, with concentration gradient; c) Extrudate surface of catalyst, 8-B, without concentration gradient and d) Powder surface of catalyst, 8-B, without concentration gradient.

across from the exterior to the interior surface of the extrudates. By contrast, the sample 8-C had a Mo concentration gradient of 17. The Co 2p XPS spectra for the samples 8-B and 8-C are presented in Fig. 4. It is evident that the sample 8-C exhibited a Co concentration gradient. The causes of high Mo

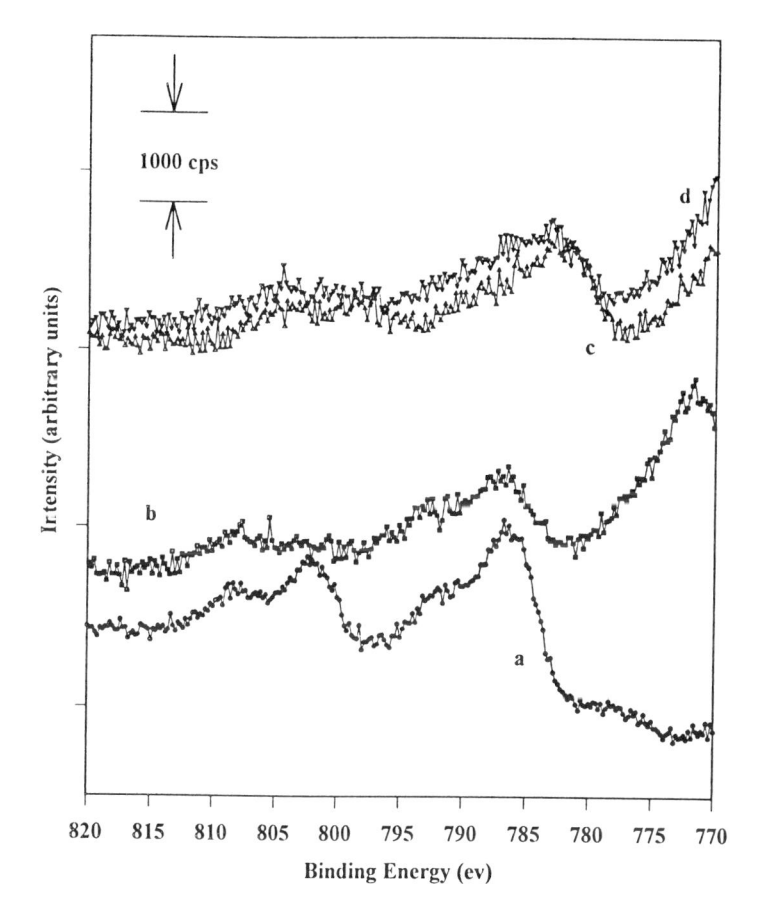

Figure 4. Co 2p XPS Spectra Showing Cobalt Concentration Gradient. a) Extrudate surface of catalyst, 8-C, with concentration gradient; b) Powder surface of catalyst, 8-C, with concentration gradient; c) Extrudate surface of catalyst, 8-B, without concentration gradient and d) Powder surface of catalyst, 8-B, without concentration gradient.

and Co concentration gradients were believed to be unstable impregnating solution and rapid drying after incipient impregnation. Fig. 5 illustrates the difference in the Mo concentration gradients between the NiMoP catalyst (sample 1-A2) and the NiMoB catalyst (sample 5-B). The addition of 6% B_2O_3

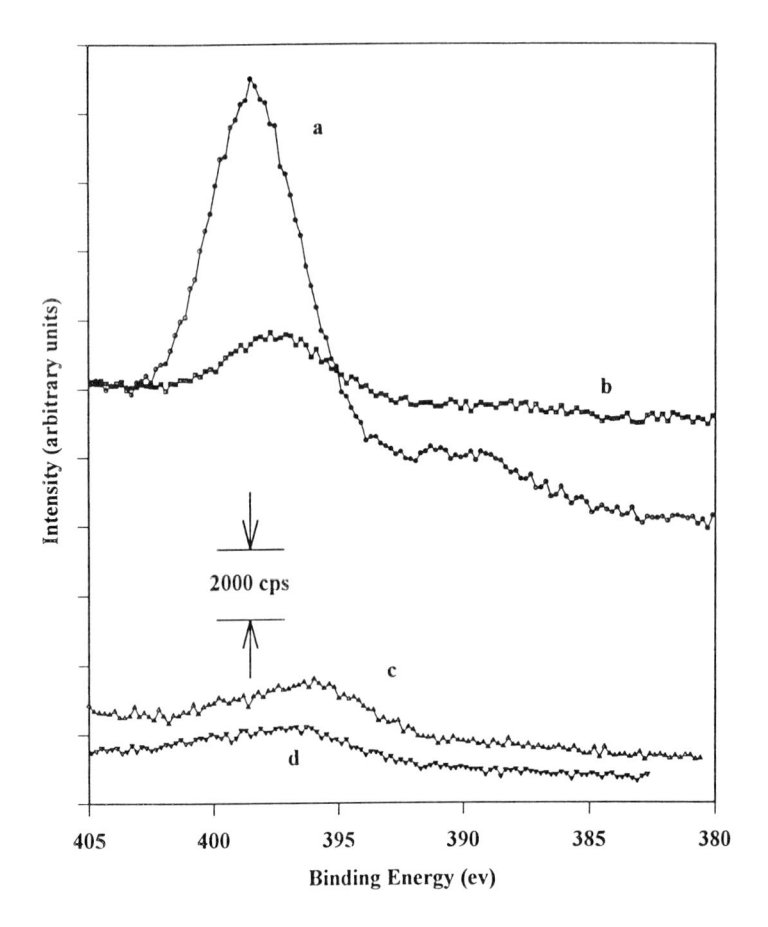

Figure 5. Mo $3p_{3/2}$ XPS Spectra for Experimental NiMo Catalysts. a) Extrudate surface of catalyst, 5-A, with concentration gradient; b) Powder surface of catalyst, 5-A, with concentration gradient; c) Extrudate surface of catalyst, 1-A2, without concentration gradient and d) Powder surface of catalyst, 1-A2, without concentration gradient.

into alumina resulted in a poor distribution of Mo across from the exterior to the interior surface of the extrudates. In contrast, Ni concentration gradient is not as pronounced as Mo gradient for the NiMoB catalyst as shown in Fig. 6 for the Ni2p XPS spetra.

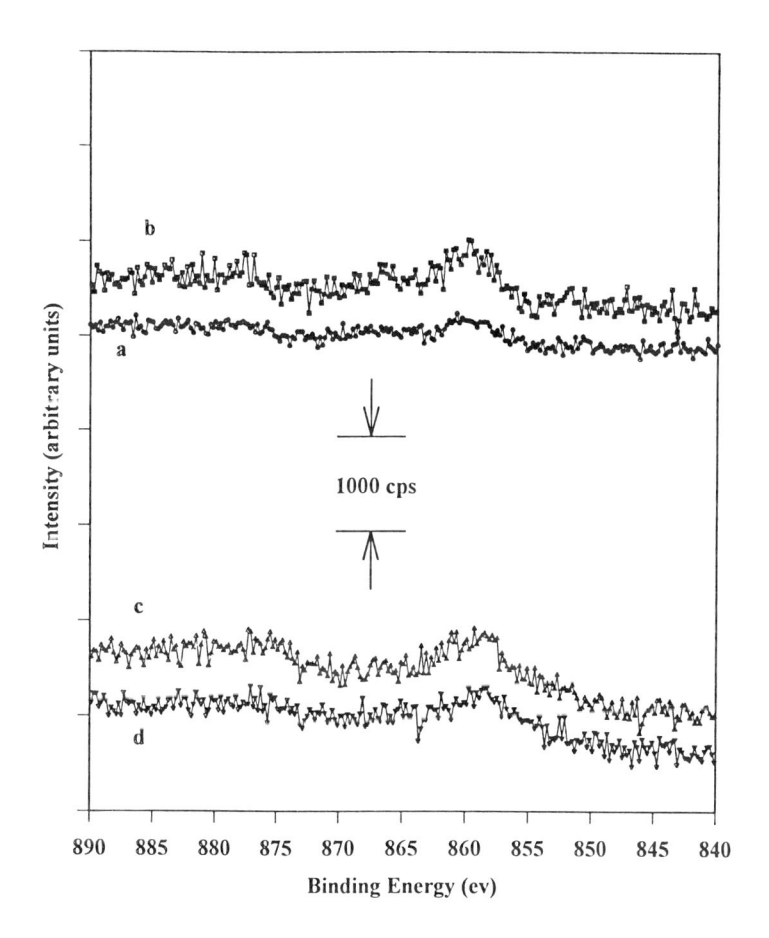

Figure 6. Ni 2p XPS Spectra Showing Nickel Concentration Gradient a) Extrudate surface of catalyst, 5-A, with concentration gradient; b) Powder surface of catalyst, 5-A, with concentration gradient; c) Extrudate surface of catalyst, 1-A2, without concentration gradient and d) Powder surface of catalyst, 1-A2, without concentration gradient

Table 5. RESULTS OF XPS ANALYSIS OF SULFIDED FORM OF THE NiMo AND CoMo CATALYSTS

Sample No.	2-A	1-B2	7-C	7-D	1-D2	8-B	8-C
Metals	NiMo	NiMoP	CoMo	CoMoP	NiMoP	CoMoP	CoMoP
(S/Mo)int	1.2	1.1	1.2	1.3	1.5	1.7	1
(S/Mo)ext	2.5	1.1	1.1	1.0	1.4	1.4	3.5
(Mo/Al)int	0.11	0.11	0.10	0.10	0.11	0.11	0.14
(Ni or Co/Al)int	0.010	0.012	0.020	0.016	0.014	0.028	0.026
Mo Gradient	3	1	2	2	1	2	9
Ni or Co Gradient	1	2	1	2	1	2	37
S Gradient	3	1	1	1	1	1	30

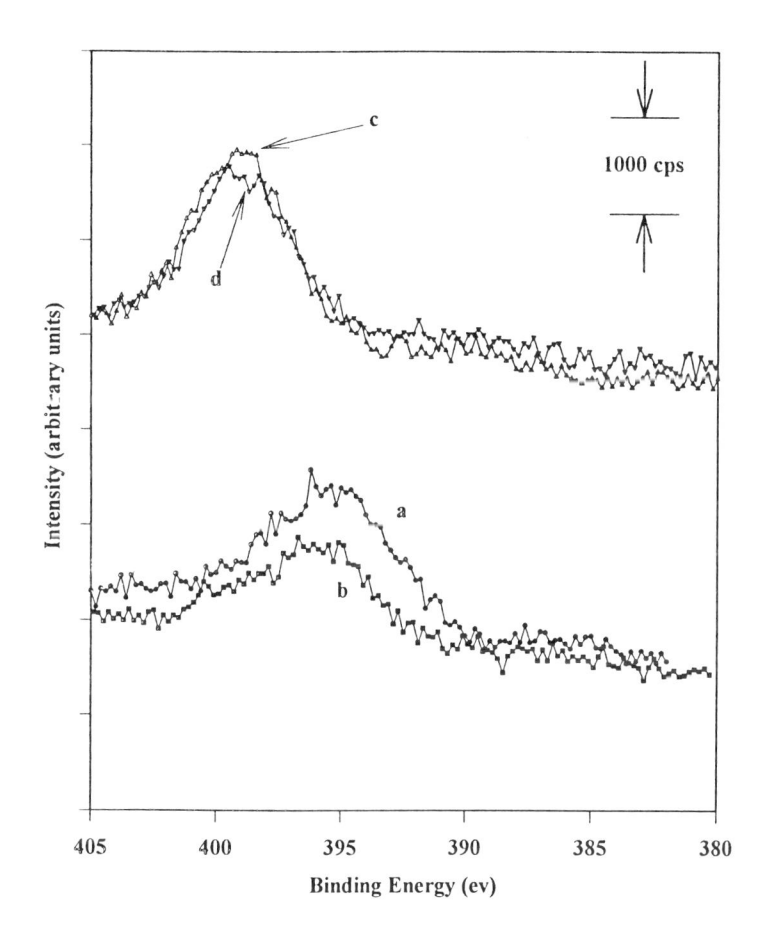

The Effect of Additives and Impregnation Stabilizers

Figure 7. Comparison of Mo $3p_{3/2}$ XPS Spectra for Sulfided and non-Sulfided CoMoP Catalysts (8-B). a) Extrudate surface of sulfided catalyst; b) Powder surface of sulfided catalyst; c) Extrudate surface of catalyst in oxide form and d) Powder surface of catalyst in oxide form.

Seven samples of sulfided NiMo and CoMo catalysts were analyzed by XPS. The results are summarized in Table 5. Fig. 7 shows the Mo$3p_{3/2}$ XPS spectra for a CoMP catalyst (sample 8-B) in both sulfided and oxide forms. The shift in binding energy of Mo$3p_{3/2}$ XPS peak to lower values indicates that the Mo in sulfided form has a lower oxidation states. In the first set of four samples including 2-A, 1-B2, 7-C, and 7-D, the effects of active metal combination and phosphorus on surface compositions of sulfur and molybdenum were examined. First of all, there is no significant rearrangement of surfaces upon sulfidation when comparing the Mo gradients between the oxide and the sulfided catalysts such as the unsulfided and sulfided forms of 2-A. In fact, it is noted that the $(Mo/Al)_{int}$ and $(Ni/Al)_{int}$ or $(Co/Al)_{int}$ remain the same after the catalysts are sulfided. These four samples had similiar $(S/Mo)_{int}$ ratios of about 1.2, suggesting that the extent of sulfidation is the same.

The results of high resolution transmission electron microscopy (HRTEM) analysis of these four samples are in agreement with Kemp et al.'s (21) results. The major difference between the sulfided NiMo and CoMo catalysts is that the Mo is dispersed in a monolayer as MoS_2-like crystallites in the CoMo catalysts, whereas, NiMo catalysts have stacking of MoS_2 layers. The presence of phosphorus in the CoMo catalysts increases the MoS_2 stacks to two or three layers thick. The addition of phosphorus in the NiMo catalysts increases the stacks of MoS_2 from 2-3 to 6-10 layers. It is conceivable that as the stacking of MoS_2 layers is increased the number of Co or Ni coordinated at the edges of MoS_2 crystallites may also be augmented, and causes an increase in the HDS activity. The increase in the stacking of MoS_2 layers is more pronounced in the NiMo catalysts than in the CoMo catalysts upon addition of phosphorus. Thus, it is easier to see the beneficial effect of phosphorus on the HDS activity in the NiMo catalysts. The positive effect of phosphorus as suggested by the previous arthors are the increase in dispersion of the catalyst components by improvement of the stability of impregnating solutions (22), and the increase of Ni^{+2} (Oh) concentration in the catalytically active phase along with an enhancement of dispersion (23).

As shown in our data, the addition of B and Si into alumina could have either beneficial or adverse effects on the HDS activity depending on the concentration and selection of the stabilizer. Cheng and Pereira (24) claimed for CoMo/alumina catalysts, that boron promotes HDS activity in the concentration range of 0.1 to 5% boria. They found that boron in the trigonal state was essential to its promoting effect. It is envisioned that the combined effect of additive and stabilizer modify the optimum composition of polymolybdate surface structures and affect the octahedral Ni^{2+} concentration in the catalytically active NiMoS phase.

CONCLUSIONS

The incorporation of Li, Mg, and B additives into alumina supports has a deleterious effect on the benzothiophene hydrodesulfurization activity of NiMo catalysts. The activities of NiMo catalysts on alumina supports containing various amounts of silica could be significantly improved through the proper choice of impregnation stabilizer. Phosphorus as an impregnation stabilizer showed a more pronounced promoting effect on the activities of NiMo catalysts than on CoMo catalysts. The combined use of phosphorus and citric acid exhibited the best enhancement in the activity of CoMo catalysts.

REFERENCES

1. J. J. Stanulonis and L. A. Pedersen, ACS, Div. Petrol. Chem. preprint 25(2), 255 (1980).
2. D. Muralidhar, F. E. Massoth and J. Shabtai, *J. Catal.* 85, 44 (1984).
3. C. W. Fitz, Jr. and H. F. Rase, *Ind. Eng. Chem. Prod. Res. Div.* 22, 40 (1983).
4. H. D. Simpson, U.S. Patent 4,255,282 to Intervep (1981).
5. A. L. Morales, and J. J. Garcia, U.S. Patent 4,743,574 to Union Oil Company (1988).
6. D. J. Sajkowski, et al., *Appl. Catal.* 62, 205 (1990).

7. D. Chadwick, D. W. Aitchison, R. Badilla-Ohlbaum and L. Josefsson, in "Studies in Surface Science and Catalysis", Vol. 16, Preparation of Catalysts III, (G. Poncelet, P. Grange and P. A. Jacobs, Eds), p. 323, Elsevier, Amsterdam (1983).
8. J. M. Lewis, R. A. Kydd and P. M. Boorman, in "Studies in Surface Science and Catalysis" (K. J. Smith, E. C. Sanford, Eds), Vol. 73, Progress in Catalysis, p. 45. Elsevier, Amsterdam (1992).
9. A. Spojakina, S. Damyanova, L. Petrov and Z. Vit, *Appl. Catal.* 56, 163 (1989).
10. P. Atanasova, T. Halachev, J. Uchytil and M. Kraus, *Appl. Catal.* 38, 235 (1988).
11. M. L. Occelli and T. P. Debies, *J. Catal.* 97, 357 (1986).
12. C. Kordulis, S. Voliotis, A. Lycourghiotis, D. Vattis and B. Delmon, *Appl. Catal.* 11, 179 (1984).
13. H. D. Simpson and P. B. Borgens, Eur. Patent 0 341 893 to Union Oil Company (1989).
14. E. P. Barrett, L. G. Joyner and P. H. Halenda, *J. Amer. Chem. Soc.* 73, 373 (1951).
15. C. N. Campbell and E. P. Dai, Paper presented at AIChE Spring Natl. Mtg., Houston, 1993, session 72B.
16. E. DeCanio and D. A. Storm, *J. Catal.* 132, 375 (1991).
17. J. C. Duchet, N. Gnofam, L. Bekakra, S. Kasztelan, J. Grimblot, J. L. Lemberton, G. Perot, C. Moreau and J. Joffre, *Catal. Today* 10(4), 593 (1991).
18. G. Muralidhar, F. E. Massoth and J. Shabtai, ACS, Div. of Petrol. Chem. preprint 27 (3), 722 (1982).
19. G. Muralidhar, F. E. Massoth and J. Shabtai, *J. Catal.*, 85, 44 (1984).
20. P. J. Mangnus, V. H. J. de Beer and J. A. Moulijn, *Appl. Catal.*, 67, 119 (1990).
21. R. A. Kemp, R. C. Ryan and J. A. Smegal, in *"Proceedings 9th International Congress on Catalysis"*, p 128, Calgary (1988).
22. H. Hilfman, U. S. Patent 3 617 528 (1971).
23. R. E. Tischer, N. R. Narain, G. J. Stiegel and D. L. Cillo, *Ind. Eng. Chem. Prod. Res. Rev.*, 26, 422 (1987).
24. W. Cheng and C. J. Pereira, U. S. Patent 4,724,226 (1988).

Molten Salt Preparation of Mixed Transition Metals on Zirconia: Application to Hydrotreating Reactions

P. AFANASIEV[1], C. GEANTET[1], M. BREYSSE[1], T. des COURIERES[2]

[1] IRC, 2 Av. A Einstein 69626 Villeurbanne, France.
[2] ELF Antar France, Centre de Recherches ELF Solaize, BP22, 69360 St Symphorien d'Ozon.

ABSTRACT

The use of molten alkali nitrates as a reaction medium is a new route for synthesizing catalysts. This method was applied to the one step preparation of molybdenum based catalysts (CoMo, NiMo, FeMo) supported on zirconia. Simultaneous reaction of the precursors in the alkali nitrate melt allows to obtain solids having high surface areas and a well dispersed oxospecies of the d-metals on the surface of zirconia. According to the data of the XRD and DRS UV vis characterizations, the solids as prepared contain the amorphous species of the molybdates of VIII group d-metals, dispersed on zirconia. By contrast to the conventional preparation methods, the species of the active components of molten salt preparations play the role of textural promoters and improve both the specific surface areas of the solids and the stability to calcination in air. After sulfidation of the samples, their activity in the hydrogenation, hydrodesulfurization and hydrodenitrogenation model reactions was measured. The results of the catalytic tests evidenced the potentiality of this new preparation method.

INTRODUCTION

In the near future, new environmental regulations will be required for refiners. In particular, more stringent specifications on diesel fuel sulfur and aromatic contents are already applied in some countries and will be generalized in the others. This opens up a great challenge in the development of new hydrotreating catalysts. Since the last decade, three main approaches to enhance the catalytic properties were used: (i) doping the alumina carrier with the elements such as phosphorus or fluorine, (ii) replacing the alumina by another support, (iii) changing the conventional CoMo, NiMo or NiW active phase by using other sulfides (1). The first point gave rise to a new generation of industrial catalysts more efficient but unable to perform the regulations now required. Numerous literature data indicate that new supports such as zirconia, titania or mixed oxide supports demonstrate the favorable interactions with the active phase and enhanced catalytic activities (2,3). New active phases such as ruthenium sulfide based catalysts supported on alumina or zeolites demonstrate also interesting properties (4). Notwithstanding all these results, the challenge remains intact. In all these cases, the catalysts were prepared by using classical aqueous impregnation method.

Unconventional preparation methods may create new types of interactions between the support and the active phase. In this way, molten salts provide an original chemistry for preparing catalysts. Molten nitrates have many peculiar features as non-aqueous reaction media, such as moderate melting points, excellent heat transfer properties, quite different solubility of chemical compounds compared to water. Several studies on molten nitrates demonstrated that oxides can be precipitated in these melts but a few works have been devoted to the preparation and characterization of the solid products. Durand et al. have shown that the molten salt method can be used for the preparation of ZrO_2 ceramic precursors and catalytic supports (5-8). Recently we have demonstrated that the reaction of the mixture of a transition metal precursor salt with zirconium oxychloride in the molten nitrate can yield three types of products: (i) a mixture of two oxides, (ii) a solid oxide solution on the base of zirconia, (iii) a dispersion of oxoanions at the surface of small zirconia particles

(9). In this latter case, an interaction occurs in the melt between the oxoanions and the growing zirconia particles. As a result, the oxoanions stabilize high surface area and generate a new kind of Mo/ZrO$_2$ catalysts (10). Since high loadings of the dispersed active phase could be achieved, enhanced catalytic activities have been found.

In the commercial hydrotreatment catalysts, large amount VIII group d-metal is always present. Co and Ni are known to act as the promoters of hydrotreating reactions and Fe was shown to have little or no effect on the HDS activity of MoS$_2$ (11). However, FeMo (or FeW) unsupported catalysts were found to provide more active sites for HDN (12).

In the present work, our purpose was to study the capability of molten salt reactions to obtain in one step the promoted hydrotreating catalysts and to explore the potential of these new materials.

EXPERIMENTAL

Preparation and characterization

Hydrated zirconium oxychloride was employed as the precursor of zirconia; ammonium heptamolybdate and nickel, cobalt or iron hydrated chlorides were applied as admixtures. The initial Mo/M (M=Ni, Co, Fe) atomic ratio in the reaction mixture was equal to 2. The salts of d-metals were tightly mixed with the 10-fold molar excess of NaNO$_3$-KNO$_3$ eutectic. Reaction was carried out in the Pyrex reactor under a flow of nitrogen. All the samples were prepared by using the following sequence of treatment: dehydration at 150°C for 2h, reaction at 500°C for 2h, extraction by distilled water and drying at 100°C in air. The solids were characterized by XRD, UV-vis and XPS spectroscopy. Chemical composition was determined using the atomic absorption method.

A second group of catalysts was prepared by the combination of the molten salt method and the conventional impregnation technique. Being prepared as described above, (M-Mo)/ZrO$_2$ molten salt catalysts were taken as supports. Additional amounts of M and (or) Mo were introduced by impregnation (M = Ni, Co, Fe). Solutions of Fe, Co or Ni nitrates and ammonium heptamolybdate were used for impregnation. These catalysts are denominated NiMo$_{im}$, CoMo$_{im}$, FeMo$_{im}$.

The activation of the catalysts was performed at 400°C under a flow of H$_2$/H$_2$S mixture (15% vol. H$_2$S) for 4 h.

Catalytic tests

The catalytic properties for the conversion of thiophene (TH), tetraline (TE) and n-pentylamine (PA) were determined in dynamic conditions, in the vapor phase, using flow microreactors. Table 1 summarizes the conditions of the catalytic tests. Specific activities per gram of catalyst, A$_s$ were measured at the pseudo-stationary state after 15 h. of reaction.

RESULTS AND DISCUSSION

XRD patterns

Reactions of the individual precursor salts of Zr, Fe, Co, Ni or Mo with molten nitrate are known to lead to the precipitation of the corresponding oxides: ZrO$_2$ (13), NiO, Co$_3$O$_4$ (14), Fe$_2$O$_3$ (15), or to the formation of alkali monomolybdates (16). We have found earlier (9) that at the loadings equal to those used in this work, simultaneous reactions of ZrOCl$_2$ with Ni, Co or Fe salts in the nitrate melt yield the mixtures of zirconia particles and the large

Table 1. Experimental conditions of the catalytic tests

Test	Total pressure (10^{5}Pa)	Temp. of reaction (°C)	Pressure of reactant (10^{2}Pa)	H$_2$S in the feed (10^{2}Pa)
HYD	41	350	TE= 67	638
HDS	1	300	TH= 18	0
HDN	1	275	PA= 3.3	40

TE=tetraline, TH= thiophene, PA= n-pentylamine

crystallites of NiO, Co_3O_4 or Fe_2O_3, visible in the XRD patterns. Precursor salts of other elements such as Mo, W, Nb, Cr react differently with $ZrOCl_2$ in molten nitrate (17). During the synthesis, an interaction between the growing zirconia particles and the oxospecies occurs. As a consequence, polyoxospecies well-dispersed on zirconia surface with enhanced surface area are obtained and no M-containing phases (M= Mo, W, Nb, Cr) appeared in the XRD patterns (17,18).

The reaction of ammonium heptamolybdate and nickel nitrate in the molten nitrate bath at 500°C led to the formation of crystalline $NiMoO_4$ (JCPDS card 33-0948). Iron and cobalt nitrates gave more complicated mixtures of crystalline products, including sometimes alkali metals : $K_3FeMo_4O_{15}$ (34-0093), $NaFe(MoO_4)_2$ (30-1195), $Fe_2(MoO_4)_2$ (35-0183) $NaCo_{2.31}(MoO_4)_3$ (17-095), and $CoMoO_4$ (254-1434). Therefore, crystalline mixed molybdates are formed in the presence of molybdate species in the nitrate melt.

Simultaneous reactions of Ni(Co,Fe), Mo and Zr precursors gave no bulk oxides of Ni, Co or Fe, nor crystalline molybdates of these elements. XRD patterns of the products of the molten salt reaction (after drying or calcination at 600°C) indicate only the presence of tetragonal (T) zirconia and traces of its monoclinic (M) form (Fig. 1, a). Even after calcination at 800°C no extra phases appeared for the NiMo sample (Fig. 1, b). Therefore, the presence of Mo species in the reaction mixture hinders the formation of the bulk oxides of Fe, Co, and Ni. On the other hand, the mixture of NiO and ZrO_2 prepared in molten nitrate reacted upon the treatment with a molybdate containing melt, resulting in the disappearance of NiO reflexes (Fig 2). Nevertheless, all the initial amount of Ni remained in the solid. Thus, the molybdate species dissolved in nitrate melt, reacted with Ni oxide. Such an interaction leads to the spreading of the Ni-Mo mixed oxospecies over the surface of the zirconia support. Due to the presence of zirconia, Ni-Mo oxospecies did not produce the crystalline nickel molybdate. Apparently, simultaneous reaction of Ni, Mo and Zr salts in molten nitrate gives small crystallites of zirconia, covered by highly dispersed NiMo oxospecies. A similar behavior has been observed for the Co and Fe containing samples.

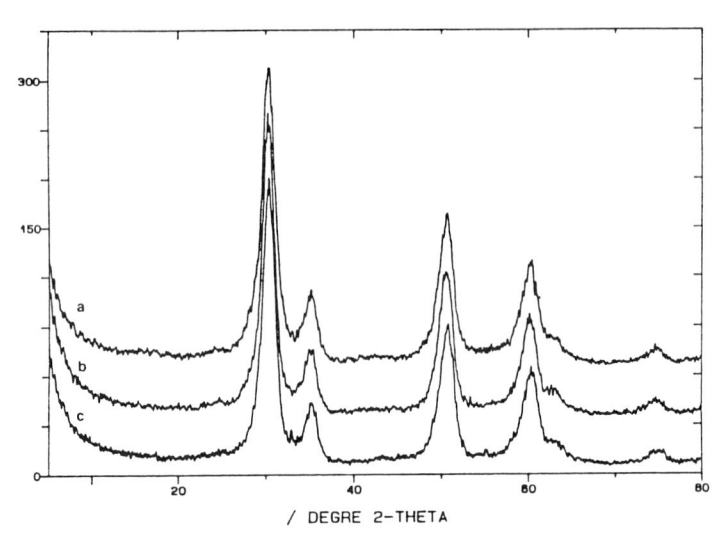

Fig. 1. XRD patterns of: a) NiMo/ZrO$_2$ molten salt catalyst, b) NiMo/ZrO$_2$, catalyst after calcination at 800 °C, c) NiMo$_{imp}$/ZrO$_2$ catalyst.
All the peaks belong to ZrO$_2$ (T).

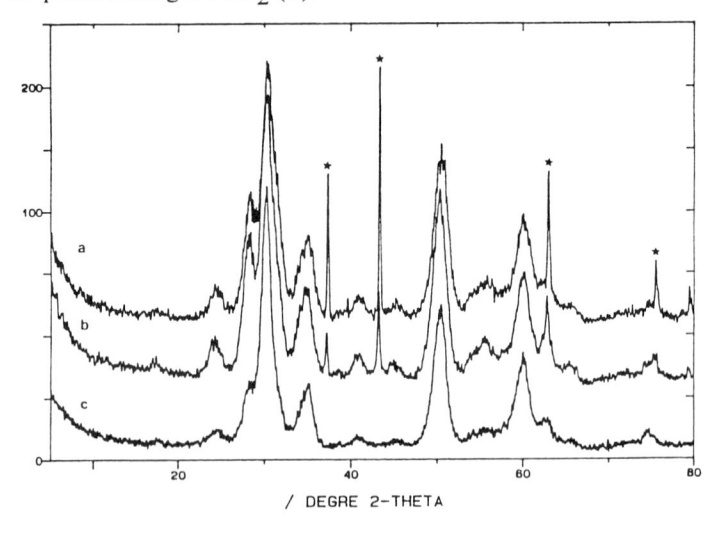

Fig. 2. XRD patterns of the product: a) of the molten salt reaction of Ni and Zr precursors, b) of treatment of the previous product with sodium molybdate at 500°C for 1h. c) Idem. 3h at 500°C.
(*) - NiO ; other peaks - ZrO$_2$ (M and T)

Table 2. Chemical composition of the samples (%wt)

Name	Mo	Fe	Co	Ni
NiMo	7.46			3.3
NiMo$_{imp}$	16.1			6.6
CoMo	7.06		3.6	
CoMo$_{imp}$	16		6.6	
FeMo	6.8	4		
FeMo$_{imp}$	15.8	6.6		
NiMo$_{com}$	9.2			2.4

Impregnation of MS catalysts with the aqueous solutions of the salts of d-metals was carried out to adjust the atomic M/Mo ratios optimal for the hydrotreating performance and to increase as much as possible the loading of the dispersed active phase on the support (denominated NiMo$_{imp}$, CoMo$_{imp}$, FeMo$_{imp}$ catalysts). A twofold increase of the loading of active phase (Table 2) did not lead to the appearance of any new features in the XRD patterns (Fig.1c). (After the impregnation FeMo$_{imp}$, CoMo$_{imp}$, NiMo$_{imp}$ samples were calcined in air at 450°C for 2 h).

BET surface areas

Table 3 gives the specific surface areas of the samples versus the calcination temperature. Compared to the classical preparations of zirconia supported systems, the surface areas are two times higher and thus indicate the determinant role of the dopants on the growth of zirconia crystallites in the melt. In the Mo/ZrO$_2$ solids, prepared from molten nitrates, molybdenum presents on the surface as polymolybdate species, which stabilize tetragonal zirconia with enhanced surface area (10, 18). NiMo, CoMo and FeMo pairs of dopants provide even more pronounced stabilizing effect than sole Mo dopant.

Table 3. Specific surface areas ($m^2.g^{-1}$) of the samples versus calcination temperatures.

	Tcalc°C		
Sample	500	600	800
CoMo/ZrO$_2$	232	123	45
NiMo/ZrO$_2$	250	199	150
FeMo/ZrO$_2$	235	151	58
Mo/ZrO$_2$	220	150	60
Ni/ZrO$_2$	130	91	4
ZrO$_2$	126	80	2.5

The NiMo catalyst was the most stable. Even after the calcination at 800°C, its XRD pattern demonstrated only the reflexes of tetragonal zirconia. Calcination of CoMo and FeMo solids leads to the formation of bulk molybdates. More rapid sintering than in the case of NiMo was observed for the Co and Fe containing samples. It can be noticed that for the conventional preparations specific surface area of the initial support is always higher than that of the catalyst, due to the plugging of the support pores by the particles of the active phase. In our case, inversely, the species of the active component serve at once as a textural promoters.

DRS UV-vis spectroscopy.

Electron d-d transition bands of Ni(II) (d^8), Co(II) (d^7) and Fe(III) (d^5) were observed in the spectra of zirconia samples, doped by these elements, as well as in the commercial alumina-supported NiMo (further NiMo$_{com}$ and CoMo$_{com}$) catalysts. DRS spectra of the catalysts and some reference solids are presented in Fig.3. Values of the 10Dq parameter for Ni are listed in Table 4.

Molten Salt Preparation of Mixed Transition Metals 243

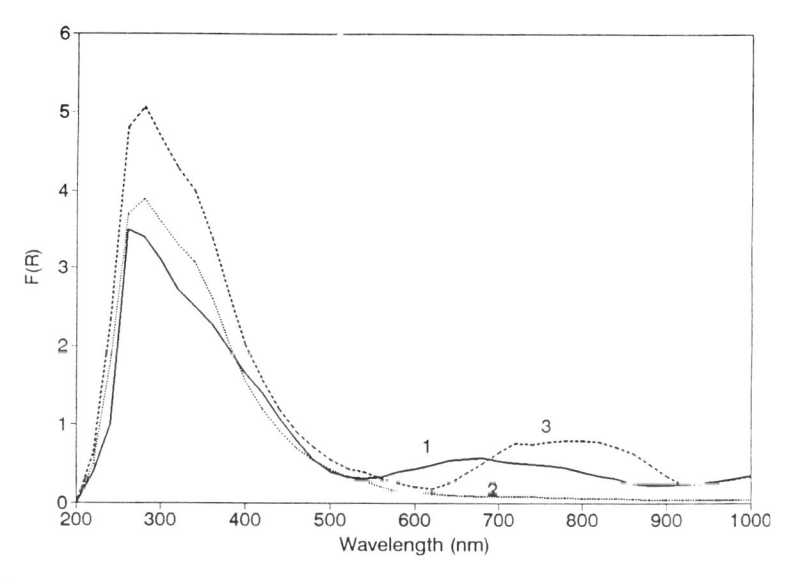

Figure 3a). UV spectra of NiMo industrial catalyst (1), NiMo molten salt catalyst (2) and nickel molybdate (3).

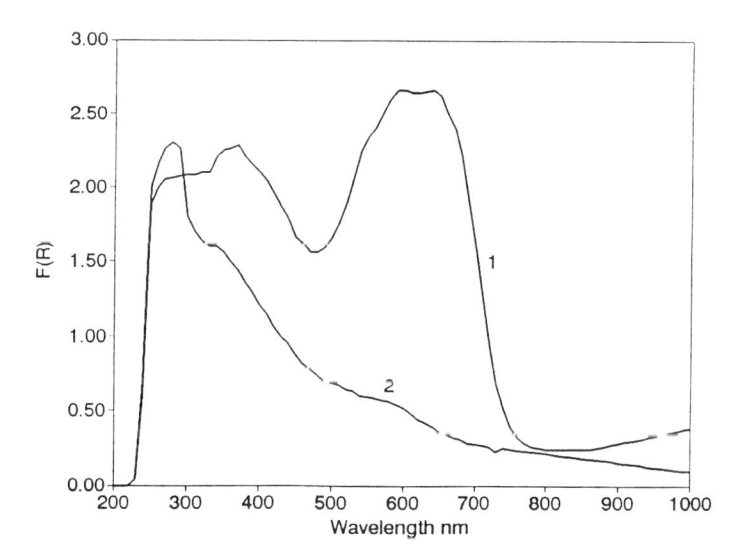

Figure 3b). UV spectra of CoMo industrial catalyst (1) and CoMo molten salt catalyst (2).

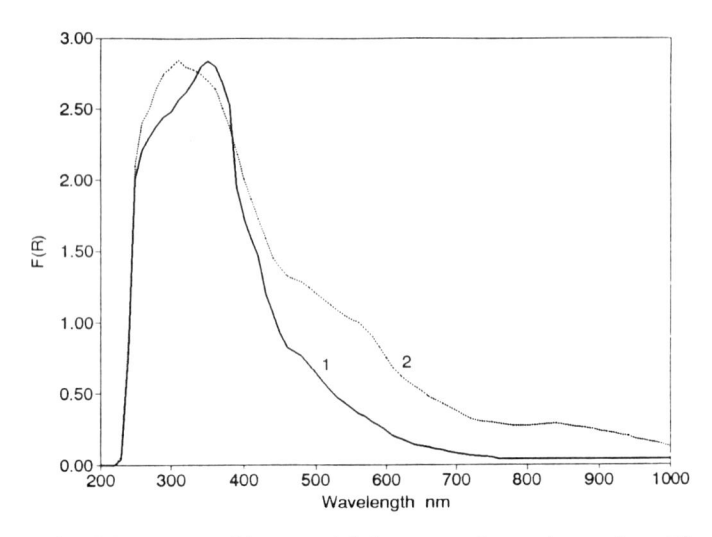

Figure 3c). UV spectra of iron molybdenum molten salt catalyst (1) and iron impregnated on zirconia (2).

Light-gray sample of the molten salt preparation 10%NiO/ZrO$_2$ demonstrated absorption bands at 33900 (shoulder at 29400), 17241, and 9090 cm^{-1}, corresponding to the charge transfer (the first band) and d-d transitions of the octahedraly coordinated Ni(II) ions in the NiO oxide (19, 20).

For the pale-green commercial NiMo - alumina catalyst, broad and intense charge transfer band of polymolybdate (35500 cm^{-1}), and d-d bands of Ni (14700, 8470, cm^{-1}), were observed. Because of superposition with intense charge transfer band, d-d transition $^3A_{2g} \rightarrow {}^1T_{1g}$ could not be observed in the NiMo$_{com}$ catalyst. From the positions of two d-d bands of Ni, we conclude that some of Ni atoms have an octahedral coordination in the NiMo$_{com}$ catalyst, calcined at 773K. No doublet of bands at 16900 and 15800 cm^{-1} characteristic

Table 4. Values of the 10Dq parameter of Ni(II) in the molten salt preparations and reference compounds.

Sample	NiAl$_2$O$_4$	NiO	NiMoO$_4$	Ni/ZrO$_2$	NiMo$_{com}$	NiMo(m.s)	NiMo$_{imp}$
10Dq	970	900	790	909	847	-	790

for Ni aluminate (21) was observed in the UV-vis spectrum of NiMo$_{com}$. The 10Dq parameter for nickel in this catalyst (847 cm^{-1}) was intermediate between that of NiAl$_2$O$_4$ (970 cm^{-1}) and NiMoO$_4$ (790 cm^{-1}). Being calcined at 800°C, NiMo$_{com}$ demonstrated the absorption bands of Ni aluminate.

By contrast, d-d transfer was not observed in DRS spectrum of mustard colored NiMo/ZrO$_2$ molten salt preparation. Only a charge transfer band extended to low wavenumbers was seen in the spectrum of this sample. Additional amount of Ni and Mo introduced in this sample by impregnation, yield weak bands corresponding to pure NiMoO$_4$ (10Dq=790 cm^{-1}, (22)). Nickel molybdate prepared in the molten salt had the same coloration as a zirconia-supported catalyst and the same absorption band of nickel. Calcination at 800°C yield no considerable changes in the UV-vis spectra of NiMo/ZrO$_2$ samples. Therefore, we can suggest that in the NiMo/ZrO$_2$ solids nickel is coordinated by oxygen of molybdate groups rather than by more basic oxygen of the support. The similar difference between DRS UV-vis spectra of alumina and zirconia supported CoMo and FeMo catalysts was observed (Fig.3). Considerable decrease of the intensity of the d-d transition bands and their bathochromic shift can be related to the coordination of VIII group metal by molybdate moieties.

XPS characterizations.

1. Binding energies.

The BE of S2p, Mo3d$_{5/2}$, and M2p$_{3/2}$ (M = Fe, Co, Ni) after sulfidation are given in Table 5. Binding energy of Zr3d$_{5/2}$ was taken as a reference at 182.0 eV. Mo 3d$_{5/2}$ and S 2p BE values are in agreement with those obtained earlier on Mo/Al$_2$O$_3$ and Mo/ZrO$_2$ sulfided catalysts (23, 24). (27).

Table 5. XPS - Binding energies (eV) of the MMo/ZrO$_2$ catalysts.

SAMPLE	S 2p	Mo 3d5/2	M 2p M=Fe,Co,Ni
CoMo	162.2	228.8	778.8
CoMo$_{imp}$	162	228.8	779
NiMo	161.8	228.6	853.9
NiMo$_{imp}$	161.8	228.8	853.9
FeMo	162.3	228.6	710.9 708.6
FeMo$_{imp}$	161.9	228.6	710.8 708.1

One type of Ni species was found in both Ni-containing samples, with the BE value corresponding either to published data for supported nickel sulfides (25) or to the "NiMoS" phase according to (24). The Co2p$_{3/2}$ peak consisted of one component at 778.8 - 779 eV, which can be assigned to cobalt in the sulfided state (26) or the "CoMoS" phase

By contrast to the results of XPS study of the sulfidation of the CoMo/Al$_2$O$_3$ catalysts (26), no peaks of oxidic cobalt were observed after sulfidation of CoMo and CoMo$_{imp}$ samples at 673 K, since the interaction of the zirconia support with the Co ions is obviously much weaker, and no mixed oxide phases of Co(II, III) and Zr(IV) is known.

Two forms of superficial Fe have been identified in XPS patterns, having Fe2p$_{3/2}$ BE peaks at 710.8 and 708.1-708.6 eV. Since both forms of Fe were observed previously in the unsupported FeMoS catalysts, prepared by homogeneous sulfide precipitation method (28), they can be assigned to the sulfided species of Fe. None of them can be ascribed to the FeS_2 sulfides (BE value 706.4-706.6). The species with BE at 710.8 and 708.1-708.5 eV has been assigned earlier to pyrrhotite $Fe_{1-x}S$ and "FeMoS" phases respectively (28).

Table 6. Atomic ratios of metals in the sulfided samples, determined by XPS and chemical analysis (CA).

	Sample	Atomic Ratio		
		Fe/Mo	Fe/Zr	Mo/Zr
	FeMo			
CA		1.14	0.11	0.1
XPS		0.64	0.058	0.09
	FeMo$_{imp}$			
CA		0.88	0.23	0.26
XPS		0.4	0.1	0.25
		Co/Mo	Co/Zr	Mo/Zr
	CoMo			
CA		0.85	0.09	0.105
XPS		0.85	0.09	0.105
	CoMo$_{imp}$			
CA		0.68	0.21	0.31
XPS		0.76	0.22	0.29
		Ni/Mo	Ni/Zr	Mo/Zr
	NiMo			
CA		0.71	0.081	0.11
XPS		0.83	0.064	0.077
	NiMo$_{imp}$			
CA		0.66	0.2	0.3
XPS		0.64	0.18	0.28

Intensities and stochiometries.

From the ratios of the integrated areas of XPS peaks, the superficial concentrations of the elements were determined and compared with the chemical analysis data (Table 6). Except the FeMo samples, superficial contents of the elements was close to the bulk values. Even at high loadings, the atomic ratios close to those calculated from the data of chemical analyses have been obtained. It indicates the high dispersion of the active components either in the molten salt preparations or in the catalysts obtained after their supplementary impregnation. In the Fe-containing samples, the total amount of Fe detected by XPS was decreased in the sulfided state, by comparison to the oxidic state, apparently due to the formation of $Fe_{1-x}S$ crystallites, as discussed in (28).

The XPS atomic ratios (S/(Mo+M)) were always higher than 2 thus indicating complete sulfidation of the samples.

Catalytic activity.

Catalytic activities of the sulfided catalysts in HDS, HYD and HDN reactions are presented in Table 7 and compared to the reference catalyst $NiMo_{com}$. After sulfidation, all molten salt catalysts are found to be active in these model reactions. $NiMo/ZrO_2$ presents an activity close to that obtained on the commercial catalyst even with a lower metal content; CoMo was found to be slightly less active and FeMo catalysts poorly active. By adding active phases to the molten salt catalyst, activities of the $NiMo_{imp}$ and $CoMo_{imp}$ catalysts are enhanced independently on the nature of the model reaction. Catalytic activity remains proportional to the metals loading even at the values as high as 16.1% wt. of Mo and 6.6% wt. of Ni. A $NiMo_{imp}$ catalyst was almost twice more active than the $NiMo_{com}$ reference in all the model reactions tested. With this respect, certain progress has been achieved compared to the previous data on the activity of NiMo on zirconia catalysts (29). Zirconia support was prepared in (29) by means of molten salt method, but the active components were introduced using the conventional aqueous impregnation. The catalysts prepared in (29) were more active than the commercial reference per atom of Mo, but not per a unit of mass of catalyst. In the present work, using simultaneous molten salt reactions we have obtained a considerable gain of the specific surface area, and as a consequence, of the specific activity per a unit of mass.

Table 7. Catalytic activities in HDS, HYD and HDN reactions per gram of catalyst (10^{-8} mol.g^{-1}.s^{-1}) and per mol of Mo, A/Mo (10^{-5} mol.g-at Mo^{-1} s^{-1}).

Test	NiMo	NiMo$_{imp}$	CoMo	CoMo$_{imp}$	FeMo	FeMo$_{imp}$	NiMo$_{com}$
	HDS						
	150	250	100	180	12	11	160
A/Mo	193	149	136	108	17	6.7	167
	HYD						
	21	53	16	28	12	12	25
A/Mo	27	31	22	16.8	17	7	27
	HDN						
	9	17.8	9	13.5	6.2	8	11
A/Mo	12	11	12	8	9	5	11

CONCLUSIONS

Molten salt chemistry opens up new perspectives in the preparation of hydrotreating catalysts. Zirconia is a promising support for hydrotreating reaction but classical preparation methods cannot achieve surface areas as high as those obtained on alumina. One step synthesis in molten nitrates provides high surface area of zirconia with well dispersed mixed phase at the surface. The thermal stability of these new catalysts is also enhanced. This preparation method generates free space for adding more active phase in a conventional manner. As the result, well-dispersed high loaded catalysts can be prepared with this new technique. The HDS, HYD and HDN activities as twice higher than those observed on a commercial catalyst can be achieved by the use of catalysts prepared from molten salts. Here the enhanced activity in the reactions of simple molecules was achieved due to the extensive increase of the number of active sites. However we can expect that these systems will demonstrate unusual catalytic activity and selectivity in the reactions of more complicated molecules or real feeds, where the modified acidobasic properties of the zirconia support can play a more important role.

ACKNOWLEDGEMENT

We gratefully acknowledge M. Cattenot for the HDN catalytic activity measurements.

REFERENCES

(1) B. Delmon, Catal. Lett. 22 , 1 (1993). (ref. therein)
(2) F. Luck, Bull. Soc. Chim. Belg., 100 (11-12) , 781 (1991).
(3) M. Breysse, J.L. Portefaix, M. Vrinat, Catal. Today, 10 , 489 (1991).
(4) M. Breysse, C. Geantet, M. Lacroix, J.L. Portefaix, and M. Vrinat, this Issue.
(5) M. Jebrouni, B. Durand and M. Roubin, Ann. Chim. Fr., 16, 569 (1991).
(6) D. Hamon, M. Vrinat, M. Breysse, B. Durand, F. Beauchesne and T. des Courieres, Bull. Soc. Chim. Belg, 100 , 933 (1991).
(7) D. Hamon, M. Vrinat, M. Breysse, B. Durand, M. Jebrouni, M. Roubin, P. Magnoux, and T. des Courieres, Catal. Today, 10 , 613 (1991).
(8) M. Jebrouni, B. Durand, and M. Roubin, Ann. Chim. Fr,. 17 , 143 (1992).
(9) P. Afanasiev, C. Geantet, and M. Breysse, *in* "Book of Abstracts, Int Symp. Chimie douce, Soft chemistry routes to new materials" 6-10 Sept 1993, Nantes, France.
(10) P. Afanasiev, C. Geantet, M. Breysse, and T. des Courieres, J. Catal., 153, 17 (1995).
(11) M. Zdrazil, Catal. Today, 3(4) , 269 (1988) (p.332)
(12) T.C. Ho, A.J. Jacobson, R.R. Chianelli, and C.R.F. Lund, J. Catal. 138 , 351 (1992).
(13) D.H. Kerridge, and J. Cancela Rey, J. Inorg. Nucl. Chem. 39 , 405 (1977).
(14) H. Frouzanfar, and D.H. Kerridge, J. Inorg. Nucl. Chem. 42, 1382 (1980).
(15) D.H. Kerridge, and A.Y.Khudhari, J. Inorg. Nucl. Chem. 37 , 1893 (1975).
(16) D.A. Habboush, and D.H. Kerridge, Thermochim. Acta, 10 : 187 (1974).
(17) C. Geantet, P. Afanasiev, and M. Breysse, in "Proceedings, 6th Int Symp: Scientific Bases for the Preparation of Heterogeneous Catalysts", Sept 5-9, 1994, Louvain-la-Neuve (Belgium).
(18) P. Afanasiev, C. Geantet, M. Breysse, G. Coudurier, J.C. Vedrine, J. Chem. Soc. Faraday Trans. 90 (1) , 193 (1994).

(19) M. Houalla, and B. Delmon, J. Phys. Chem. 84, 2194 (1980).
(20) B. Sheffer, J.J. Heijenga, and J.A. Moulijn, J. Phys. Chem. 91, 4752 (1987).
(21) G.L.M. Souza, A.C.B. Santos, D.A. Lovate, and A.C. Faro Jr, Catal. Today, 5, 451 (1989).
(22) P. Courtine, and J.C. Daumas, C. R. Acad. Sci. Paris. 268, 1568 (1969).
(23) J. Grimblot, P. Dufresne, L. Gengembre, J.P. Bonnelle, Bull. Soc. Chim. Belg. 90, 1261 (1981).
(24) F. Maugé, J.C. Duchet, J.C. Lavalley, S. Houssenbay, E. Payen, J. Grimblot, S. Kasztelan, Catal. Today 10, 561 (1991).
(25) H. van der Heide, R. Hemmel, F. van Bruggen, and C. Haas, J. Solid State Chem. 33, 17 (1980).
(26) R. Prada Silvy, J.M. Beuken, P. Bertrand, B.K. Hodnett, F. Delannay, B. Delmon, B., Bull. Soc. Chim. Belg. 93 (8-9), 775 (1984).
(27) I. Alstrup, I. Chorkendorff, R. Candia, B.S. Clausen, and H. Topsoe, H., J. Catal 77, 397-409 (1982).
(28) M. Karroua, J. Ladrière, H. Matralis, P. Grange, B. Delmon, J. Catal. 138, 640 (1992).
(29) M. Vrinat, D. Hamon, M. Breysse, B. Durand, T. des Courieres, Catal. Today 20, 273 (1994).

The Role of Hydrogen in the Hydrogenation and Hydrogenolysis of Aniline on the Nickel Single Crystal Surfaces: Its Implication on the Mechanisms of HDN Reactions

Sean X. Huang, Daniel A. Fischer,* and John L. Gland

Department of Chemistry, University of Michigan, Ann Arbor, MI 48109
*Materials Science and Engineering Laboratory, National Institute of Standards and Technology, Gaithersburg, MD 20899

ABSTRACT

The adsorption and reaction of aniline on nickel single crystal surfaces have been studied both in the presence and absence of hydrogen. The hydrogen reactive environment is created in a high pressure cell filled with up to 0.001 Torr of flowing hydrogen. A combination of temperature programmed reaction spectroscopy and near edge x-ray absorption fine structure measurements was utilized to probe the structure of adsorbed species and reaction mechanism *in situ*. In vacuum, aniline dehydrogenation dominates and produces a tenacious surface carbonaceous deposit which fragments and desorbs at 800 K. In contrast, hydrogenolysis reaction is increasingly favored on both nickel surfaces studied with increasing amount of external hydrogen. The Ni(100) surface exhibits significantly higher hydrogenolysis activity compared to the closed packed Ni(111) surface, and produces benzene at 470 K. Spectroscopic study of the reaction precursors reveals the formation of hydrogenated aniline which aromaticity decreases in the presence of hydrogen. We believe that the hydrogenated aniline is readily subject to hydrogen-induced C-N bond activation, in a way similar to that in catalytic HDN reactions.

I. INTRODUCTION

The selectivity of hydrogenation and hydrogenolysis reactions for organonitrogen compounds on transition metal surfaces depends heavily on the availability of surface hydrogen under reaction conditions. The surface hydrogen produced during dehydrogenation of adsorbed aniline upon thermal activation does not significantly modify hydrogenolysis reactions because it desorbs below the reaction temperatures. A series of experiments which uses external hydrogen to increase surface hydrogen concentration at reaction temperatures are reported here. *In situ* kinetic measurements in the presence of reactive hydrogen environments have been used to probe the details of the adsorbed species and reaction mechanisms. Nickel single crystal have been used as well-defined substrate for aniline hydrogenation and hydrogenolysis studies.

Previously, this group has reported the effect of external hydrogen on aniline hydrogenolysis on the Pt (111) surface [1]. Carbon-nitrogen bond activation is substantially enhanced on the Pt(111) surface in the presence of hydrogen. The increased C-N bond cleavage in hydrogen is facilitated by a parallel configuration of the aromatic ring at reaction temperature. While in the absence of surface hydrogen, the adsorbed intermediate tilts away from surface because of partial dehydrogenation with increasing temperature at about 400 K. On evaporated nickel film, aniline adsorbs both molecularly via the π-electrons and dissociatively via an anion formed by the release of a proton from the amino group [3]. On both the Ni(100) and Ni(111) surfaces, adsorbed aniline was reported to form a strongly bound surface species at relatively high temperatures, postulated to be polyaniline [4, 5]. Polymerization occurs when the positively charged ring carbon atoms, upon electrophilic attack from nickel, react with the nitrogen on adjacent adsorbed aniline molecules. This polymerized species is very stable on the nickel surfaces, and does not decompose till above 600 K. This paper will present our recent study of aniline reactions on the Ni(100) and Ni(111) surfaces both in the presence and absence of hydrogen. Reactivity comparisons will also be made for these two nickel surfaces towards adsorbed aniline.

The reactions were studied both under ultrahigh vacuum (UHV) conditions and in hydrogen pressures up to 0.001 Torr. The gas phase reaction products were monitored by multiplexed mass spectrometer during temperature programmed reaction studies (TPRS) performed at the University of Michigan. The structure and orientation of adsorbed surface intermediates were characterized using the angular dependence of Near Edge X-ray Absorption Fine Structure (NEXAFS) spectra. The NEXAFS experiments, as described elsewhere [2], were performed on U-1 beamline at the National Synchrotron Light Source. Real time monitoring of the evolution of surface species was achieved by measuring the intensity change of key NEXAFS resonances during temperature programmed experiments. Fluorescence yield detection technique allows us to study the surface reaction kinetics even in the presence of high reactive gases. Aniline was adsorbed onto clean, annealed nickel surfaces at 100 K and annealed to 210 K to remove multilayers formed at low temperature.

II. RESULTS AND DISCUSSION

The existence of a highly aromatic adsorbate at elevated temperatures is illustrated by temperature programmed fluorescence near edge spectroscopy (TP FYNES) for a monolayer of aniline on both Ni(100) and Ni(111) surfaces (solid

Hydrogenation and Hydrogenolysis of Aniline

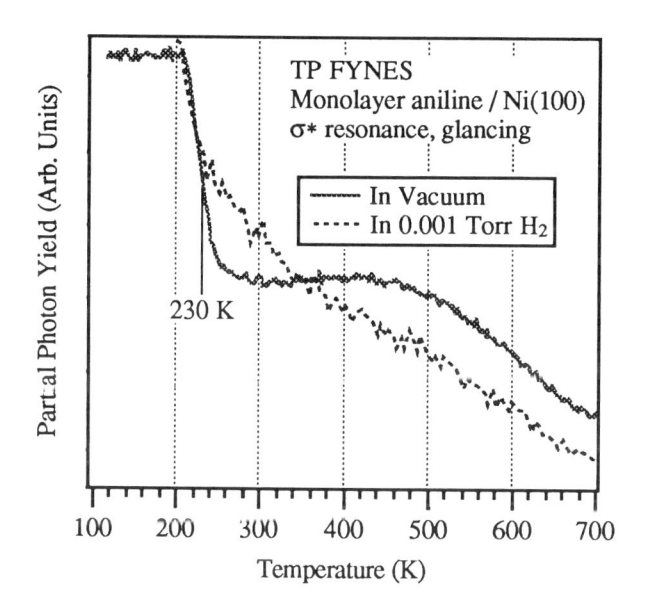

Figure 1: Temperature Programmed FYNES spectra for monolayer aniline on the Ni(100) surface. The heating rate for the TP FYNES is 2 K/s.

curves, Figure 1 and Figure 2). In these spectra, the fluorescence yield intensity of the σ* resonance at 293 eV was monitored as the sample temperature increased. This σ* resonance corresponds to a transition from carbon core level into one of the antibonding C-C molecular orbitals [6]. After physisorbed aniline desorbs at 220 K from the Ni(111) surface (230 K from the Ni(100) surface), this σ* resonance remains constant till about 500 K. Therefore, the plateau in the TP FYNES spectra between 220 K and 500 K indicates that the surface intermediate at 500 K has an carbon ring structure rather similar to molecular aniline at 230 K. The TPRS results reveal that no decomposition product desorbs till 700 K (*vida infra*). This chemical stability far exceeds what is expected for molecular aniline. These data are consistent with an earlier proposal by Benziger et al that polymerization occurs during thermal dehydrogenation of aniline on nickel [4, 5].

By increasing the amount of surface hydrogen available, we significantly altered the dominant reaction pathways for aniline on both the Ni(100) and Ni(111) surfaces. Two methods have been used to increase the concentration of hydrogen on the metal surfaces: (1) heating adsorbed aniline in hydrogen flow, as was done in the synchrotron experiments (All figures except Fig. 4); and (2) preadsorbing hydrogen before aniline dosing, as in the TPRS experiments on the Ni(111) surface (Fig. 4). Various degrees of C-N bond

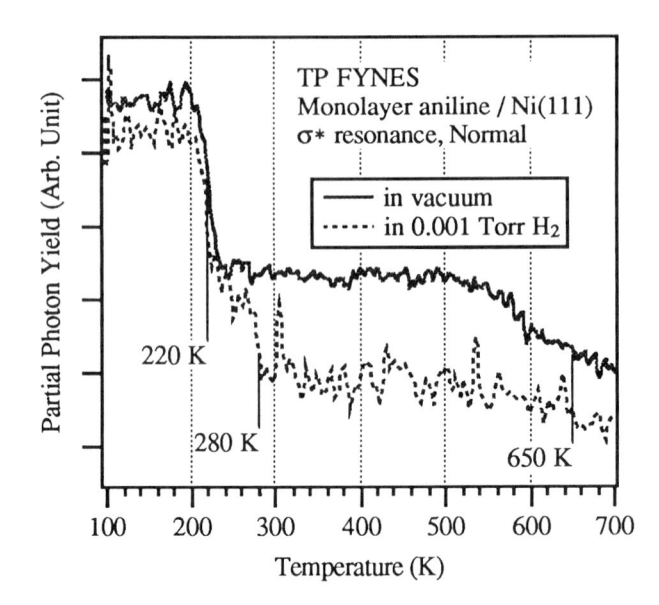

Figure 2: Temperature Programmed FYNES spectra for monolayer aniline on the Ni(111) surface. The temperature tags indicate the monolayer desorption temperature (210 K) and aniline (polyaniline) fragmentation temperatures. The heating rate for the TP FYNES is 2 K/s.

activation were observed by using these two methods. Nevertheless, both methods result in enhancement in hydrogenation/hydrogenolysis rates and retardation of aniline polymerization. However, the most remarkable changes caused by hydrogen in the TP FYNES spectra are the dramatically diminishing σ* resonances at the temperatures far below those seen in vacuum annealed cases. On the Ni(100) surface (dashed curve, Fig. 1), we observe that the C-C σ* resonance decrease steadily with increasing temperature after desorption of the physisorbed layer at ~ 200 K. On the Ni(111) surface (dashed curve, Fig. 2), we observed a rapid drop in the σ_{C-C}* resonance at 280 K, in contrast to 600 K in vacuum. We propose that the increased availability of surface hydrogen due to hydrogen flow results in ring hydrogenation before dehydrogenation of the amino group. The hydrogenation of the phenyl ring appears to prevent polymerization of aniline and stabilize a monomeric aniline derived species. Our carbon K-edge NEXAFS spectrum (*vida infra*) for aniline annealed in 0.001 Torr hydrogen at 320 K on the Ni(111) surface does not contain sharp π* resonances, indicating that the structure of adsorbed aniline has been substantially modified.

Hydrogenation and Hydrogenolysis of Aniline 257

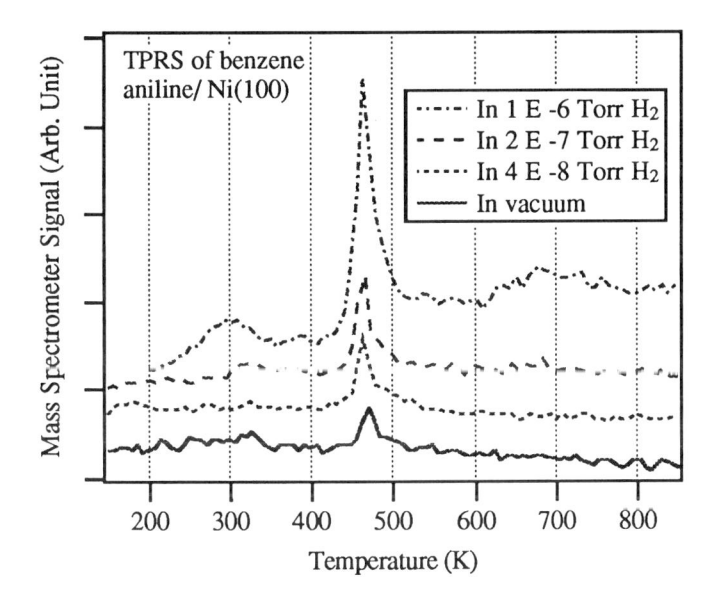

Figure 3: Temperature Programmed Reaction Spectra of the benzene formation from 0.66 L adsorbed aniline on the Ni(111) surface. The dashed curves show progressively higher pressure of hydrogen in which the heating takes place. The heating rate during the TPRS is 5 K/s.

In the presence of external hydrogen, the aniline derived surface species has a substantial propensity to undergo carbon-nitrogen bond cleavage. In the temperature programmed reaction experiments on the Ni(100) surface, benzene is observed as an aniline hydrogenolysis product at 470 K, even in the absence of external hydrogen (Fig. 3). The reaction pathway is enabled by the disproportionation of adsorbed aniline. Hydrogen removed from aniline could join the hydrogen addition to the C-N bond, as the side effect of the massive dehydrogenation which is observed as a coexisting reaction pathway at hydrogenolysis temperature. External hydrogen significantly enhances the production of benzene. With relatively little amount (1×10^{-6} Torr) of hydrogen, benzene production is increased by more than an order of magnitude.

In the temperature programmed reaction experiments with coadsorbed hydrogen on the Ni(111) (Figure 4), we observed ammonia production at 380 K which does not appear in the vacuum TPRS spectrum. The ammonia production remains relatively small even in ~ 10^{-7} Torr hydrogen flow. We also performed the aniline hydrogenolysis reaction in 10^{-3} Torr hydrogen pressure using fluorescence yield as the detection method. Up to 20 % of a monolayer aniline

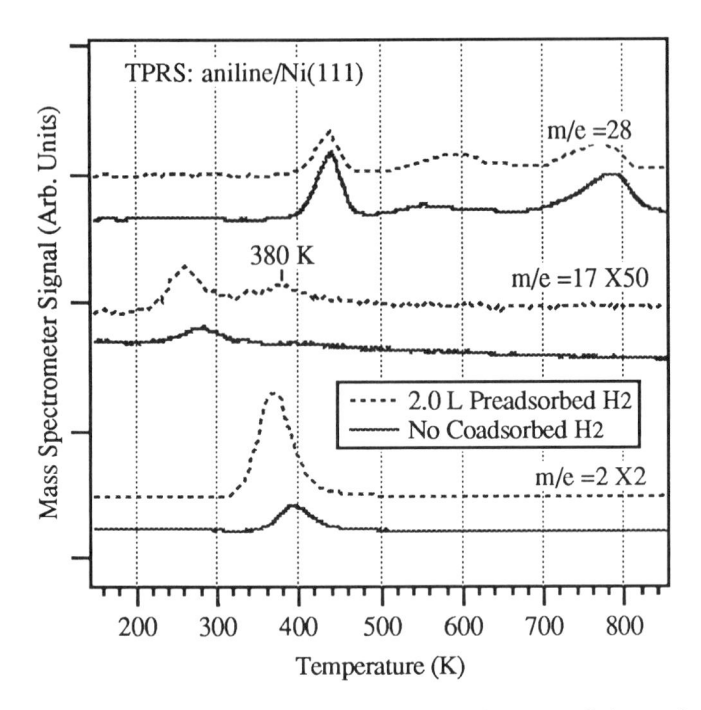

Figure 4: Temperature Programmed Reaction Spectra of the surface reaction products aniline adsorbed with an exposure of 0.01 L on the Ni(111) surface at 100 K. The dashed lines are the spectra taken for aniline with 2.0 L of preadsorbed hydrogen. The heating rate during the TPRS is 2 K/s.

is hydrogenolyzed at 0.001 Torr H_2 pressure, the threshold of our technique. Obviously, higher degree of hydrogenolysis could be achieved at higher pressure of reactive gases. Therefore, we believe the selectivity of reaction pathways depends heavily on the availability of hydrogen.

Carbon K edge spectra of monolayer aniline on the Ni(100) surface are shown in Figure 5. In the absence of hydrogen, adsorbed aniline remains aromatic, undisturbed from the structure similar to molecular aniline. Hydrogen treatment causes dramatic changes in the NEXAFS spectra compared to that in vacuum. The hydrogen spectrum (lower panel, Fig. 5) is typified by a reduction in π^* intensities, together with the increase in σ_{C-H}^* at 289 eV. At 0.001 Torr of hydrogen, σ_{C-H}^* becomes the dominant feature in the NEXAFS spectrum, indicating adsorbed aniline has been converted into hydrogenated form. The hydrogenated and/or partially hydrogenated surface aniline are proposed to be the primary precursor for C-N bond cleavage on the Ni(100) surface. This result

Hydrogenation and Hydrogenolysis of Aniline

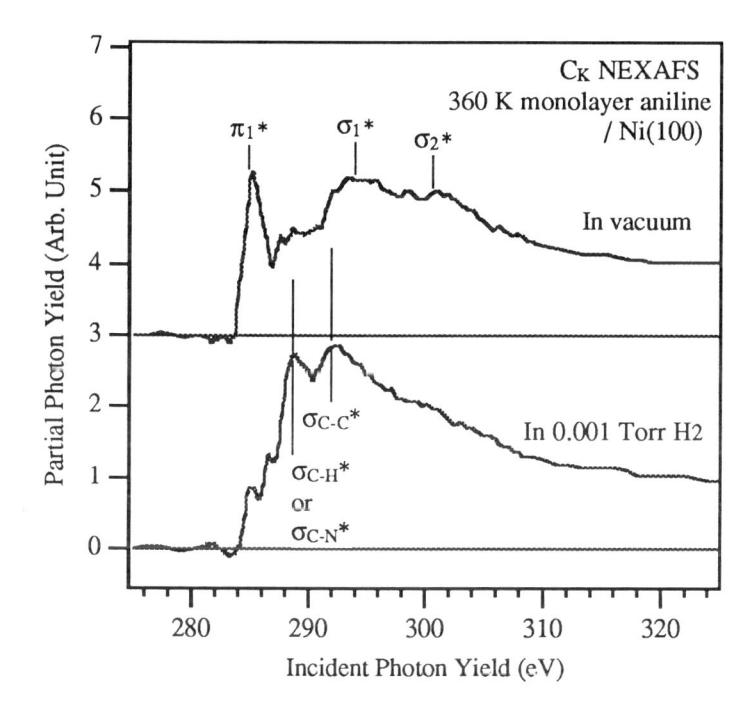

Figure 5: Carbon K-edge NEXAFS spectra for monolayer aniline on Ni(100) annealed to 360 K in vacuum and in 0.001 Torr hydrogen. The incident x-ray was at glancing incidence with respect to the Ni(100) surface. The upper and lower panels show the spectra taken in the absence and presence of external hydrogen, respectively.

clearly indicates that partial aniline hydrogenation enhances hydrogen-induced C-N bond activation. This observation is also in agreement with the widespread catalytic observation that aromatic amines are saturated before undergoing hydrodenitrogenation [7].

Carbon K edge spectra of monolayer aniline on Ni(111) annealed to 320 K are shown in Figure 6. When annealed in vacuum, the adsorbate remains strongly aromatic, with a sharp π^* resonance at 285.1 eV (upper panel, Fig. 6). The spectrum is similar to the characteristic NEXAFS spectra for molecular aniline [1], suggesting that the adsorbate remains undissociated at 320 K. The changes observed in the NEXAFS spectrum appear to be caused by electron transfer to the nickel surface, and the reduction in C-C and C-N bond order, as suggested by Myers et al in their aniline adsorption model calculation using the INDO technique [4]. Partial dehydrogenation of the adsorbed intermediates on

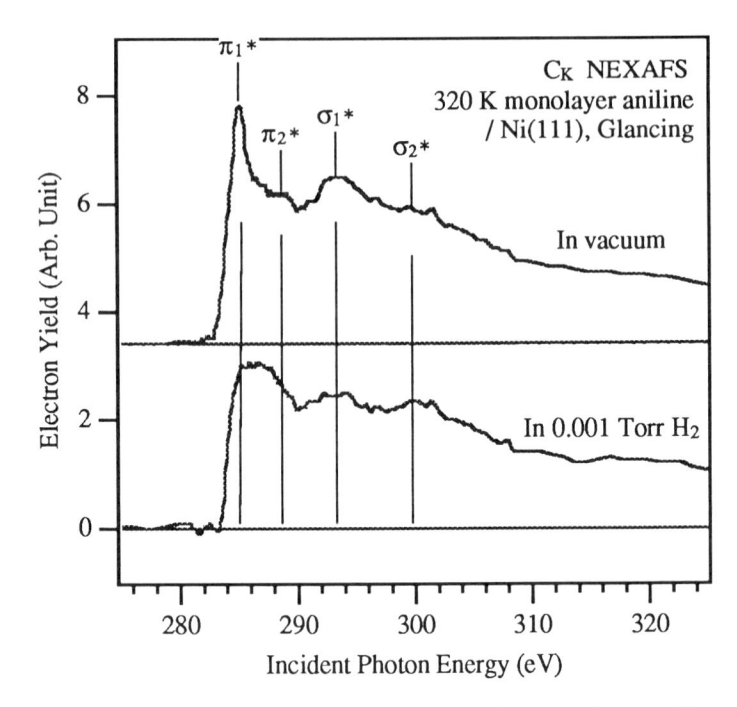

Figure 6: Carbon K-edge NEXAFS spectra for monolayer aniline on Ni(111) annealed to 320 K in vacuum and in 0.001 Torr hydrogen. The incident x-ray was at glancing incidence with respect to the Ni(111) surface. The upper and lower panels show the spectra taken in the absence and presence of external hydrogen, respectively.

the Ni(111) happens at 400 K, producing a relatively small hydrogen desorption peak in the TPRS spectrum (Figure 4). We propose that this is the temperature range where aniline starts to polymerize. Polymer formation stabilizes the surface carbon species from further decomposition till 800 K. The NEXAFS spectra of chemisorbed aniline are also marked by a strong orientational dependence of the resonant transitions, indicating the aromatic ring is parallel or nearly parallel to the surface. This parallel configuration is kept in the presence of hydrogen (lower panel, Fig. 6) enables the adsorbates to interact with the nickel surface through the delocalized π resonance, as illustrated by the clear broadening of the π^* resonances.

Juxtaposing the results on the two surfaces, we found that the Ni(100) surface is more reactive for hydrodenitrogenation of aniline than the more closed Ni(111) surface. Even in the absence of hydrogen, hydrogenolysis product benzene is observed on the Ni(100) surface. As mentioned above, the Ni(111) surface exhibits stronger surface-substrate interaction and parallel adsorption of the ring system. With hydrogen pressures, the π^* resonances for adsorbed aniline on the Ni(111) surface are substantially more broadened, indicating more π interaction with the substrate. The carbon-ring's interaction with the Ni(111) surface proposedly weaken the surface's bonding with the amine functionality, and therefore the propensity for hydrogenolysis. In contrast, on the Ni(100) surface, the surface precursor maintains the uptilt configuration as a result of hydrogenation, as indicated by the orientational dependence of resonance intensities (data not shown due to limited space). Assuming a chair configuration of cyclohexyl functional group, the hydrogenated carbon ring could position an axial C-N bond parallel to the surface when the ring itself is at large angle. With the C-N bonds parallel bonded to the surface, the probability of its activation is optimized on the Ni(100) surface. To summarize, aniline hydrogenolysis on nickel model catalysts is a structure sensitive reaction. The hydrogenolysis activity depends heavily on a its orientation (parallel) and a hydrogenated reaction precursor.

III. CONCLUSION

The adsorption and desorption of aniline on the Ni(100) and Ni(111) surfaces have been studied both in the presence and absence of hydrogen. A combination of TPRS and NEXAFS measurements was utilized to probe the structure of adsorbed species and reaction mechanism. In vacuum, the aromatic structure of adsorbed aniline is stable up to 600 K, as indicated by *in situ* kinetic measurements. In contrast, the adsorbate's aromatic character decreases significantly at 270 K in the presence of 0.001 Torr hydrogen. The adsorbed aniline also exhibits substantially higher hydrogenation and hydrogenolysis reactivity on the Ni(100) surface, as compared to the close-packed Ni(111) surface. The external hydrogen enhances the hydrogenolysis of aniline on both nickel surfaces. Partially hydrogenated aniline is observed spectroscopically, and is an important precursor for hydrogen-induced C-N bond activation. Our results suggest that during the HDN of aromatic amines on nickel, the activation of the C-N bond is often achieved after hydrogenating the organonitrogens into more saturated amines. This observation is consistent with that in catalytic HDN on Ni-Mo-S catalysts. We believe that this research will elucidate the role of multiple coordinated nickel sites in the nitrogen removal process in the presence of hydrogen.

IV. ACKNOWLEDGMENT

Acknowledgment is made to the Office of Basic Energy Sciences, the Department of Energy (Grant no. DE-FGO2-91ER14-190), for the support of this research.

V. LITERATURE CITED

1. S. X. Huang, D. A. Fischer, and J. L. Gland, Preprints of ACS Petroleum Chemistry Division, 38(3): 692 (1993).
2. S. X. Huang, D. A. Fischer, and J. L. Gland, J. Vac. Sc. Technol., A12(4): 2164 (1994).
3. K. Kishi, K. Chinomi, Y. Inoue, and S. Ikeda, J. Catal., 60: 228 (1979).
4. A. K. Myers, and J. B. Benziger, Langmuir, 5: 1270 (1989).
5. G. R. Schoofs, and J. B. Benziger, J. Phys. Chem., 92: 741 (1988).
6. For the assignment of the aniline NEXAFS resonances, see J. L. Solomon, R. J. Madix, and J. Stöhr, Surf. Sc., 255: 12 (1991) and reference (1).
7. T. C. Ho, Catal. Rev. -Sci. Eng., 30(1): 117 (1988).
8. S. X. Huang, D. A. Fischer, and J. L. Gland, J. Phys. Chem., *in preparation.*

Hydrodenitrogenation of Carbazole on an Alumina-Supported Molybdenum Nitride Catalyst

Masatoshi Nagai, Toshihiro Miyao, and Shinzo Omi
Department of Chemical Engineering
Tokyo University of Agriculture and Technology
Koganei, Tokyo 184, Japan

I. INTRODUCTION

Hydrodenitrogenation (HDN) on molybdenum–containing catalysts has received considerable attention during the past decades. The kinetics and mechanisms have been investigated with single nitrogen-containing compounds in laboratory reactors. Mechanistically, the HDN of unsaturated nitrogen-containing compounds proceeds as follows: first, hydrogenation of the nitrogen-containing ring takes place, followed by the breaking of secondary C-N bonds to give hydrocarbons and ammonia. Normally, ring hydrogenation is the rate-determining step. Pyridine [1,2] and quinoline [3–5] were often selected to do this type of study. Sonnemans et al. [1] studied the kinetics of pyridine hydrogenation over a CoMo/Al$_2$O$_3$ catalyst under atmospheric pressure. They reported that the Langmuir-Hinshelwood equation described the reaction well with adsorption of hydrogen and nitrogen bases on different sites. Recently, Massoth and his coworkers [5] studied the kinetics of indole hydrogenolysis at 350°C and 3.55 MPa total pressure over a sulfided CoMo/Al$_2$O$_3$ catalyst. The reported that the rate of the first C-N bond-breaking step to o-ethylaniline depended on the square root of the H$_2$S partial pressure and was inhibited by indole and dihydroindole. Ring hydrogenation reactions were similarly inhibited, but only slightly affected by H$_2$S. Kinetic analysis supported the concept that the Langmuir-Hinshelwood equation with different catalytic sites involved the C-N hydrogenolysis and ring hydrogenation reactions. Nagai et al. [6] studied the kinetics of acridine HDN on a reduced Mo/Al$_2$O$_3$ catalyst and reported that the C-N hydrogenolysis of perhydroacridine in the HDN reaction could be described by a Langmuir-Hinshelwood type of equation with a competition term for hydrogen and perhydroacridine.

Molybdenum nitride is known to be an active catalyst that emulates the catalytic properties of noble metals for a variety of reactions. Oyama et al. [7] first reported high activity and selectivity of molybdenum nitrides for quinoline HDN in the application of transition metal nitrides for the industrial hydrotreatment of heavy oil. Furthermore, Nagai and Miyao [8] found that the alumina-supported molybdenum nitride catalyst was extremely active in the HDN of carbazole compared with sulfided and reduced molybdenum catalysts. In this study, the kinetics of C-N hydrogenolysis of perhydrocarbazole on the Mo/Al_2O_3 nitrided at 500°C was observed to obtain rate expressions for C-N hydrogenolysis in which it was the rate-determining step with equilibrium of the hydrogenation. The object of this study is to present a suitable rate expression for the C-N bond hydrogenolysis step involved in the HDN of carbazole at high temperature and high hydrogen pressure.

II. Experimental

The molybdena catalyst (11.7% MoO_3 supported on alumina) was prepared from a mixture of ammonium paramolybdate and γ-alumina. The oxide catalyst was crushed and sieved to 10–20 mesh granules and calcined in air at 450°C for 24 hours. The catalyst was nitrided according to the procedure Vople and Boudart employed [9]. After the catalyst was heated to 450°C under an air flow, 2.0 g sample of the catalyst was contacted with pure flowing ammonia at a rate of 4 l/h under atmospheric pressure at 300°C. The nitriding temperature was increased from 300 to 500°C at a rate of 1°C/min, and held at 500°C for 3 hours. The catalyst was then cooled to the desired temperature in flowing NH_3. The liquid feed consisted of 0.25 wt% carbazole dissolved in xylene. The feed was introduced at 20 ml/h with a 6 l/h hydrogen flow at 10.1 MPa total pressure. Product samples were quantitatively analyzed by gas chromatography. The kinetic study of carbazole HDN was performed at 340, 350, and 360°C under vapor-phase conditions. Partial pressures of carbazole and hydrogen varied in the following ranges: carbazole, 0.05–0.45 wt%; H_2, 3.40–9.12 MPa. Repeated experiments demonstrated that conversion under standard conditions remained constant within 5%. The HDN activity of the catalyst represents the rate of formation of bicyclohexyl per catalyst weight.

The surface area of the catalysts was measured by nitrogen adsorption using a standard BET apparatus. The value was 226 and 245 m^2/g for the fresh and nitrided molybdena-alumina, respectively. Nitriding of the molybdena-alumina at 500°C decreased the surface area by only 8%. Nitrogen content analysis was carried out using a Perkin-Elmer CHN elemental analyzer. The nitrogen content of the nitrided catalyst was 1.46 and 0.28 wt% upon evacuation at 10^{-2} Pa and at 200°C and 700°C, respectively. The N_2 desorbed from the catalyst amounted to 0.42 mmol/g upon heating from 200°C to 700°C during evacuation.

III. Results and Discussion

In the product distribution of carbazole HDN, the major hydrocarbon product was bicyclohexyl. Other hydrocarbons (cyclohexylbenzene, hexylcyclohexane, hexylcyclohexene and 2-ethylbicyclo[4.4.0]decane) were also formed in smaller amounts above 340°C. The hydrogenated carbazole compounds [such as hexahydrocarbazole (0.003 wt%), octahydrocarbazole (0.007 wt%), and perhydrocarbazole (0.003 wt%)] at 350°C for the 0.25 wt% carbazole feed were barely observed in the reaction products except for a relatively high amount of tetrahydrocarbazole. Furthermore, side reactions occurred above 330°C during the isomerization of bicyclohexyl to 2-ethylbicyclo[4.4.0]decane and the decomposition to cyclohexylhexane.

A. Contact Time

The change in the product distribution of the HDN of carbazole on the nitrided catalyst as a function of contact time at 300°C and 340°C is shown in Figs. 1 and 2. At 300°C (Fig. 1), the concentrations of bicyclohexyl, tetrahydrocarbazole, and perhydrocarbazole were increased with increasing contact time, while the carbazole concentration decreased. Furthermore, the concentrations of hexahydro- and octahydrocarbazole were increased, reached a maximum, and then decreased with contact time as shown in Fig. 1b. From these results, the hydrogenation of carbazole to perhydrocarbazole was successive in the HDN of carbazole as shown in Fig. 3. Bicyclohexyl was produced from the C-N bond scission of perhydrocarbazole into bicyclohexyl and ammonia.

The hydrogenation of tetrahydrocarbazole to octahydrocarbazole was in equilibrium but the hydrogenations of carbazole to tetrahydrocarbazole and of octahydrocarbazole to perhydrocarbazole were not yet equilibrated at 300°C. On the other hand, at 340°C (Fig. 2), carbazole and the hydrogenated compounds disappeared and only perhydrocarbazole was produced above the contact time of 0.1 [g-cat h/ml]. The perhydrocarbazole concentration was almost constant above the contact time of 0.1 [g-cat h/ml]. This result suggested that the hydrogenation of carbazole to perhydrocarbazole was in equilibrium around 340°C.

The C-N bond of the nitrogen compounds such as pyridine and quinoline [1–4] was first broken after the complete saturation of the heteroaromatic ring (piperidine and decahydroquinoline). Hexylcyclohexene was observed in the reaction products of carbazole HDN. This compound was possibly formed during the C-N hydrogenolysis of decahydrocarbazole which was not a completely hydrogenated carbazole compound. This result indicates that the nitride catalyst is very active during the highly selective C-N hydrogenolysis of partially hydrogenated nitrogen compounds. Therefore, nitriding the MoO_3/Al_2O_3 catalyst offers the development of a catalyst preparation method for an effective C-N hydrogenolysis process with less hydrogen consumption.

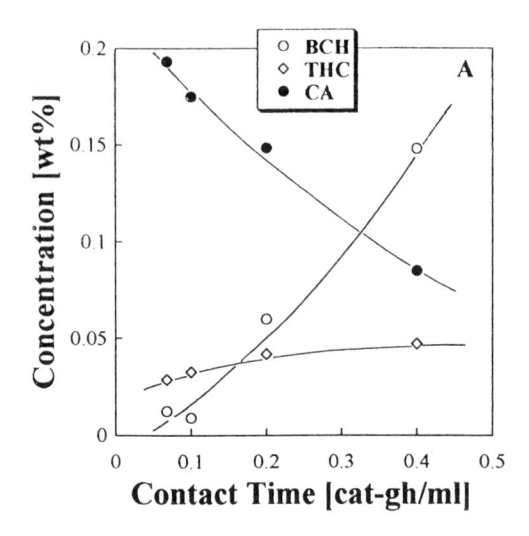

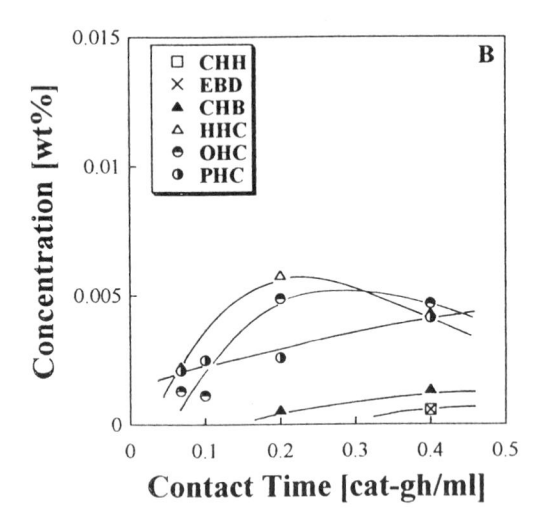

Fig. 1 Product distribution as a function of contact time at 300°C. (CA) carbazole, (THC) tetrahydrocarbazole, (HHC) hexahydrocarbazole, (OHC) octahydrocarbazole, (PHC) perhydrocarbazole, (BCH) bicyclohexyl, (CHB) cyclohexylbenzene, (EBD) 2-ethylbicyclo[4.4.0]decane, (CHH) cyclohexylhexene.

Hydrodenitrogenation of Carbazole

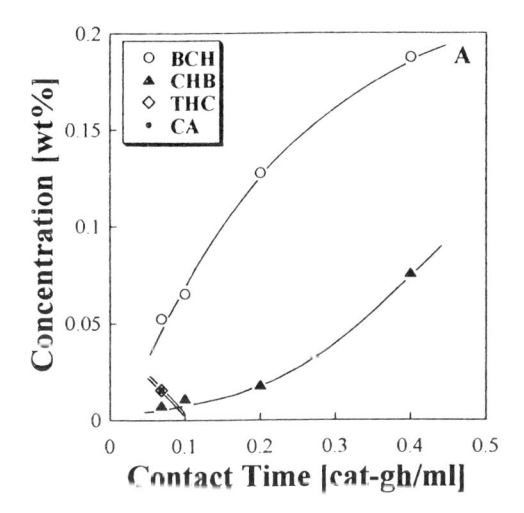

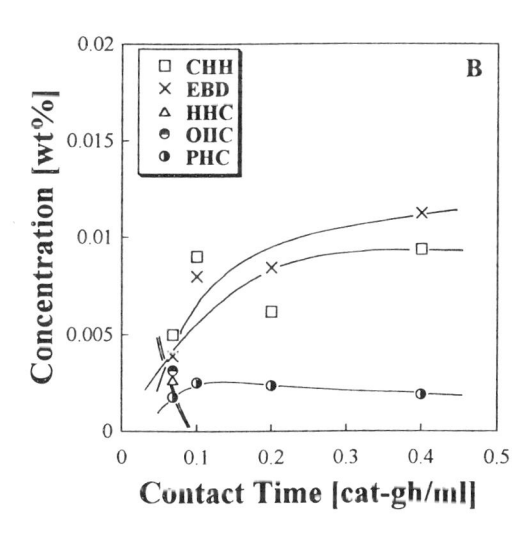

Fig. 2 Product distribution as a function of contact time at 340°C. See Figure 1 for the symbols.

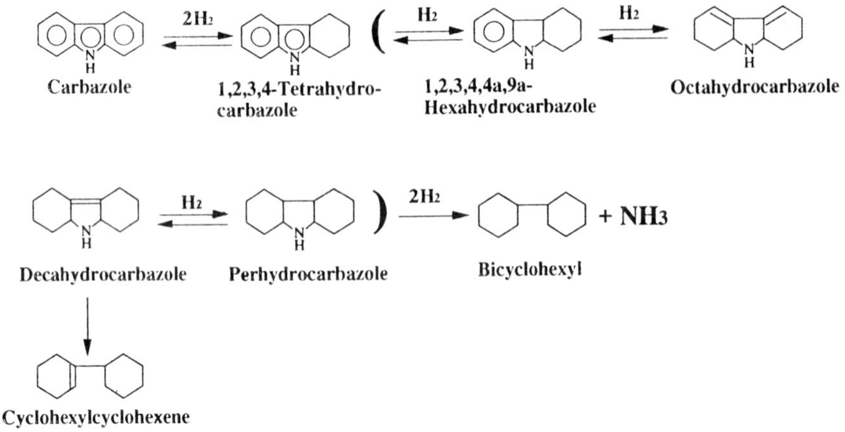

Fig. 3 Reaction scheme for carbazole HDN.

B. Mass Transfer Limitation

The carbazole HDN data were collected in two separate experiments using the nitrided catalyst with two particle sizes (d_i) of 0.85–1.70 mm (average d_{a1} = 1.27 mm) and 0.5–0.71 mm (average d_{a2} = 0.605 mm) at temperatures of 280–360°C and a constant partial pressure of carbazole and hydrogen. When the catalyst has a finer particle size, the pressure drop through the catalyst bed is probably increased. When the reaction was limited by interpolate mass transfer, the C-N hydrogenolysis rate varied inversely with particle size; $r \propto 1/d$. Because the average particle sizes of the two catalysts are 1.27 and 0.605, d_{a2}/d_{a1} (= r_{a1}/r_{a2}) becomes 0.48. However, C-N hydrogenolysis rates, r_1 and r_2, were obtained for the two particle distributions of the catalyst at the reaction temperatures. The ratios of r_1/r_2 were 1.01 at 300°C, 1.00 at 340°C, and 1.02 at 360°C. The ratios of r_{a1}/r_{a2} were independent for the catalyst particle sizes that were significantly higher than 0.48 in this experiment. From these results, the effect of external mass transport was expected to be negligible.

C. Kinetics

Typical kinetic data are shown in Figs. 4 and 5. Reactivity measurements were carried out by varying the partial pressures of carbazole and hydrogen at different temperatures. Representative plots of some of the data are also shown in Fig. 6 to indicate the trends and comparisons with the equations. Curve

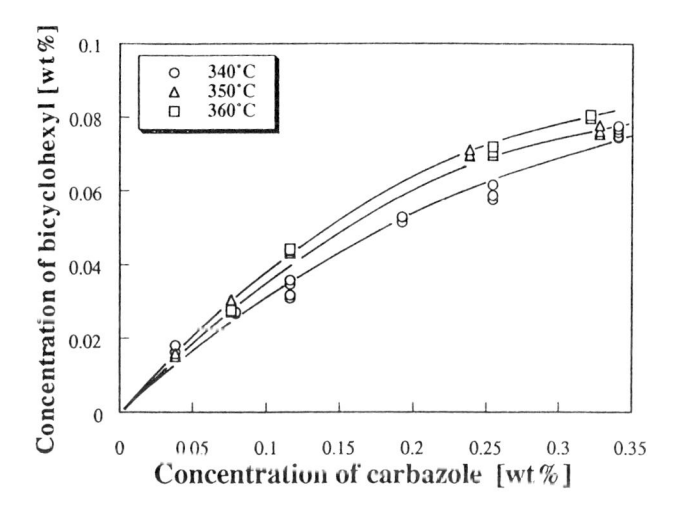

Fig. 4 Dependence of initial carbazole concentration on the formation of bicyclohexyl.

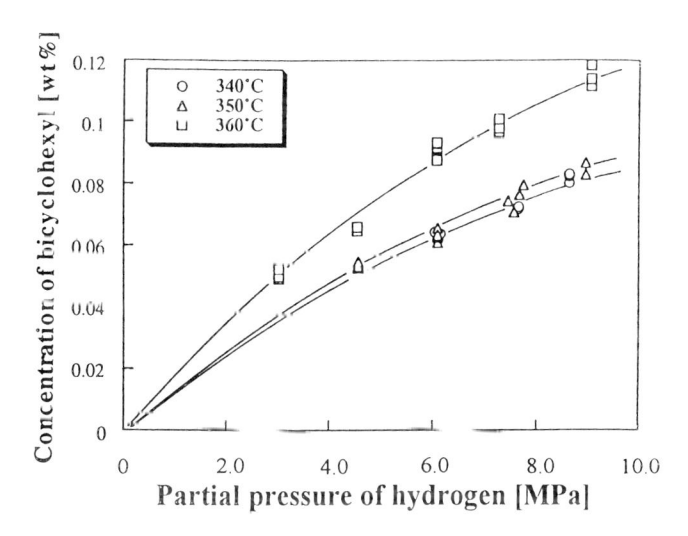

Fig. 5 Dependence of hydrogen pressure on the formation of bicyclohexyl.

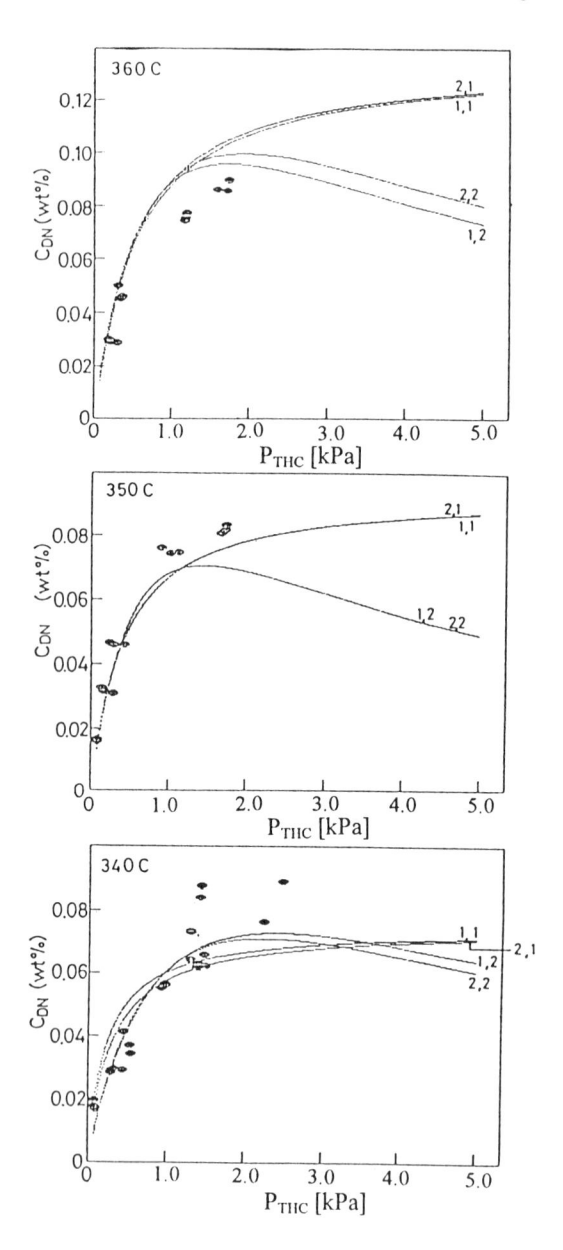

Fig. 6 Carbazole HDN kinetics at three temperatures: comparison of data with prediction of Model A.

fittings are based on the rate constants and parameters for Eqs. 1 and 2. Most studies find that nitrogen removal is first order with respect to feed-nitrogen concentration. Miller and Hineman [4] reported that the pseudo-first order rate constant was observed to decrease with increasing initial nitrogen concentration. Glola and Lee [10] reported that the rates of most hydrogenation and dehydrogenation reactions obeyed the Langmuir-Hinshelwood mechanism with different adsorption sites of hydrogen from organic compounds in quinoline HDN. The existence of two different types of sites on Mo catalysts under hydrodesulfurization and HDN conditions has been evidenced by Hadjiloizou et al [2] and Massoth et al. [5]. The present study showed that a Langmuir-Hinshelwood-type equation was adopted to account for the experimental data. The formation of several hydrocarbons except for bicyclohexyl only negligibly occurred on the C-N hydrogenolysis sites since these compounds were formed in small amounts during the reaction. Ammonia was also negligibly adsorbed on the C-N hydrogenolysis sites, because the strength of adsorption of the nitrogen compounds is influenced by its aromaticity. Ammonia was one of the nitrogen compounds with weak gas-phase basicity and had much lower basicity than pyrrole and aniline [11]. Because the hexahydro-, octahydro-, and perhydrocarbazoles were barely observed in the reaction products, they must be highly reactive intermediates. The hydrogenation sequence was likely to be equilibrated at 330°C, therefore, the partial pressure of perhydrocarbazole was substituted for that of tetrahydrocarbazole. The rate equation for the C-N hydrogenolysis was calculated by the formation of bicyclohexyl from perhydrocarbazole as follows:

$$r_{BCH} = \frac{k (K_H P_H)^l}{(1 + (K_H P_H)^l)^m} \frac{K_{THC} P_{THC}}{(1 + K_{THC} P_{THC})^n} \quad \text{(Model A)}$$

$$r_{BCH} = \frac{k(K_H P_H)^l K_{THC} P_{THC}}{(1+(K_H P_H)^l K_{THC} P_{THC})^q} \quad \text{(Model B)}$$

where m=0, 1, 2; l=0.5, 1, and n=0, 1; q=0, 1. r_{BCH}, k, K_i and P_i are the rate of formation of bicyclohexyl, reaction rate coefficient, adsorption coefficient, and partial pressure, respectively. The subscripts, H and THC, are hydrogen and tetrahydrocarbazole. A number of plausible rate equations of the Langmuir-Hinshelwood type are considered in attempting to correlate the data of Fig. 6. We empirically proceeded, selecting equations on the basis of good fit to the data and deferring the discussion of mechanistic implications until the final paragraph.

A nonlinear least-squares regression analysis was used with Eqs. A-1 and A-2 to determine the best equations to describe the data in Fig. 6. The output of the parameters and criteria with which to judge the goodness of fit with the particular equation: (i) σ_{min}, the minimized sum of the squares of the differences between the observed and predicted rates for each of the observations and (ii) the parameters are omitted when the analysis does not converge during the calculation. Several other criteria were applied to measure the physical meaningfulness of the parameters; (iii) a plot of the logarithm of each adsorption equilibrium constant versus the reciprocal absolute temperature (van't Hoff plot) should produce positive values within their error bounds; (iv) a plot of the logarithm of the formation rate versus reciprocal absolute temperature (Arrhenius plot) should be linear with a negative slope; and (v) each heat of adsorption (ΔH_i) should be linear with a positive slope, except when chemisorption is endothermic [12].

We have found no kinetic evidence for the dissociative adsorption of hydrogen. When the model was calculated at l=0.5 (dissociative adsorption of hydrogen), the solution became divergent. The half power of the term $K_H P_H$ in the denominator was omitted. The temperature dependences of K_i, K_H, and K_{THC} were calculated and are given in Table 1. Eqs. A-2, A-4, and B-8 show negative heats of adsorption of hydrogen and tetrahydrocarbazole. These values are not physically meaningful of the parameters [12]. Therefore, Eqs. A-2,

Table 1 Activation Energies and Adsorption Constants for Model A and B

Equation	m	n	q	l	Activation energy [kJ/mol]	Adsorption heat [kJ/mol] tetrahydro-carbazole	hydrogen	Variance $\Sigma \sigma_{ij}^2 \times 10^{-4}$
Model A								
A-1	1	1	—	1	68.2	101.7	135.1	2.95
A-2	1	2	—	1	125.1	115.5	−52.7	2.16
A-3	2	1	—	1	139.7	61.1	90.4	2.86
A-4	2	2	—	1	96.7	60.2	−23.0	2.44
Model B								
B-1	—	—	1	1	200.4	111.7	110.5	7.07
B-2	—	—	2	1	104.6	72.0	−46.0	2.35

(Model A) $r_{BCH} = \dfrac{k (K_H P_H)^l}{(1 + (K_H P_H)^l)^m} \dfrac{K_{THC} P_{THC}}{(1 + K_{THC} P_{THC})^n}$

(Model B) $r_{BCH} = \dfrac{k(K_H P_H)^l K_{THC} P_{THC}}{(1+(K_H P_H)^l K_{THC} P_{THC})^q}$

A-4, and B-2 were not suitable for the equation. From the results, Eqs. A-1, A-3, and B-1 were the only ones giving meaningful values of the heats of adsorption; the values for hydrogen are 135.1, 90.4, and 110.5 kJ/mol, respectively. Equation B-1 gave a poorer fit, as indicated by the higher variance values of Eqs. A-1 and A-3. Variance in Eq. A-1 is slightly higher than that in Eq. A-3. The values of the heat of adsorption of hydrogen (135.1 and 90.4 kJ/mol) and activation energy (68.2 and 139.7 kJ/mol) for Eqs. A-1 and A-3 should be estimated, respectively, although none for tetrahydrocarbazole have been reported yet. The heat of adsorption for hydrogen in Eq. A-1 is 135.1 kJ/mol, which is in agreement with the value (138.5 kJ/mol) reported by O'Brien et al. [13] for the hydrodesulfurization of dibenzothiophene. The activation energy for bicyclohexyl (139.7 kJ/mol) for Eq. A-3 is also in agreement with the activation energy of 140.2 kJ/mol which was reported by Nagai et al. [14] from studies of carbazole HDN on a reduced Mo/Al_2O_3 catalyst at 280–360°C. The activation energy for the hydrogenolysis of decahydroquinoline to propylcyclohexyl aniline was 138 kJ/mol, reported for the overall HDN of quinoline at 280–360°C by Satterfield and Cocchetto [3]. This value is in agreement with that in Eq. A-3.

Equation A-3 is recommended for its good fit, its simple form, and its low parameter correlation and error bounds relative to the other equations. In this case, the mechanism of carbazole HDN on the nitrided Mo/Al_2O_3 catalyst is best correlated by a dual-site model. This model is different from the single-site model in a previous study on the HDN of acridine [6] using the reduced Mo/Al_2O_3 catalyst. The single-site model assumes that perhydrocarbazole and hydrogen are competitively adsorbed on the same site. The nitrided Mo/Al_2O_3 catalyst is much more active than the reduced catalyst [8]. The kinetic results demonstrate that the creation of two independent active sites by nitriding the catalyst gives rise to higher activity of the nitrided molybdenum catalyst than that of the reduced catalyst.

The XRD study of the as-prepared 97.1% MoO_3/Al_2O_3 at 500°C showed MoO_2 with a small amount of MoO_3 [15]. This result indicated no formation of molybdenum nitrides in the catalyst, although the catalyst has a lot of NH_3 and N_2 from the nitrogen analysis. When the as-prepared 500°C-nitrided catalyst was heated from room temperature to 700°C in flowing He, the catalyst consisted of a mixture of γ-Mo_2N, β-Mo_2N, and MoO_2. From these results, the MoO_2 was nitrided by adsorbed NH_3 and N_2 which remained on the catalyst, because the catalyst nitrided at 500°C was cooled to room temperature from 500°C in flowing NH_3. These results for the 97.1% MoO_3/Al_2O_3 catalyst nitrided at 500°C were also observed like those for the nitrided 11.7% MoO_3/Al_2O_3 catalyst [15]. Thus, even though the catalyst was treated at 500°C in flowing NH_3, NH_x (x=0–3) adsorbed on the Mo/Al_2O_3 catalyst slightly nitrided MoO_2 to form γ-Mo_2N and β-Mo_2N on the surface, consequently bringing about the creation of new active Mo species on the catalyst. This formation of new Mo species as well as MoO_2 possibly leads to a dual site mechanism with a high activity for carbazole HDN.

IV. Conclusions

The kinetics of the C-N hydrogenolysis of perhydrocarbazole in the HDN of carbazole on the nitrided catalyst was studied at 340–350°C and 3.40–9.12 MPa hydrogen pressure. Rate equations of the Langmuir-Hinshelwood type were compared with the rate data using a nonlinear least-squares regression technique. The kinetic results demonstrate the adsorption of perhydrocarbazole on one kind of catalytic site and of hydrogen on another.

$$r_{BCH} = \frac{k_i K_H P_H}{(1 + K_H P_H)^2} \frac{K_{PHC} P_{PHC}}{(1 + K_{PHC} P_{PHC})}$$

V. Acknowledgments

The authors thank the Scientific Research Foundation of the Ministry of Education of the Japanese Government for the partial financial support through Contract (03805093).

References

1. J. Sonnemans and P. Mars. J. Catal. 31:220 (1973).
2. G. C. Hadjiloizou, J. B. Butt, and J. S. Dranoff. J. Catal. 131:545 (1991).
3. C. N. Satterfield and J. F. Cocchetto. Ind. Eng. Chem. Proc. Des. Dev. 20:53 (1981).
4. J. T. Miller and M. F. Hineman. J. Catal. 85:117 (1984).
5. F. E. Massoth, K. Balusami, and J. Shabtai. J. Catal. 122:256 (1990).
6. M. Nagai, T. Hanaoka, and T. Masunaga. J. Catal. 101:284 (1986).
7. S. T. Oyama, J. C. Schlatter, J. E. Metcalfe, and J. M. Lambert, Jr. Ind. Eng. Chem. Res. 27:1648 (1988).
8. M. Nagai and T. Miyao, Catal. Letters 15:105 (1992).
9. L. Vople and M. Boudart. J. Phys. Chem. 90:4874 (1986).
10. F. Glola and V. Lee. Ind. Eng. Chem. Process Des. Dev. 25:918 (1986).
11. J. E. Bartmess, J. A. Scott, and R. T. McIver, Jr. J. Am. Chem. Soc. 79: 6046 (1979).
12. G. A. Somorjai. In *Chemistry in Two Dimensions; Surfaces*, Cornell University Press, New York, 1981, p. 27.

13. W. S. O'Brien, J. W. Chen, R. V. Nayak, and G. S. Carr. Ind. Eng. Chem. Process Des. Dev. 25:221 (1986).
14. M. Nagai, T. Hanaoka, and T. Masunaga. Fuel & Energy 2:645 (1988).
15. M. Nagai, A. Miyata, T. Kusagaya, and S. Omi. Bull. Soc. Chim. Belg. 104:311 (1995).

Aromatics Hydrogenation over Alumina Supported Mo and NiMo Hydrotreating Catalysts

N. Marchal, D. Guillaume, S. Mignard and S. Kasztelan

Kinetics and Catalysis Division, Institut Français du Pétrole
B.P. 311, 92506 Rueil-Malmaison Cedex, France

Abstract

The effect of H_2S on the toluene hydrogenation activities of sulfided Mo, NiMo and NiMoP/alumina catalysts has been determined at 6 MPa, 350°C over a large range of H_2S partial pressure. Similar trends are found for the three catalysts with in particular no effect of H_2S at high partial pressure. Comparison of the hydrogenation activities of the unpromoted and promoted catalysts shows that H_2S has an influence on the promotion effect of Ni with a maximum at medium H_2S partial pressure.

I INTRODUCTION

Hydrotreating catalysts are basically composed of sulfides of molybdenum or tungsten promoted by nickel or cobalt and supported on alumina. Their active phase is usually described as nickel (or cobalt) ions located around the edges of MoS_2 (WS_2) crystallites and the promotion effect on the activity for reactions such as hydrodesulfurization, hydrogenation and hydrodenitrogenation due to nickel (or cobalt) is explained either by electronic considerations, or by production of new catalytic sites [1-7].

There is a renewed interest in the properties of hydrotreating catalysts for aromatics hydrogenation in light of future specifications on aromatics content in gas-oil [8-11]. Industrially, aromatics hydrogenation in gas-oil may be obtained in a single stage process where hydrogenation and hydrodesulfurization take place together under high H_2S partial pressure. Another possibility is a two-stage process where deep hydrodesulfurization is performed in a first reactor followed by stripping of the effluent to remove H_2S and NH_3 and hydrogenation is performed in a second reactor on a catalyst working under low H_2S partial pressure. NiMo or NiW hydrotreating catalysts may be used in both types of processes and an adjustement of the H$_2$S partial pressure may be needed to maximize their hydrogenation activities. Therefore it is of interest to study the influence of H$_2$S on the hydrogenation activity of these catalysts.

Many studies on the influence of H$_2$S on the activity of hydrotreating catalysts in reactions such as hydrodesulfurization, hydrodenitrogenation or aromatic compounds hydrogenation have been reported in the literature. These studies have shown that the effect of H$_2$S is complex with reports of promotional, inhibiting or absence of effect depending on the reaction, the reactant and the experimental conditions considered [12-22]. Conversely, there is little information on the effect of H$_2$S on the promotion effect of Ni or Co in mixed sulfide catalysts.

In view of the interest of getting a better knowledge of the hydrogenation properties of sulfide catalysts, we have started an investigation of the effect of H$_2$S on the toluene hydrogenation activities of hydrotreating catalysts and we report in this work preliminary results on the influence of H$_2$S on the promotion effect of nickel in NiMo/Al$_2$O$_3$ catalysts over a large range of H$_2$S partial pressure.

II EXPERIMENTAL

All of the catalysts tested in this work were prepared from a γ-alumina in form of 1.2 mm cylindrical extrudates (Rhône-Poulenc, 240 m^2/g^{-1}, 0.57 cm^3/g) by pore-filling impregnation with aqueous solutions of ammonium heptamolybdate, nickel nitrate and/or phosphoric acid. The wet samples were then dried at 120°C overnight and calcined at 500°C for 4h. The metals contents were measured by X-ray fluorescence and are reported in Table 1.

Table 1. Composition of Catalysts Tested

Catalyst	Mo (wt %)	Ni (wt %)	P (wt %)
Mo/Al$_2$O$_3$	10.2	/	/
NiMo/Al$_2$O$_3$	9.3	2.6	/
NiMoP/Al$_2$O$_3$	11.4	2.4	2.4

Aromatics Hydrogenation over Hydrotreating Catalysts

Toluene hydrogenation tests were performed in a high pressure fixed bed continuous flow "Catatest" unit from Vinci Technologies. The experimental conditions used were a total pressure between 1 and 6 MPa, a reaction temperature between 220 and 380°C, a liquid hourly space velocity (LHSV) between 0.5 and 6 h^{-1}, a H$_2$/HC ration equal to 450 l/l and 40 cm^3 of catalyst.

The liquid feed was composed of toluene (20 wt %), thiophene (between 0 and 13 wt %) and cyclohexane (for balance). Thiophene was used as a H$_2$S generator and was found completely hydrogenated into butane and H$_2$S in the experimental conditions employed as checked by gas chromatographic analysis. The sequence of tests at different H$_2$S partial pressure was performed by changing the amount of thiophene in the feed starting from the higher thiophene content, i.e. higher H$_2$S partial pressure. The test performed with the feed containing no thiophene is assumed to be a test at 0 Pa H$_2$S partial pressure. Toluene conversion, ranging between 5 and 60 %, was determined in a steady state reached generally after 4 h. on stream.

The liquid products of the reaction were analyzed by gas chromatography using a 50 m PONA column at 60°C and a flamme ionization detector.

Prior to catalytic tests, the samples were sulfided in situ by passing a feed containing 2 wt % dimethyldisulfide in cyclohexane over the catalyst at 6 MPa and 350°C during 4 hours.

From the conversion of toluene the first order rate constant of hydrogenation in mol/(kg.h) is computed by :

$$r = -\frac{F}{w} * \ln(1-x)$$

with F : molar flow of toluene (mol/h) and w : weight of catalyst. A correction of the rate of hydrogenation was performed assuming first orders of reaction relatively to H$_2$ and to toluene to account for the variation of hydrogen and toluene partial pressures due to the consumption of H$_2$ by the thiophene decomposition reaction. Then all of the rates of hydrogenation reported in this work are referred to the H$_2$ and toluene partial pressure of the test with no H$_2$S, i.e. 4 MPa H$_2$ and 0.37 MPa toluene, respectively.

III RESULTS

A Influence of H$_2$S partial pressure

The influence of H$_2$S on the rates of toluene hydrogenation of sulfided Mo, NiMo and NiMoP catalysts has been measured from 300,000 to 0 Pa of H$_2$S partial pressure. The results for the Mo catalyst are reported in figure 1.

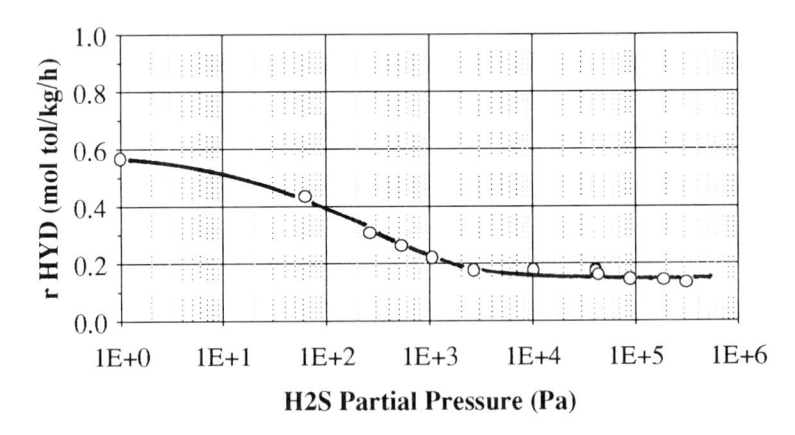

Figure 1. Effect of H$_2$S on the rate of toluene hydrogenation of the Mo/Al$_2$O$_3$ catalyst at 6 MPa and 350°C.

Figure 1 shows the complex effect of H$_2$S on the rate of toluene hydrogenation of the Mo catalyst. Three domains of H$_2$S partial pressure can be distinguished. In the low H$_2$S partial pressure domain, 0 to 50 Pa, the rate of toluene hydrogenation is maximum, stable and independent of the H$_2$S partial pressure. In the medium H$_2$S partial pressure domain, H$_2$S has an inhibiting effect on the rate of hydrogenation and at high H$_2$S partial pressure, beyond 2,700 Pa, H$_2$S has surprisingly no more effect.

The effect of H$_2$S on the rates of hydrogenation of the Mo, NiMo and NiMoP catalysts are reported in figure 2. The NiMo and NiMoP catalysts are clearly more active than the Mo catalyst whatever the H$_2$S partial pressure is. This is the well known promoting effect of nickel on molybdenum. In addition, the NiMoP formula is more active than the NiMo formula, showing the also well known promoting effect of phosphorus on the rate of hydrogenation of the NiMo catalyst.

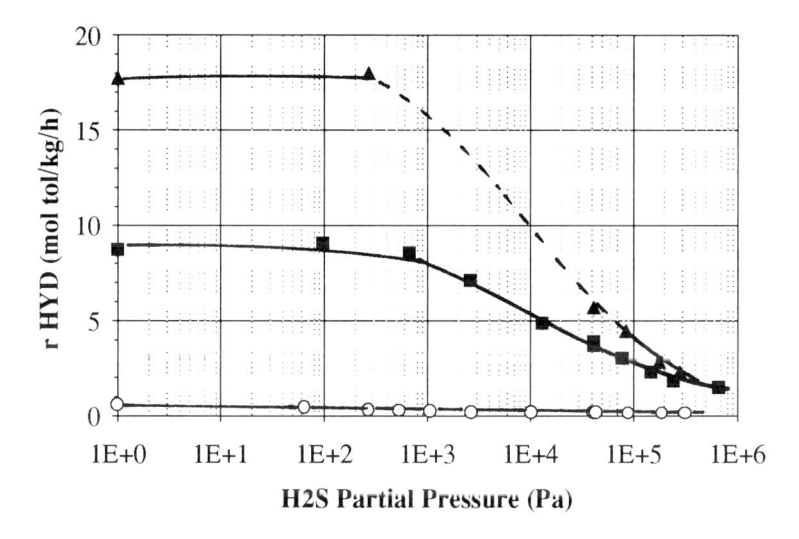

Aromatics Hydrogenation over Hydrotreating Catalysts

Figure 2. Effect of H$_2$S on the rate of toluene hydrogenation at 6 MPa and 350°C for (○) Mo/Al$_2$O$_3$, (■) NiMo/Al$_2$O$_3$ and (▲) NiMoP/Al$_2$O$_3$ catalysts.

The variation of the rate of hydrogenation of the three catalysts studied in this work shows qualitatively the same behavior towards H$_2$S partial pressure. The three domains of H$_2$S partial pressure defined for the Mo catalyst seem however to be shifted towards higher H$_2$S partial pressures for the NiMo and NiMoP catalysts.

A log-log plot of the rate of hydrogenation versus H$_2$S partial pressure (figure 3) allows to clearly put in evidence the three domains aforementioned for each catalyst. Reaction orders relative to H$_2$S partial pressure have also been determined for each domain and, for the sake of clarity, have been reported in table 2. For the NiMo catalyst, the medium H$_2$S partial pressure domain has been divided into two subdomains.

At low and high H$_2$S partial pressure, a zero order of reaction relative to H$_2$S is found for the three catalysts. In these two domains, H$_2$S has no effect on the rate of toluene hydrogenation. At medium H$_2$S partial pressure, H$_2$S has an inhibiting effect with a slightly negative order of reaction, between - 0.25 and - 0.5.

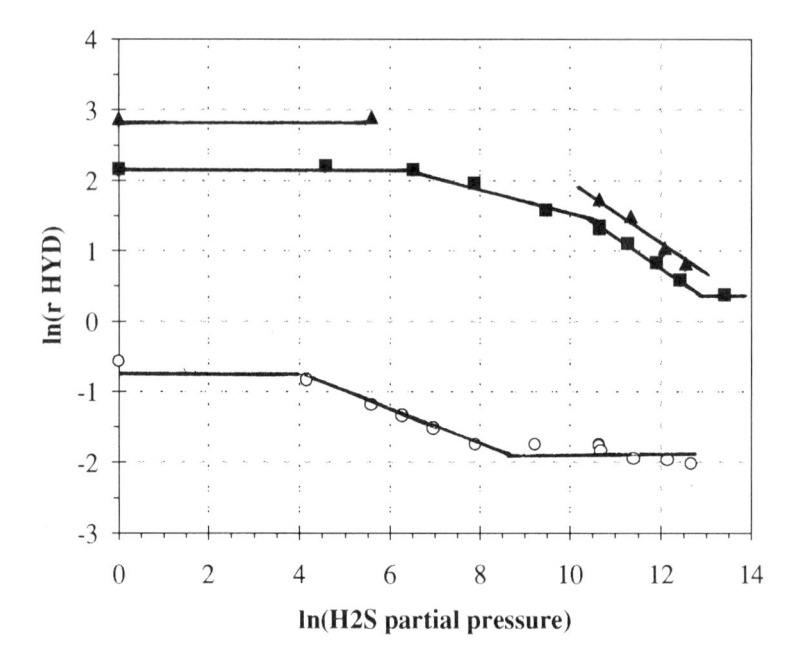

Figure 3. Log-log plot of the rate of toluene hydrogenation versus H_2S partial pressure at 6 MPa and 350°C for (○) Mo/Al_2O_3, (■) $NiMo/Al_2O_3$ and (▲) $NiMoP/Al_2O_3$ catalysts.

Table 2. Reaction orders relative to H_2S partial pressure at 6 MPa total pressure, 350°C for Mo, NiMo and $NiMoP/Al_2O_3$ catalysts.

	\multicolumn{3}{c}{H_2S Partial Pressure Domain}	
	Low H_2S	Medium H_2S	High H_2S
Mo/Al_2O_3	0	- 0.25	0
$NiMo/Al_2O_3$	0	- 0.22 - 0.42	0
$NiMoP/Al_2O_3$	0	- 0.50	no data

B Kinetic studies

To investigate further the differences between the three catalysts studied in this work, some kinetic parameters such as the apparent activation energy, the orders of reaction relative to H_2 and toluene were determined.

Apparent activation energies were determined at 42,000 Pa H_2S partial pressure, between 280 and 350°C and are reported in table 3. In addition, for the NiMoP catalyst, the apparent activation energy was determined at 300 Pa H_2S, between 220 and 280°C and at 85,000 Pa H_2S, between 280 and 350°C. For the Mo catalyst a determination of the apparent activation energy was also done at 300 Pa H_2S, between 320 and 350°C. The values reported in table 3 are given at ± 2 kcal/mol.

Table 3. Apparent activation energies at various H_2S Partial Pressure for Mo, NiMo and NiMoP/Al_2O_3 catalysts.

	H_2S Partial Pressure		
	300 Pa	42,000 Pa	85,000 Pa
Mo/Al_2O_3	17 kcal/mol	17 kcal/mol	
NiMo/Al_2O_3		17 kcal/mol	
NiMoP/Al_2O_3	16 kcal/mol	17 kcal/mol	17 kcal/mol

Within the uncertainty of the measurements, the apparent activation energies measured appear remarquably similar. This suggests that no major modification of kinetic behavior occurs when changing the catalyst formulation in the NiMo system and whatever the range of H_2S partial pressure.

Measurements of the reaction orders with respect to H_2 and toluene were done only for the NiMoP catalyst. The reaction order relative to toluene was determined by changing the contact time (1/LHSV) in three different sets of experimental conditions. 85,000 Pa H_2S, 350°C; 42,000 Pa H_2S, 350°C and 300 Pa H_2S, 240°C. Figure 4 shows that in all three cases, the first-order rates of hydrogenation vary linearly with contact time. So, in our experimental conditions, the toluene hydrogenation reaction follows a first-order reaction kinetic in good agreement with literature data [19,21].

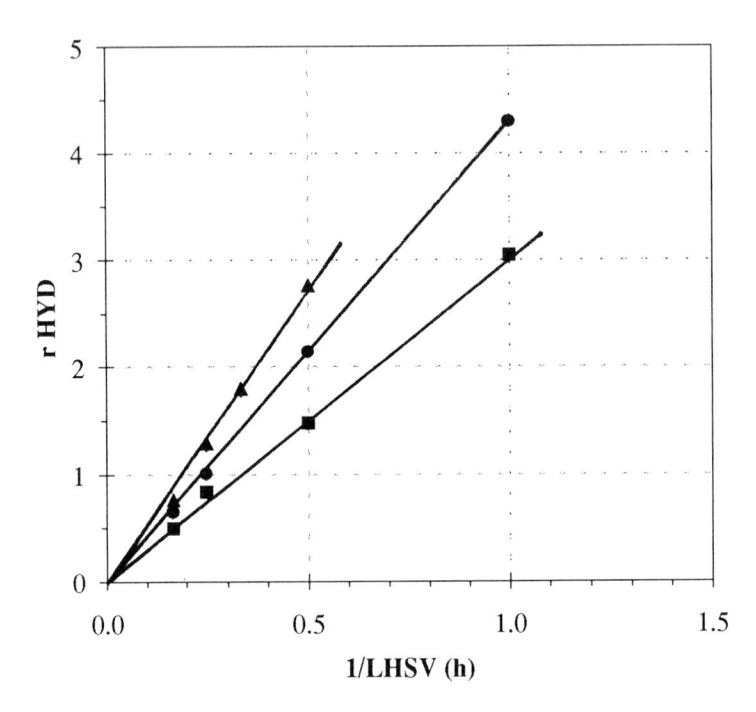

Figure 4. Rate of toluene hydrogenation versus contact time for the NiMoP/Al$_2$O$_3$ catalyst at 6 MPa and (●) 350°C, 85,000 Pa H$_2$S, (▲) 350°C, 42,000 Pa H$_2$S, (■) 240°C, 300 Pa H$_2$S.

The order of reaction with respect to hydrogen partial pressure was determined for two experimental conditions : 85,000 Pa H$_2$S, 350°C, and 300 Pa H$_2$S, 240°C. The log-log plot of the rate of toluene hydrogenation versus H$_2$ partial pressure, reported in figure 5, gives linear relationships with an order relative to H$_2$ equal to 1.2 ± 0.2 at 85,000 Pa H$_2$S and equal to 0.1 at 300 Pa H$_2$S.

In the literature, the order with respect to the hydrogen partial pressure is often reported equal to 1 [13,21]. Our results at 85,000 Pa H$_2$S partial pressure is in quite good agreement with these data. More surprising, however, is the small order of reaction found at very low H_2S partial pressure which may indicate a change of kinetic regime as previously suggested for a Mo catalyst [17].

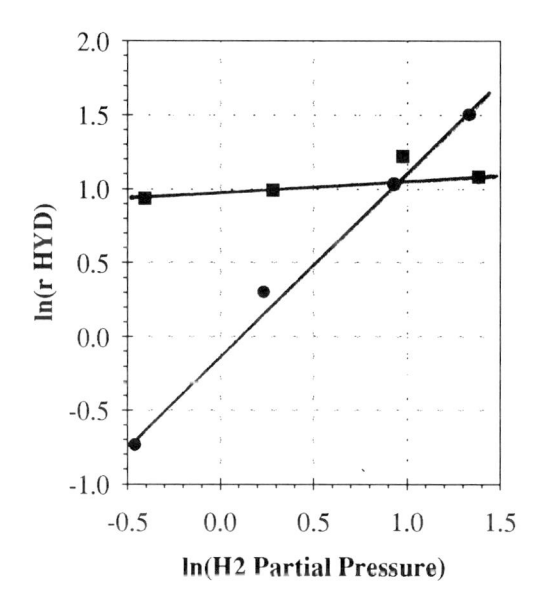

Figure 5. Log-log plot of the rate of toluene hydrogenation versus hydrogen partial pressure at (●) 350°C and 85,000 Pa H_2S, (■) 240°C and 300 Pa H_2S.

IV DISCUSSION

It is usually assumed that H_2S has an inhibiting effect on the hydrogenation activity of hydrotreating catalysts [21, 22]. However, many studies can be found in literature on the effect of H_2S partial pressure in hydrotreating reactions and their conclusions differ widely. According to the considered reaction, H_2S partial pressure may have :

- a promoting effect, as described in the case of hydrodenitrogenation of indole over a sulfided $CoMo/Al_2O_3$ catalyst [14] and in the case of hydrogenation of ortho-xylene over a sulfided $CoMo/Al_2O_3$ catalyst [18].

- an inhibiting effect, as observed during the hydrogenation of propylbenzene over a sulfided NiMo/Al$_2$O$_3$ catalyst [19] or during the hydrogenation of benzene over a sulfided CoMo/Al$_2$O$_3$ catalyst [18].

- no effect, as for example during the hydrogenation of 1-hexene over a sulfided CoMo/Al$_2$O$_3$ catalyst [20] or during the hydrogenation of toluene over a sulfided CoMo/Al$_2$O$_3$ catalyst [18].

- a complex effect as for instance during the hydrodenitrogenation of quinoline where at low H$_2$S partial pressure the rate is accelerated and reaches a plateau at high H$_2$S partial pressure [15]. On the other hand, during the first hydrogenation step of 2.6-dimethylaniline hydrodenitrogenation, an inhibiting effect of H$_2$S partial pressure followed by no effect is observed [16]. For thiophene hydrodesulfurization over a sulfided CoMo/Al$_2$O$_3$ catalyst, the curve activity versus H$_2$S partial pressure passes through a maximum, which means a promoting effect followed by an inhibiting effect of H$_2$S [12].

While different reaction mechanisms can be invoked to explain the various effects of H$_2$S on different reactions such as hydrodenitrogenation and hydrogenation (22), it is likely that the discrepancies reported for the same catalytic function, hydrogenation for example, arise from different experimental conditions and procedures. In particular experiments have seldom been done over a very large range of H$_2$S partial pressure and therefore the range of H$_2$S partial pressure may not be always comparable.

The results reported in this work indicate that H$_2$S has an inhibiting effect on the aromatics hydrogenation reaction in a certain range of H$_2$S partial pressure whatever the Mo, NiMo or NiMoP catalyst. However at high H$_2$S partial pressure, H$_2$S has no more effect on the hydrogenation activity of both the Mo and NiMo catalyst.

This result confirms a previous observation made for a Mo/Al$_2$O$_3$ catalyst following a different testing procedure. The three domains of H$_2$S partial pressure were observed at almost the same values of H$_2$S partial pressure and were interpreted by kinetic rate laws deduced from heterolytic reaction mechanisms [17]. According to this work, the low and medium H$_2$S partial pressure ranges correspond to the rate limiting addition of the first hydrogen to toluene according to the reaction

$$\ast - R + \ast - H \longrightarrow \ast - RH + \ast - V \qquad (1)$$

with $\ast - V$: free surface site.

Aromatics Hydrogenation over Hydrotreating Catalysts

The high H$_2$S partial pressure range corresponds to the rate limiting addition of the second hydrogen according to :

$$\ast - RH + \bullet - SH \longrightarrow \ast - V + \bullet - S + RH_2 \qquad (2)$$

with both H$_2$ and H$_2$S being adsorbed heterolitically according to :

$$H_2S + \ast - V + \bullet - S \longrightarrow \ast - SH + \bullet - SH \qquad (3)$$

$$H_2 + \ast - V + \bullet - S \longrightarrow \ast - H + \bullet - SH \qquad (4)$$

The similarity between the curves rate of hydrogenation versus the H$_2$S partial pressure for the three catalysts studied suggests that the same kind of reaction mechanism prevails for the NiMo and NiMoP catalysts.

In this work, the effect of H$_2$S on the rate of toluene hydrogenation is assumed to be kinetic in nature rather than being due to structural modifications of the active phase. The three catalysts studied in this work provide an example of the difficulty to compare the rate of hydrogenation of different catalysts because the curves rate of hydrogenation versus H$_2$S partial pressure of NiMo and NiMoP catalysts are shifted compared to the curve for the Mo catalyst.

The shift of one curve relatively to another can be assigned to different heat of adsorption of H$_2$S on these catalysts. The shift of the rate of hydrogenation curve of the NiMo catalyst towards higher H$_2$S partial pressure compared to the Mo catalyst suggests that H$_2$S is less strongly adsorbed on NiMo sulfide than on the Mo sulfide catalyst. Similarly the shift towards even higher H$_2$S partial pressure for the NiMoP catalyst suggests that H$_2$S is less strongly adsorbed on the P doped NiMo sulfide phase than on the undoped NiMo sulfide phase.

Although the determination of the level of H$_2$S partial pressure needed to observe the start of the inhibiting effect of H$_2$S in figure 2 is not very precise, this value can be used to establish a preliminary ranking of the catalysts according to the strength of H$_2$S adsorption namely :

$$NiMoP \leq NiMo \ll Mo.$$

It is interesting to note that the inhibiting effect of H$_2$S is different between the Mo and NiMo catalyst. The ratio of the rate of hydrogenation of the NiMo catalyst over the rate of hydrogenation of the Mo catalyst is usually defined as the promotion factor. Then and according to figure 1 the promotion factor must vary with the H_2S partial pressure.

The promotion factor calculated from the data of figure 2 has been plotted versus the H_2S partial pressure in figure 6 for the NiMo and NiMoP catalysts. Both curves in figure 6 show that the promotion factor varies with the H_2S partial pressure from 15 up to 40 for the NiMo catalyst and from 15 up to 70 for the NiMoP catalyst. In addition, the promotion factor passes through a maximum at about 3,000 Pa for the NiMo catalyst and about 2,000 Pa H_2S partial pressure for the NiMoP catalyst. In the latter case, a lack of experimental data makes it difficult to determine the precise position of the maximum.

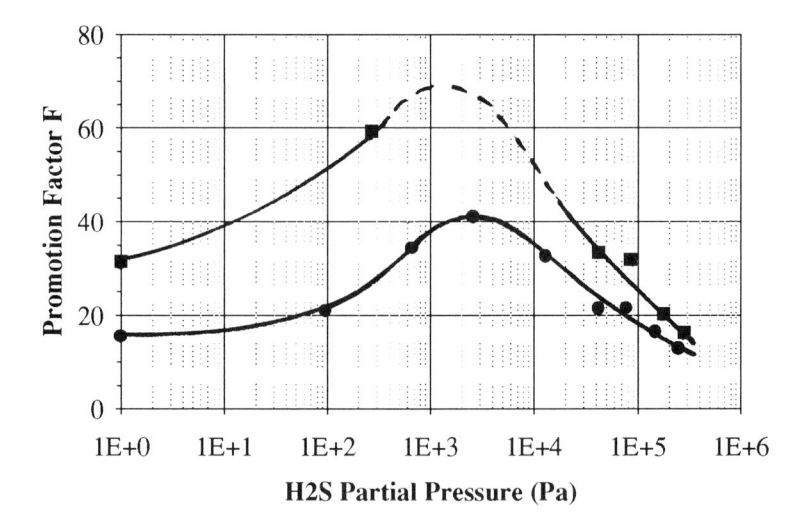

Figure 6. Promotion factor due to Ni (F) in NiMo and NiMoP catalysts versus H_2S partial pressure at 6 MPa and 350°C. (●) NiMo ; (■) NiMoP.

This result shows that the value of the promotion factor is dependent on kinetic factors of which the H_2S partial pressure is an important one. However the value of the promotion factor obtained in our conditions for toluene hydrogenation is always large, between 15 and 70, and this indicates that the kinetic effect is of minor importance compared to more fundamental causes of the promotion effect such as an electronic effect.

V CONCLUSION

This work has shown that the effect of H_2S on the rate of toluene hydrogenation over Mo, NiMo and $NiMoP/Al_2O_3$ catalysts is complex. Three domains can be distinguished for the three catalysts : a low H_2S partial pressure domain, where H_2S partial pressure has no effect on the rate of toluene hydrogenation, a medium H_2S partial pressure domain, where H_2S partial pressure has an inhibiting effect on the rate of hydrogenation and a high H_2S partial pressure domain, where H_2S partial pressure has no more effect on the rate of hydrogenation.

The position of each three domains as well as the levels of the rate of hydrogenation are different for each of the catalyst. The three domains are shifted towards higher H_2S partial pressures for NiMo and NiMoP catalysts compared to Mo catalyst suggesting that H_2S is less strongly adsorbed on the promoted catalysts. Noteworthy, the promoting effect of nickel on molybdenum is also dependent on the H_2S partial pressure and shows a maximum at medium H_2S partial pressure for the toluene hydrogenation reaction.

Acknowledgement

We are indebted to B. Bétro and H. Deschamps for their experimental assistance.

REFERENCES

1. R. Prins, V.H.J. De Beer and G.A. Somorjai, Catal. Rev. Sci. Eng., 31 : 1 (1989).
2. S. Harris and R.R. Chianelli, J. Catal., 98 : 17 (1986).
3. H. Topsøe, R. Candia, N.Y. Topsøe and B.S. Clausen, Bull. Soc. Chim. Belg., 93 : 783 (1984).
4. B. Delmon, Stud. Surf. Sci. Catal., 53 : 1 (1989)1.
5. H. Topsøe, B.S. Clausen, N.Y. Topsøe and P. Zeuthen, Stud. Surf. Sci. Catal., 53 : 77 (1989).
6. H. Knozinger, in Proc. 9th Int. Cong. Catal., Calgary, 1 : 20 (1988).
7. S. Kasztelan, H. Toulhoat, J. Grimblot and J.P. Bonnelle, Applied Catal., 13 : 127 (1984).
8. P. Crow and B. Williams, Oil and Gas J., January 23 (1989) 15.

9. A.J. Suchanek, Oil and Gas J., May 7 (1990) 109.
10. S.L. Lee and M. de Wind, Preprints, Div. Petr. Chem., ACS, 37 : 718 (1992)8.
11. A. Stanislaus and B.H. Cooper, Catal. Rev. Sci. Eng., 36 : 75 (1994).
12. J. Leglise, J. van Gestel and J.C. Duchet, J. Chem. Soc., Chem. Commun., 611 (1994).
13. M.L. Vrinat, Applied Catal., 6 : 137 (1983).
14. F.E. Massoth, K. Balusami and J. Shabtai, J. Catal., 122 : 256 (1990).
15. C.N. Satterfield and S. Gültekin, Ind. Eng. Chem. Process Des. Dev., 20 : 62 (1981).
16. J. van Gestel, J. Leglise and J.C. Duchet, Applied Catal. A: General, 92 : 143 (1992).
17. S. Kasztelan and D. Guillaume, Ind. Eng. Chem. Res., 33 : 203 (1994).
18. M. Yamada, T. Obara, J.W. Yan and S. Hayakeyama, Sekiyu Gakkaishi, 31 : 118 (1988).
19. S. Gültekin, S.A. Ali and C.N. Satterfield, Ind. Eng. Chem. Process Des. Dev., 23 : 179 (1984).
20. R. Ramachandran and F.E. Massoth, J. Catal., 67 : 248 (1981).
21. J.F. Le Page et al., in Applied Heterogeneous Catalysis; (Technip, ed), Paris, 1987 p 370.
22. M.J. Girgis and B.C. Gates, Ind. Eng. Chem. Res., 30 : 2021 (1991).

Applications of Pillared and Delaminated Interlayered Clays as Supports for Hydrotreating Catalysts

M.F. WILSON[1], J.-P. CHARLAND[1], E. YAMAGUCHI[2] AND T. SUZUKI[2]

[1]CANMET Energy Research Laboratories, Natural Resources Canada, 555 Booth Street, Ottawa, Ontario, K1A 0G1

[2]Sumitomo Metal Mining Company Ltd., Catalyst Group, Central Research Laboratory, 3-8-5, Nakakokubun, Ichikawa, Chiba, Japan

ABSTRACT

A new concept is presented for preparation of hydrotreating catalysts and involves supporting MoS_2 on pillared and delaminated interlayered clays dispersed in γ-alumina. The modified composite supports were designed to counteract fouling and coking by problematical feedstock components. Details of characterization and testing of Ni-Mo catalysts prepared using synthetic and naturally occurring smectite clays are presented. Characterization of support material and finished catalyst using transmission electron microscopy (TEM) showed unique surface chemical structures. Dispersion of MoS_2 appeared to be controlled by interaction with delaminated clay particles dispersed within the support. TEM results also suggest that the dispersed clay induced stacking of MoS_2 crystallites. Pyridine/FTIR adsorption studies also confirmed a modified surface acidity. Catalyst resistance to deactivation by adsorption of foulants was demonstrated through time-on-stream monitoring of nitrogen conversion during hydrotreating of coal-derived liquid. Catalyst activity for heteroatom conversion was evaluated by hydrotreating heavy gas oil derived from coprocessing coal/bitumen.

INTRODUCTION

In many hydrotreating operations, serious problems are encountered due to catalyst fouling and coking by problematical feedstock components. Deactivation may occur due to irreversible adsorption of basic compounds at Brønsted or Lewis acid sites located on the surface of the catalyst support (1,2,3). The adsorbed species are believed to act as precursors for the formation of carbonaceous deposits which result in loss of surface area and plugging of catalyst pores. A new concept is demonstrated for supporting MoS_2 catalysts in hydrotreating problematical feedstocks. The method involves dispersing pillared and delaminated interlayered clay particles in γ-alumina on which MoS_2 crystallites may then be deposited. The dispersed clay particles provide protection against adsorption and fouling thus producing catalysts with enhanced resistance to deactivation. In the present work the supported MoS_2 crystallites were also promoted by Ni.

Stohl and Stephens demonstrated that in hydrotreating coal-derived liquid fractions, loss of catalyst activity and occurrence of carbon deposition correlated with the base strength of nitrogen compounds in the feedstock with some contribution from phenolic compounds (2). In their work the catalyst support used was γ-alumina oxide, a material showing a well defined Lewis acidity and used extensively in the production of commercial catalysts. Similarly, polynuclear aromatic compounds and asphaltenic materials may also adsorb on and foul hydrotreating catalysts leading to formation of carbonaceous deposits (4). Other feedstocks containing high concentrations of catalyst foulants include synthetic crudes from oil sands bitumen, heavy oils and liquids derived from coprocessing coal/bitumen.

In recent years strategies for improving hydrotreating catalyst resistance to deactivation have included development of supports with optimum bimodal pore structures (5,6) and utilization of porous carbons as support materials (7,8,9). Among the problems encountered using carbon is its low crush strength. Previous work using pillared clays for hydrotreating catalyst supports was undertaken by Occelli and Rennard (10). Delaminated clays have also been used in the development of FCC catalysts by Occelli et al. (11,12). The strategy in the present work was to combine the properties of smectite clays with those of γ-alumina and utilize composite support materials. It is well established that γ-alumina displays superior properties for dispersion of MoS_2 crystallites and was equally useful in this work as a medium for dispersion of clay particles. By this means the physical and chemical properties of the support were significantly modified to counteract deactivation.

EXPERIMENTAL

Preparation of Pillared Clays

Pillared clays (PILCs) were prepared using Polargel, a commercially processed montmorillonite, and Volclay HPM-20, a naturally occurring montmorillonite, both supplied by American Colloid Company. Materials were pillared by ion-exchange using published procedures (13). The pillaring agent used was the $[Al_{13}O_4(OH)_{24}(H_2O)_{12}]^{7+}$ cation (ACH) contained in 50 wt % aqueous Chlorohydrol solution supplied by Reheis Chemical Company. The PILCs were washed until free of chloride ion and air dried. A reference PILC material used in previous work (14) and prepared from a naturally occurring montmorillonite, Accofloc 350, was also used in BET N_2 adsorption studies.

Preparation of Delaminated Clays

Delaminated clays were formed using Polargel and Laponite, a synthetic hectorite supplied by Laporte Industries. Clay samples were delaminated in aqueous suspension (approximately 2.0 wt %) and allowed to flocculate. For Polargel, ACH pillaring agent was used in the delamination procedure (13), and for Laponite, $[Al_8(OH)_{20}ZrO]^{6+}$ (ZACH), which was supplied as Rezal 36G by Reheis Chemical Company. After flocculation and centrifuging, materials were washed continuously to remove excess pillaring agent and freeze dried to produce final delaminated clay products.

Preparation of Ni-Mo Catalyst Extrudates

Samples of pillared and delaminated clays were wet mixed with γ-alumina, and after peptizing with nitric acid, formed into 1/32 in. extrudates. The support materials were air dried and calcined in air at 500°C for 5 h. Extrudates contained between 15-30 wt % PILC or delaminated clay material. Final catalysts were prepared using the pore filling method. Solutions of ammonium paramolybdate and nickel nitrate were introduced sequentially to give the approximate theoretical Mo/Ni loadings, i.e., 15% MoO_3, 4% NiO. Impregnated catalysts were oven dried, calcined in air as above and reduced to 14-20 mesh prior to screening. A standard Ni-Mo/γ-alumina catalyst was prepared using the same procedure, and a commercial Ni-Mo hydrotreating catalyst, AKZO KF-153S, was also used in activity tests.

Catalyst Characterization

Elemental analyses of final catalyst extrudates was carried out by Element Analysis Corporation, Tallahassee, Fl. using proton-induced X-ray emission spectrometry (PIXE). BET N_2 adsorption measurements were performed to determine Ni-Mo catalyst surface areas and pore volumes using a Quantachrome Autosorb unit. X-ray diffraction (XRD) analysis was carried out on pillared and delaminated clays using a Siemens D500TT automated diffractometer. Data were collected over the angular range 2-70° (2θ) using Cu Kα radiation. Clay crystallite sizes were calculated using Scherrer's equation applied to diffraction peak profiles corrected for instrumental broadening (15). The average number of clay layers per crystallite was derived by dividing crystallite size by the basal spacing. Pyridne adsorption/Fourier transform infrared (FTIR) measurements were undertaken to evaluate the propensity for catalyst materials to adsorb basic nitrogen compounds. FTIR spectra of adsorbed pyridine were measured using a Nicolet 60SX spectrometer at 2 cm^{-1} resolution. A temperature- and pressure-controlled optical cell was used in the transmittance mode. Sample wafers of 15 mg were evacuated at $\sim 10^{-5}$ torr and degassed by heating for 16 h at 400°C prior to carrying out adsorption/desorption measurements. To probe the catalyst surface acidity, the sample temperature was maintained at 100°C and ~ 25 torr of pyridine vapour was introduced. The sample was then evacuated and spectra were recorded.

Transmission electron microscopy (TEM) analysis was carried out using a JEM-2000EX microscope operating at a voltage of 200 keV. Magnifications were in the range X 100,000-300,000. Analysis was done on support material containing delaminated Laponite and on a corresponding used Ni-Mo catalyst. The used catalyst was obtained from hydrotreating heavy gas oil feedstock derived from coprocessing coal/bitumen and was run time-on-stream for approximately 240 h. Catalyst extrudates were soxhlet extracted for 12 h with tetrahydrofuran. Fresh and used catalyst materials were prepared for TEM analysis by milling to a fine powder <400 mesh and dispersing in ethanol by ultrasonic agitation. Specimens were made by dropping the suspension on a holey carbon film from which micrographs were obtained from parts of the specimen over holes in the film.

Catalyst Deactivation Studies Using Coal-Derived Liquid

Deactivation of Ni-Mo catalysts was evaluated by hydrotreating a coal-derived spent donor solvent from the Japanese NEDOL process. The boiling range was approximately 200°C-550°C. An automated microreactor system equipped with a fixed-bed stainless steel tubular reactor 0.305 m long, and 0.635 cm ID was operated in the continuous upflow mode. The bed contained catalyst particles of 14-20 mesh, was 0.14 m long, and had a volume of 4.50 cm^3. Pre-heating and post-heating zones were filled with quartz particles of 20-48 mesh. The catalyst was presulfided in situ with a mixture of 10% H$_2$S/H$_2$ using a standard procedure (14). Hydrotreating conditions were: 380°C, hydrogen pressure 10.3 MPa, LHSV 1.00 and gas flowrate 1000 L H$_2$/L feed. Experimental time-on-stream runs were performed for 140-180 h and the product nitrogen content was determined at regular intervals to monitor the rate of catalyst deactivation. Elemental analysis was undertaken on liquid feedstocks and products using a Perkin Elmer 240 analyzer for C, H and N, and a Dohrmann analyzer for S.

Catalyst Activity Studies Using Coprocessed Heavy Gas Oil

Experimental and commercial Ni-Mo catalyst activities were compared by determining nitrogen and sulfur conversions in hydrotreating coprocessed heavy gas oil. This material gave minimal catalyst deactivation and therefore allowed reliable determinations. Experiments were carried out using a continuous flow Robinson-Mahoney gradientless stirred tank microreactor of approximate internal volume 50 cm^3 equipped with an annular catalyst basket and internal recycle impeller (16). The catalyst bed was approximately 10 cm^3 and contained particles of 14-20 mesh. Presulfiding was carried out with 3 wt % butanethiol in diesel oil as described previously (14). Hydrotreating of diesel oil was then performed for approximately 72 h until steady-state conditions were achieved. Conversions of nitrogen and sulfur in coprocessed heavy gas oil were measured by hydrotreating at WHSV 0.75 and 2500 rpm. Temperatures were 300°C-380°C with hydrogen pressure and flowrates as above.

RESULTS AND DISCUSSION

Physical Properties of Catalyst Materials

Pillaring and delamination of smectite clays were investigated extensively by Pinnavaia et al. (13) who demonstrated the dependence of pore size distribution on method of drying and showed that the extent of delamination, i.e., layer separation, is strongly related to clay particle size and morphology. Thus it was shown that freeze drying of flocculated clay particles having a lathlike morphology, as opposed to air drying, preserved their delaminated structures. Delaminated clays have disordered, random platelet orientations when compared to the ordered stacking arrangement or face-to-face associations in pillared clays. The disordered platelet orientations have been referred to as a house-of-cards type structure (12), and consist of edge-to-face and edge-to-edge layer associations which are retained during the freeze drying process.

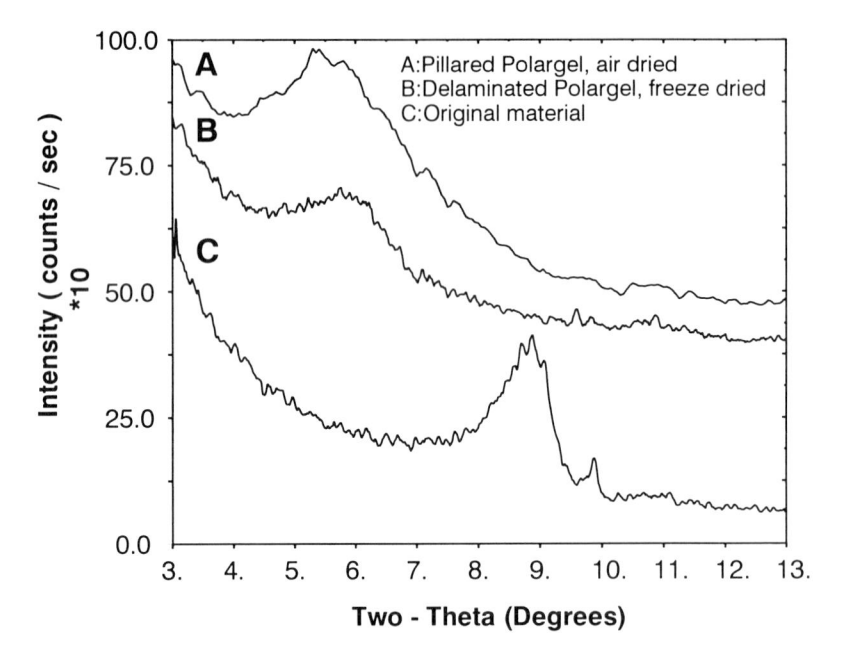

Fig. 1. XRD spectra of pillared and delaminated Polargel compared with original material.

In this work, XRD was used to determine pillared clay crystallinity and extent of delamination in freeze dried materials. Basal spacing measurements were made by monitoring the first order or (001) reflection. Occelli et al. have shown that delaminated Laponite displays no (001) reflection (12). In the present study, flocculation in aqueous suspension and freeze drying of Polargel was investigated using XRD. Figure 1 shows X-ray diffractograms for three Polargel samples: (A) pillared and air dried, (B) delaminated and freeze dried and (C) original material. Spectra (A) and (C) show basal spacing peaks characteristic of laminated structures, the peak displacement being typical of pillared clay formation. For freeze dried Polargel (spectrum B), basal spacing peak intensity is considerably less and indicates loss of crystallinity due to delamination. Table 1 presents basal spacings and crystallite sizes for pillared and delaminated Polargel and Laponite. The extent of delamination of freeze dried materials is indicated by a reduction in the number of clay layers. It is shown that on average delaminated Polargel was reduced to 8-9 layers, whereas delaminated Laponite contained 4. Increases in clay crystallite size were observed in PILC materials as indicated for pillared Polargel which showed an approximate increase from 165 to 249 Å.

Table 1 - Basal Spacings and Crystallite Sizes for Pillared and Delaminated Clays

Clay	Pillaring agent	Angle (°,2θ)	Basal spacing (Å)	Crystallite size (Å)	Average No. of layers
Polargel	---	6.99	12.6	165	13
Polargel PILC[a]	ACH	4.52	19.5	249	13
Polargel Delam.[b]	ACH	4.88	18.1	165	9
Polargel Delam.[b]	ZACH	4.20	21.0	156	8
Laponite	---	6.74	13.1	99	7
Laponite Delam.[b]	ZACH	5.77	15.2	56	4

[a]Air dried
[b]Freeze dried

To elucidate the effects of clay particle size and morphology on pore diameter, normalized incremental pore volume distributions for freeze dried delaminated Laponite and Polargel determined from BET N_2 adsorption isotherms are presented in Fig. 2. The data are compared with the corresponding pore volume distribution for a purified pillared clay, Accofloc 350, used in previous work (14). It is observed that the pore volume distributions for the three materials are markedly different. Delaminated Laponite has about 50% of its pore volume distribution in pores of > 20 Å, whereas delaminated Polargel has about 75% and pillared Accofloc 350 about 25%. Hence delaminated Polargel has the lowest percentage distribution of microporosity and contains a wide range of mesopore diameters between 20-300 Å. In contrast the range of mesopore size in delaminated Laponite is much narrower and lies between 20-80 Å. It is apparent that differences in pore size

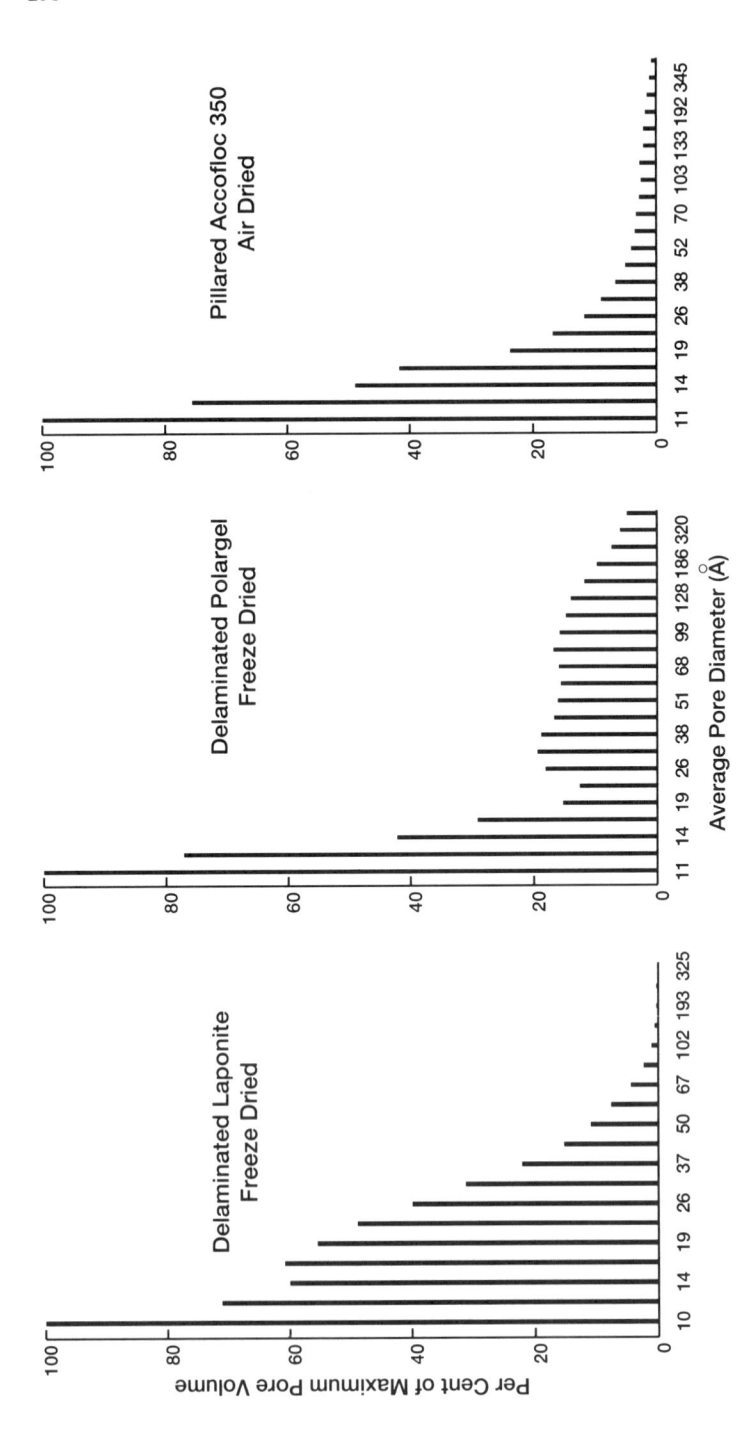

Fig. 2. Normalized incremental pore volume distributions of delaminated and pillared clays determined from BET N_2 adsorption data.

Applications of Interlayered Clays as Supports

distribution are a function of crystallite size which is significantly greater for delaminated Polargel as shown in Table 1. Figure 2 also shows that pillared Accofloc 350 has relatively few mesopores which is expected of a typical layered PILC structure. The significance of these results is realized when considering application of the materials in fabricating catalyst supports. Thus microporosity is of little importance in processing large gas oil molecules which are subject to diffusion limitations and are too large to penetrate. In this work it is shown that by dispersing delaminated Laponite and Polargel in γ-alumina, a combination of micropores and mesopores may be achieved. The result is a unique catalyst support structure.

Table 2 shows that surface areas of metal loaded catalysts prepared with varying amounts of γ-alumina were somewhat modified depending on the clay material added. In general, surface areas moderately lower than for γ-alumina were found using pillared and delaminated montmorillonites. A marginally higher surface area was achieved using 15% delaminated Laponite. Catalyst metal loadings are also presented in Table 2. Molybdenum content varied in the range 8.4-9.1 wt % and nickel was 2.8-3.1 wt %.

Determination of Surface Acidity by Pyridine/FTIR Adsorption

Investigation of surface acidity using pyridine/FTIR was focused on adsorption at Lewis acid sites. After calcination at 500°C, composite materials containing PILCs and delaminated clays showed only weak Brønsted acidity. Figure 3 shows FTIR spectra (1550-1650 cm^{-1}, 100°C) of pyridine adsorbed on various calcined support materials including some containing oxides of Ni and Mo. Spectra of the metal loaded supports are compared with those of pure γ-alumina and delaminated Polargel, and a progression between the latter materials is observed from strong to weak Lewis acidity. The characteristic γ-alumina spectrum, **A**, displays two major types of Lewis acidity, a strong band at ~ 1615 cm^{-1} (Lewis I) and a shoulder at ~ 1620 cm^{-1} (Lewis II). A third weak Lewis acidity band is also observed at ~ 1580 cm^{-1}. These results are in good agreement with literature values (17,18). Addition of oxides of Ni and Mo suppresses the Lewis II shoulder, as shown by **B**, the spectrum of pyridine adsorbed on calcined Ni-Mo/15% delaminated Laponite in γ-alumina. In this case, the spectrum is dominated by a strong Lewis I band at ~ 1610 cm^{-1} attributed to excess γ-alumina. For the Ni-Mo catalyst **C**, containing 30% Polargel PILC in γ-alumina, a similar spectrum is shown but with loss of intensity and broadening of bands indicating a modified overall surface acidity.

Spectra **D** and **E** in Fig. 3 are for pyridine adsorbed respectively on pure delaminated Polargel support and on a corresponding material loaded with oxides of Ni and Mo. Spectrum **D** for the pure material shows much weaker Lewis

Table 2 - Chemical and Physical Properties of Ni-Mo Catalysts

Ni-Mo Catalyst Support Composition	Ni Metal[1] Loading (wt %)	Mo Metal[2] Loading (wt %)	BET Surface Area (m^2/g)	Total Pore Volume (cm^3/g)	Average Pore Diameter (Å)
100% γ-alumina	2.9	8.7	195	0.24	50
15% Delaminated Laponite/ 85% γ-alumina	2.9	8.4	232	0.27	47
30% Delaminated Polargel/ 70% γ-alumina	3.1	8.9	166	0.21	51
30% Pillared Polargel/ 70% γ-alumina	2.8	9.1	151	0.21	54
30% Pillared Volclay HPM-20/ 70% γ-alumina	2.8	8.8	142	0.19	54
Commercial Ni-Mo catalyst AKZO KF-153S	2.1	7.7	248	0.42	61

[1] Analytical error in nickel metal content determined by PIXE was ± 0.19 wt %
[2] Analytical error in molybdenum metal content determined by PIXE was ± 0.58 wt %

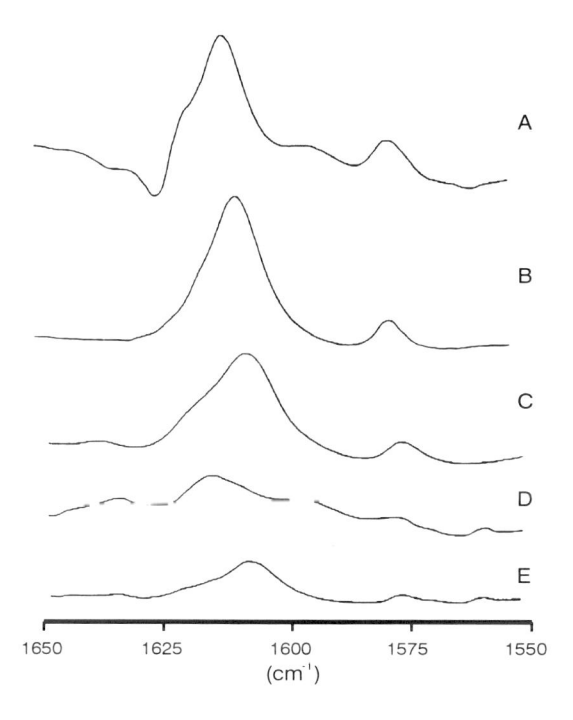

Applications of Interlayered Clays as Supports

Fig. 3. Infrared spectral regions (1550-1650 cm⁻¹) of:
    A: γ-alumina support
    B: Ni-Mo/15% delaminated Laponite/85% γ-alumina
    C: Ni-Mo/30% Pillared Polargel/70% γ-alumina
    D: 100% delaminated Polargel support
    E: Ni-Mo/100% delaminated Polargel

acidity than γ-alumina, with a Lewis II band at ~1616 cm⁻¹, and a weaker Lewis I shoulder at ~1600 cm⁻¹. Upon addition of Ni and Mo oxides, in spectrum **E**, the Lewis I shoulder becomes a well defined band and the Lewis II band is observed as a very weak shoulder. For the series of materials studied, delaminated Polargel is the only support to show this type of behaviour with Lewis II sites being more abundant than Lewis I sites. Delaminated Polargel also reveals a weak band at ~1580 cm⁻¹ indicative of hydrogen bonded pyridine (spectrum **D**) in good agreement with the results of Bodoardo et al. (19). FTIR spectra were also recorded for sulfided catalysts. A bar chart showing surface area-normalized absorbance values for pyridine adsorption on sulfided Ni-Mo catalysts containing varying amounts of delaminated clay is presented in Fig. 4. The Lewis acidity sensitive band used was located at

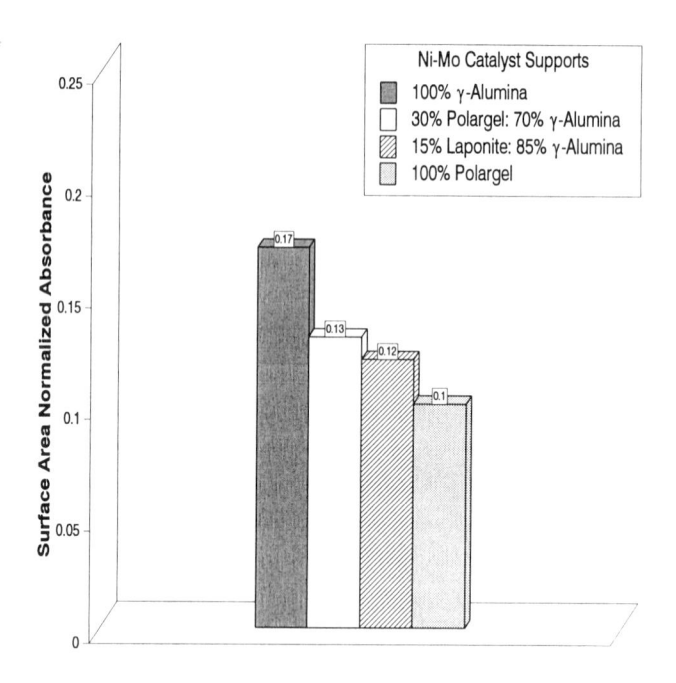

Fig. 4. Surface area-normalized Lewis acidity (~ 1450 cm^{-1}) at 100°C for sulfided Ni-Mo catalysts supported on γ-alumina and delaminated clay composite materials determined by pyridine/FTIR adsorption.

~ 1450 cm^{-1} (100°C) (20). It is observed that Ni-Mo supported on 100% γ-alumina shows a stronger pyridine adsorption than Ni-Mo supported partially or fully on delaminated Polargel. Similarly, Ni-Mo supported on 15% delaminated Laponite in γ-alumina gave a weaker absorbance. The general trend observed in Fig. 4 shows that the Lewis acidity of sulfided Ni-Mo catalysts containing delaminated clays is significantly reduced compared with 100% γ-alumina. In conclusion, the FTIR results indicate that the catalyst materials studied may be grouped according to their Lewis acidity. It is demonstrated that Ni-Mo supported on 100% delaminated Polargel had the lowest propensity to adsorb basic nitrogen components.

Transmission Electron Microscopy Studies

Figure 5 presents a TEM micrograph obtained from the extruded catalyst support material comprising 15% freeze dried delaminated Laponite dispersed in γ-alumina. The extruded material was calcined in air at 500°C and contained no metals. The micrograph shows house-of-cards type structures consisting of

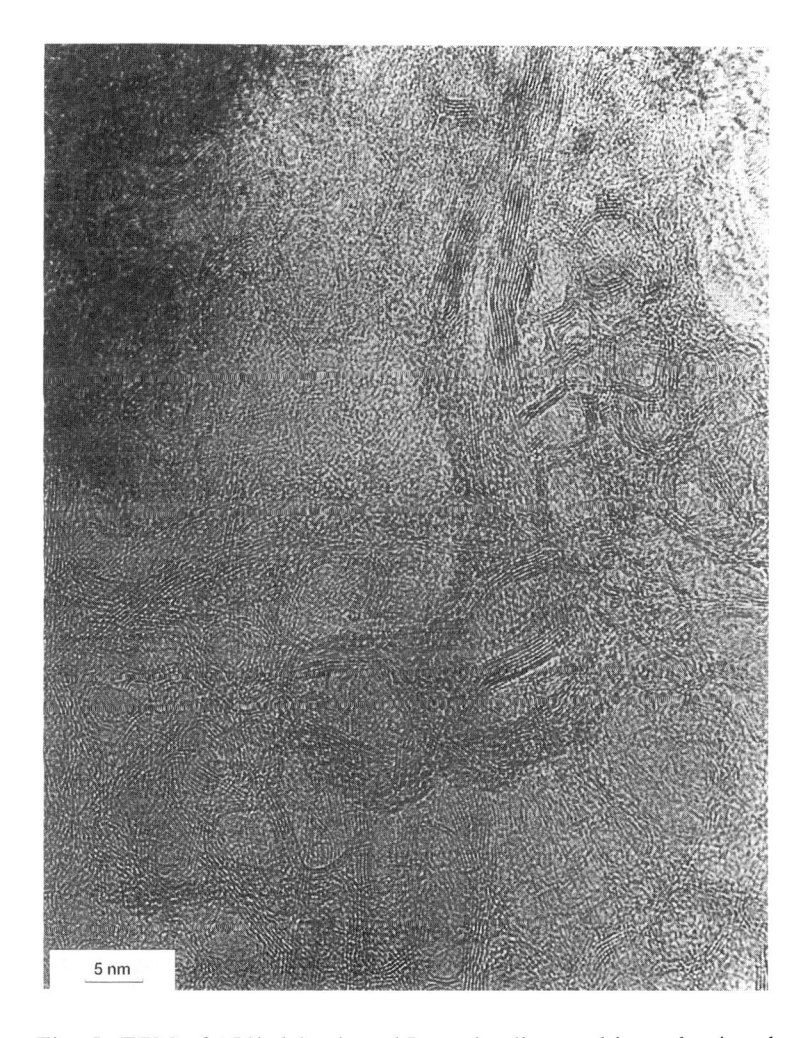

Fig. 5. TEM of 15% delaminated Laponite dispersed in γ-alumina showing mesopores formed from house-of-cards type structures.

delaminated layer aggregates of Laponite dispersed throughout the support material and largely arranged in edge-to-edge and edge-to-face connections between blocks forming mesopores in the γ-alumina matrix. In some regions almost symmetrical mesopores were obtained. It is also shown that the particle size and lath-like morphology of the synthetic hectorite is a key factor in forming the house-of-cards type structures which are 4-7 nm in diameter and constructed of short block-like aggregates containing between 3 to 8 layers. The

Fig. 6. TEM of used Ni-Mo catalyst supported on 15% delaminated Laponite dispersed in γ-alumina showing stacking of MoS_2 layers.

average dimensions of mesopores so formed are confirmed by the pore volume distribution data determined from BET N_2 adsorption isotherms and presented as a line chart for delaminated Laponite in Fig. 2.

TEM analysis was also performed on a used catalyst extrudate consisting of sulfided Ni-Mo supported on 15% delaminated Laponite in γ-alumina. The catalyst was run for approximately 240 h on oil using the Robinson-Mahoney reactor in hydrotreating coprocessed heavy gas oil. A TEM micrograph of the used catalyst is presented in Fig. 6., and shows characteristic filament-like structures of MoS_2 dispersed on the surface of the support. The layered structures were identified with MoS_2 crystallites by electron diffraction. The micrograph reveals edge planes of stacked MoS_2 layers which are between 1 and 8 and a layer spacing of 6.2 Å which is very close to the reported value of 6.15. The length of the MoS_2 crystallites varied considerably between 5 to 10 nm.

The characteristic house-of-cards type structure of delaminated Laponite is not visible in Fig. 6, but the presence of the synthetic hectorite in the used catalyst was confirmed by XRD analysis. It therefore appears that the MoS_2 structures are formed by a process of nucleation on the delaminated clay surfaces which is likely to be facilitated by dispersion across the γ-alumina component of the support. This suggests that, during its formation, MoS_2 migrates preferentially onto the delaminated clay where stacking occurs. Stacking of MoS_2 catalysts supported on γ-alumina was investigated by Ryan et al. who showed that formation of stacks is promoted by Ni and phosphorus (21). These workers used Mo metal loadings between 10-16 wt %. In the present work optimal loadings of Mo between 8-9 wt % were used and it is concluded that stacking was promoted by overloading of the clay surfaces which would also encourage lengthening of the MoS_2 crystallites.

Catalyst Deactivation Studies

Deactivation studies to assess the effects of modified supports containing both pillared and delaminated clays on Ni-Mo catalyst performance were carried out by hydrotreating a coal-derived liquid from the Japanese NEDOL process. Properties of the coal liquid are presented in Table 3. It is shown that the material contained significant amounts of nitrogen (0.65 wt %) and oxygen (1.60 wt %) as well as hexane and toluene insolubles (4.15 and 0.86 wt % respectively). Since the sulfur content was low, the material was spiked with 3 wt % butanethiol to maintain the catalyst metal ingredients in a sulfided state.

Results from time-on-stream monitoring of per cent nitrogen conversion in spent donor solvent are presented in Fig. 7, and differences in catalyst performance are apparent. For the standard Ni-Mo catalyst supported on 100% γ-alumina deactivation occurred rapidly and the conversion decreased by 30% over approximately 140 h time-on-stream. In coal liquid feedstocks nitrogen is largely contained in aromatic-type structures and deactivation is attributed to strong adsorption of foulants at surface Lewis acid sites leading to restricted access of large molecules, as indicated by the steady decline in activity. The two catalysts containing pillared clays, i.e., 30% Polargel PILC and 30% Volclay HPM-20 PILC, showed some initial deactivation but demonstrated an improved overall performance. However, the Ni-Mo catalyst giving the best performance was that containing delaminated clay, i.e., the support comprising

Table 3 - Properties of Feedstocks

Feedstock Property	NEDOL Process Spent Donor Solvent[1]	CANMET Coprocessed Heavy Gas Oil[2]
Density at 15°C (g/cm^3)	1.03	0.973
Elemental Analysis (wt %)		
Carbon	88.9	85.6
Hydrogen	8.60	10.8
Nitrogen	0.65	0.70
Sulfur	0.069	2.29
Oxygen	1.60	0.80
D-1160 Distillation (wt %)	IBP = 102°C	IBP = 171°C
IBP - 200°C	0.90	0.0
200 - 350°C	60.4	32.1
350 - 525°C	36.2	66.3
+ 525°C	2.20	1.4
Hexane insolubles (wt %)	4.15	1.68
Toluene insolubles (wt %)	0.86	0.11
Aromatic carbon (^{13}C NMR) %	58.0	40.0

[1] Coal-derived liquid feedstock prior to spiking with 3 wt % butanethiol
[2] Product was derived from coprocessing 70% Cold Lake bitumen with 30% Forestburg coal

15% delaminated Laponite in γ-alumina. Although this catalyst showed a lower initial nitrogen conversion than the standard Ni-Mo/γ-alumina catalyst, the per cent conversion remained virtually constant over 160 h of time-on-stream and the catalyst showed almost no deactivation. These results suggest that delaminated clays of suitable morphology giving well defined mesoporous structures are more effective as local carriers of MoS$_2$ than pillared clays. While both types of materials are expected to modify the chemical properties of composite supports the pore structures formed may be a critical factor. The differences may be observed in the TEM micrograph shown in Fig. 5 which reveals domains of both types of materials dispersed in γ-alumina.

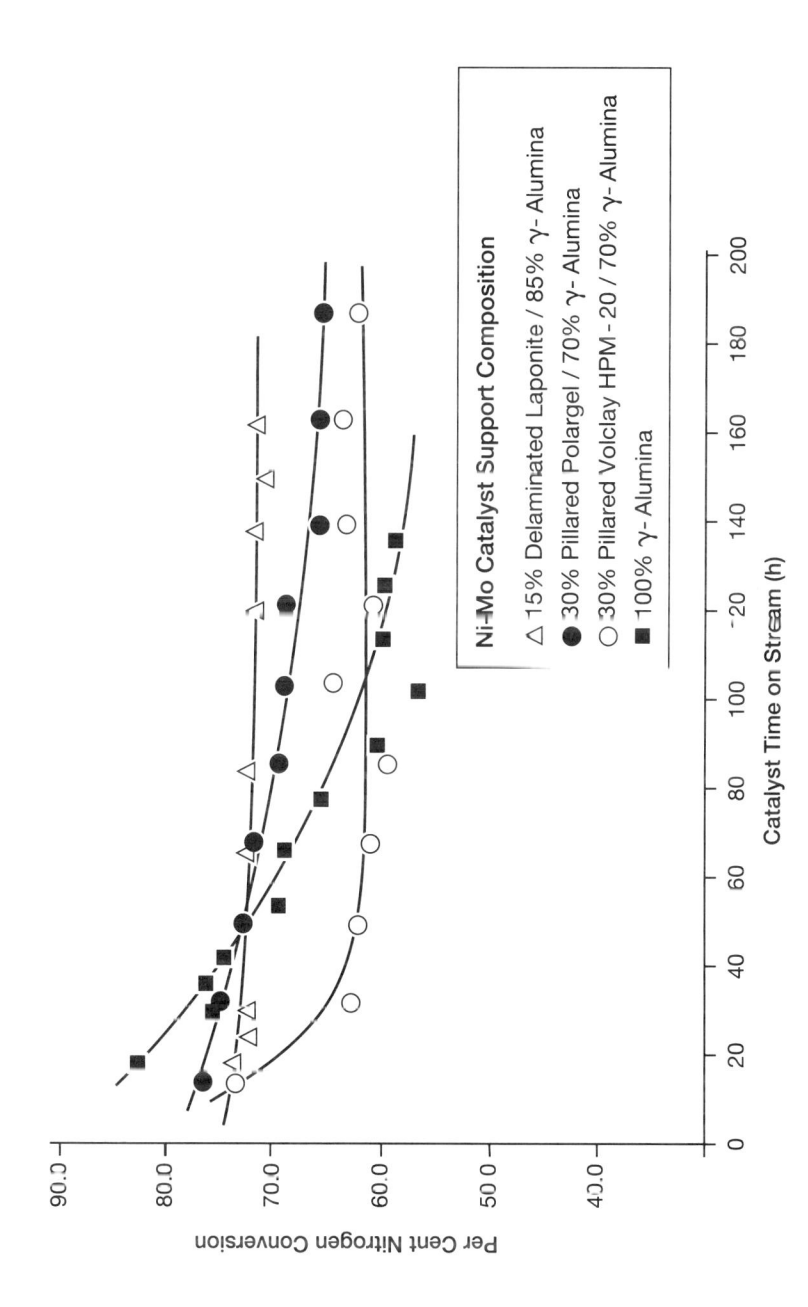

Applications of Interlayered Clays as Supports 307

Fig. 7. Effect of catalyst support composition on rate of deactivation during nitrogen conversion in hydrotreating NEDOL process donor solvent.

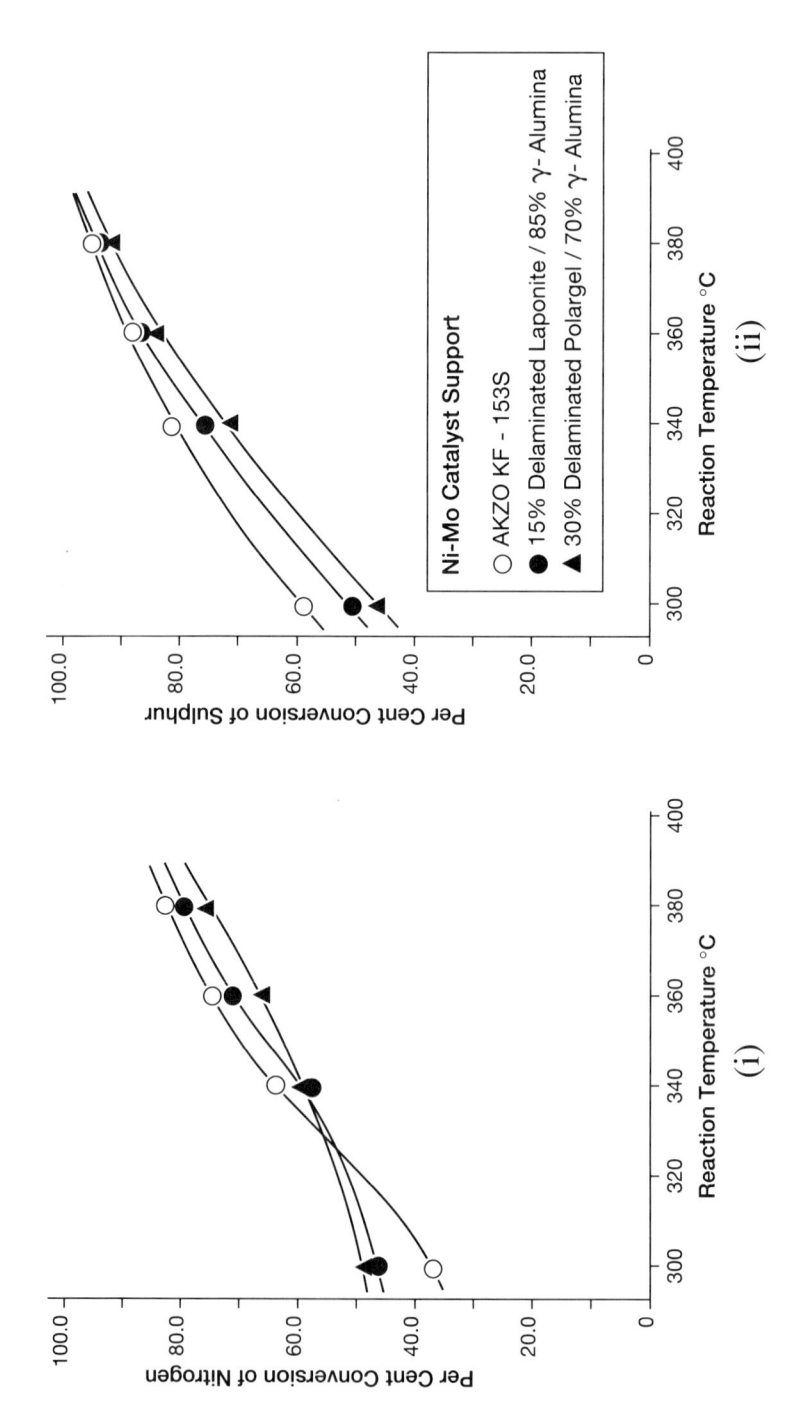

Fig. 8. Effect of catalyst support composition on conversion of nitrogen and sulfur in hydrotreating coprocessed heavy gas oil.

Catalyst Activity Studies for Heteroatom Conversion

Ni-Mo hydrotreating catalysts were tested for hydrodenitrogenation (HDN) and hydrodesulfurization (HDS) activity in heavy gas oil from coprocessed coal/bitumen using a Robinson-Mahoney gradientless stirred tank microreactor (16). The gradientless reactor allows determinations under isothermal conditions with uniform reactant concentrations. Properties of the feedstock are presented in Table 3. The activity performance of experimental catalysts with supports containing 15% delaminated Laponite and 30% delaminated Polargel was compared with that of the high surface area commercial Ni-Mo catalyst AKZO KF-153S.

Figure 8(i) presents plots of per cent nitrogen conversion versus reaction temperature. Reasonably good agreement is observed for nitrogen removal over the commercial catalyst and that containing 15% delaminated Laponite. At the maximum operating temperature, 380°C, the nitrogen conversion was approximately 80%. The catalyst containing 30% delaminated Polargel showed the lowest activity and its lower surface area may be a contributing factor. It is also noteworthy that at 300°C, the pattern for the catalysts containing delaminated clays is reversed and they gave somewhat higher conversions than the commercial catalyst.

Plots of per cent sulfur conversion versus reaction temperature are presented in Fig. 8 (ii). At 300°C, it is shown that significant differences were obtained and catalyst AKZO KF-153S gave the highest conversion. These results suggest a different reaction mechanism than for nitrogen removal. However, as the temperature is increased, the plots are seen to converge and at 380°C the performance of all three catalysts is in good agreement. For the HDS reaction high conversions, approximately 94%, were obtained at the maximum operating temperature.

CONCLUSIONS

Pillared and delaminated clays were used to fabricate composite catalyst supports by combination with γ-alumina. Resistance to deactivation by fouling and coke formation was related to surface Lewis acidity and improved performance was demonstrated by some catalysts containing clay materials. Synthetic hectorite has the most suitable size and morphology to achieve almost complete delamination, and results confirm that optimum catalyst performance was obtained using a composite support containing 15% delaminated Laponite dispersed in γ-alumina. During catalyst preparation, MoS_2 crystallites appeared to develop through a process of migration and nucleation at clay surfaces with the formation of multilayers superimposed on delaminated clay structures. In addition to promoting MoS_2 stacking, delaminated clays formed house-of-cards type structures in composite supports which may have enhanced catalytic properties. The materials provide well defined mesopores which are likely to improve access to catalytic sites. It is concluded that the combined effect of improved chemical and physical properties of composite supports containing delaminated clays enhances catalyst performance.

ACKNOWLEDGEMENTS

The authors acknowledge the contribution of heavy gas oil feedstock from CANMET Coprocessing Consortium. The contributions of CANMET staff from the Synthetic Fuels and Fuels Characterization Research Laboratories are also gratefully acknowledged.

REFERENCES

(1) Stohl, F.V. and Stephens, H.P., ACS, Div. Fuel Chem. Prep., 30 (4), 148 (1985).
(2) Stohl, F.V. and Stephens, H.P., ACS, Div. Fuel Chem. Prep., 34 (4), 251 (1986).
(3) Nishijima, A., Shimada, H., Yoshimura, Y., Sato, T. and Matsubayashi, N., Proc. 4th Intl. Symp. on Catal. Deact., Antwerp, Belgium (1987).
(4) Derbyshire, F.J. "Catalysis in coal liquefaction: new directions for research", IEA Coal Research, London, England, June 1988.
(5) Tischer, R.E., Narain, N.K., Stiegel, G.J. and Cillo, D.L., J. Catal. 95, 406 (1985).
(6) Nishijima, A., Shimada, H., Sato, T. and Yoshimura, Y., Proc. Intl. Conf. on Coal Science, Sydney, Australia, Pergamon Press, 201 (1985).
(7) Beer, V.H.J. de, Derbyshire, F.J., Groot, C.K., Prins R., Scaroni, A.W. and Solar, J.M., Fuel, 63, 1095, (1984).
(8) Scaroni, A.W., Jenkins, R.G. and Walker, P.L., Appl. Catal. 14, 173 (1985).
(9) Groot, C.K., Beer, V.H.J. de, Prins, R., Stolarski, M. and Niedzwiedz, W.S., Ind. Eng. Chem. Prod. Res. and Dev. 25, 522 (1986).
(10) Occelli, M.L. and Rennard, R.J., Catal. Today, 2, 309 (1988).
(11) Occelli, M.L., Catal. Today, 2, 339 (1988).
(12) Occelli, M.L., Landau, S.D. and Pinnavaia, T.J., J. Catal. 104, 331 (1987).
(13) Pinnavaia, T.J., Tzou, M., Landau, S.D. and Raythatha, H., J. Mol. Catal. 27, 195 (1984).
(14) Monnier, J., Charland, J.-P., Brown, J.R. and Wilson, M.F., "New Frontiers in Catalysis", L. Guczi, F. Solymosi, P. Tétéyi, eds., Akadémiai Kiadó, Budapest, and Elsevier, Amsterdam, 1943 (1993).

Applications of Interlayered Clays as Supports

(15) Klug, H.P. and Alexander, L.E, " X-ray diffraction procedures for polycrystalline and amorphous materials", John Wiley and Sons Eds., New York, 1974.
(16) Mahoney, J.A., J. Catal. 32, 247 (1974).
(17) Kiviat, F.E. and Petrakis, L., J. Phys. Chem. 77 (10), 1232 (1973).
(18) Occelli, M.L. and Lester, J.E., Ind. Eng. Chem. Prod. Res. and Dev. 24, (1) 27 (1985).
(19) Bodoardo, S., Figueras, F. and Garrone, E., J. Catal. 147, 223 (1994).
(20) Valyon, J., Schneider, R.L. and Hall, W.K., J. Catal. 85, 277 (1984).
(21) Ryan, R.C., Kemp, R.A., Smegal, J.A., Denley, D.R. and Spinnler, G.E., "Advances in Hydrotreating Catalysts", M.L. Occelli and R.G. Anthony, eds., Elsevier Science Publishers B.V., Amsterdam, 21 (1989).

Catalytic Hydrotreating with Pillared Synthetic Clays

Ramesh K. Sharma and Edwin S. Olson
University of North Dakota
Energy & Environmental Research Center
Grand Forks, ND 58202

I. ABSTRACT

Pillared clays have large micropores that can be exploited for the hydrotreatment of petroleum- or coal-derived materials. Nickel-substituted synthetic mica montmorillonite (NiSMM) was prepared and subsequently pillared by the intercalation of polynuclear hydroxyaluminum and hydroxyzirconium cations. The pillared NiSMMs were then impregnated with molybdenum and sulfided. These pillared clay catalysts have higher acidity and surface area than NiSMM. Bibenzyl and pyrene were hydrotreated with these catalysts to investigate the mechanism(s) of hydrocracking and hydrogenation. The major pathway for pillared clay is a Brønsted acid-catalyzed, but some Lewis acid-catalyzed reactions are also observed. The combined effect of clay acidity and hydrogen activation by nickel and molybdenum sulfides resulted in higher conversions in hydrocracking and hydrogenation tests.

II. INTRODUCTION

New and better catalysts are constantly sought for improving the overall economics of fossil fuel refining. Natural clays were used as catalysts in petroleum cracking until replaced by more active and selective zeolites (1). Pillared clays have been investigated in the last two decades. In the pillared clays, intercalation of hydroxylated or complexed metal cations maintains the clay layer structure after loss of water and generates large pore sizes (2). Pillared-clay hydrocracking catalysts have been prepared using several natural clays (3–5). The pillared clays are stable to 600°C (6) and even 800°C, when pillaring is performed with mixed colloidal clusters (7). Pillared clays containing molybdenum sulfide were very active hydrocracking catalysts for the aromatic structures found in coal-derived liquids (8).

Synthetic clays (SMM) containing a transition metal component were highly active for hydrocracking and hydroisomerization (9–11). Incorporation of cobalt and especially nickel into the SMM structure dramatically improves the hydrocracking and hydroisomerization of light n-paraffins (12). Pillared clay catalysts were also prepared from synthetic clays (13).

The objective of the investigations reported here was to combine a strong hydrogen activating site with a strongly acidic pillared synthetic clay. Pillared synthetic clays with and without incorporated hydrogenation catalysts were prepared and tested with model compounds representative of fossil fuel-derived components. The catalysis of aryl–alkyl bond breakage, as well as arene hydrogenation, has been investigated.

III. EXPERIMENTAL

A. Catalyst Preparation

Nickel-substituted mica montmorillonite (NiSMM) was prepared by the procedure of Heinerman (14). Sodium silicate solution was exchanged with Amberlite IR 120-H cation resin, and a solution of nickel acetate was added and stirred with aluminum isopropylate and ammonium fluoride. After evaporation, aqueous ammonia was added, and the slurry was heated in a pressure vessel at 300°C for 16 hr. The product was separated by centrifugation and washing with deionized water, dried at 120°C, and calcined in air at 540°C for 6 hr.

Pillaring of NiSMM was carried out according to the procedure described by Gaaf and coworkers (13). A slurry of the NiSMM (5 g in 1000 mL) was stirred at 70°C for 20 hr with an aged (3 day) polyoxyaluminum solution prepared from aluminum chloride (30 mmoles) and sodium hydroxide 60 mmoles). The pillared clay was separated by centrifugation, washed with deionized water several times, and freeze-dried. The dried solid was calcined in air at 350°C for 6 hrs. Zirconia-pillared NiSMM was prepared by adding the zirconia pillaring solution, prepared from zirconyl chloride (75 mmole), to 5 g NiSMM in 1000 mL water and heating at 70°C for 20 hr.

The molybdenum sulfide-supported pillared clay catalysts were prepared by adding 0.5 g of synthetic clay to 50 mL of aqueous ammonium molybdate solution and stirring overnight (15). The solvent was removed by evaporation, and the resulting product was calcined in air at 350°C for 6 hr to produce a catalyst with 5 wt% loading of molybdenum.

B. Catalyst Activation

1. Sulfidation

The catalysts were activated in a rocking microreactor with 1000 psi of hydrogen containing 10% hydrogen sulfide at 400°C for 2 hr. The catalyst was recovered and stored in airtight bottles.

2. Prereduction

Pillared NiSMM catalysts were reduced in hydrogen (1000 psi) at 450°C for 16 hr in a rocking microreactor as described above.

C. Catalyst Characterization

1. Total Acidity

The acidity of the solid acid catalysts was determined by pyridine adsorption and desorption studies. A small amount of sample (100 mg) was placed in a glass chamber attached to a vacuum pump, a gas inlet, and a gas outlet. The chamber was evacuated, and argon saturated with pyridine was introduced into the chamber until the weight increase ceased. At this stage, the chamber was evacuated until the physisorbed pyridine was removed, as indicated by the constant weight of the base absorbed sample. The amount of pyridine chemisorbed was used to determine total acidity of the catalyst.

2. Surface Area Measurements

Surface area measurements were performed with Micromeritics AccuSorb 2100E static unit for nitrogen physisorption at 77K (BET method).

D. Catalytic Reactions

In a typical run, 0.5 g of model compound and 0.25 g of the catalyst were placed in a tubing bomb (15-mL microreactor). The microreactor was evacuated, pressurized with 1000 psig of hydrogen, and placed in a rocking autoclave heated to desired temperature. The heating was continued for 3 hr. At the end of the reaction period, the microreactor was cooled in a dry ice–acetone slurry, degassed, and opened. The desired amount of the internal standards, isooctane and n-octadecane, was added to the product slurry; the product slurry was transferred into a centrifugation tube by washing with methylene chloride; and the solid catalyst removed by centrifugation. The solid was dried in vacuum at 110°C for 3 hr.

Quantitative gas chromatography–flame ionization detector (GC–FID) analyses of the liquid samples were performed with a Hewlett-Packard 5890A gas chromatograph equipped with a Petrocol capillary column and calibrated with the internal standards. The instrument was not calibrated for all the hydropyrene isomers; therefore, hydropyrene data are derived from flame ionization peak areas, assuming the response factors of pyrene and hydropyrenes to be the same.

IV. RESULTS AND DISCUSSION

Pillared synthetic clays were prepared and used as supports for molybdenum sulfide hydrogenation catalysts. Both the pillared clay supports and molybdenum sulfide-loaded pillared clays were tested for hydrocracking and hydrogenation activities using model compounds. The results were compared with commercial Ni–Moly or Co–Moly catalysts.

The total acidity of the NiSMM as measured by the amount of chemisorbed pyridine was found to be 0.84 mmol/g of catalyst. These sites may be Lewis as well Brønsted acids. Pillaring the NiSMM with polyoxyaluminum cations increased the total acidity to 1.02 mmol/g of catalyst. The increase in acidity is due to the additional acidic sites created by the dehydroxylation of the hydroxylated aluminum cations. When zirconia-pillared NiSMM was impregnated with molybdate and sulfided, the acidity was significantly reduced (0.20 mmol/g of catalyst). Similar effects were noted for sulfided chromia-pillared montmorillonites (8). Presumably, sulfidation converts acidic metal sites into metal sulfides.

The surface area of the NiSMM was found to be 240 m^2/g. Pillaring the clay with polyoxy zirconium cations resulted in an increase in the surface area (274 m^2/g). The increase in surface areas may be due to the micropores created by the dehydroxylation of the intercalated hydroxylated metal cations (pillars). The alumina-pillared NiSMM had a little lower surface area (230 m^2/g) as reported earlier (13). The repeat distance of the layers in the alumina-pillared NiSMM was previously reported to be 1.73 nm (13).

A. Hydrocracking

The hydrocracking activities of several pillared NiSMM catalysts were determined for bibenzyl (1,2-diphenylethane), a model for the types of bridged aromatic compounds present in fossil fuel feedstocks. The hydrocracking activities of pillared synthetic clays were determined, as a basis for subsequent studies with molybdenum-impregnated synthetic clays. The reaction of bibenzyl with NiSMM was shown earlier to convert 67% of the bibenzyl into lighter products (15). The activities of the pillared NiSMMs were expected to exceed that of the nonpillared NiSMM, owing to the introduction of additional acidic functionalities.

Both polyoxyaluminum- and zirconium-pillared NiSMM exhibited higher hydrocracking activity than NiSMM for cleavages of the aryl–methylene bonds in bibenzyl (Table 1). The hydrocracking activity of alumina-pillared NiSMM (77%) was about the same as that of zirconia-pillared NiSMM (74%). Benzene and ethylbenzene were the major products from both pillared and nonpillared clays. A small portion of the cracking products were hydrogenated to give cycloalkanes. Formation of these products is indicative of a carbonium ion mechanism catalyzed by Brønsted acid sites (16). This mechanism proceeds via *ipso* protonation followed by cleavage of the aryl–alkyl bond to form the phenylethyl carbonium ion intermediate (Figure 1a). The carbonium ion is then reduced to the alkyl group. Ethylbenzene is further cracked to benzene via the same steps, the extent depending upon the activity of the catalyst. The formation of toluene may have been Lewis acid-catalyzed (Figure 1b). In addition to these products, a very large number of other products in much smaller concentrations were observed. These products were formed via a variety of hydrogenation, hydroisomerization, and ring-opening reactions. The increase in the activity of the NiSMM (from 67% to 77%) is due to the higher acidity of the pillared catalysts. Although the hydrocracking activity of pillared and nonpillared synthetic clays compares very well with acid catalysts and zeolites, there is no evidence for formation of coke or oligomers (from the GC analysis) as a result of retrograde condensation reactions.

Reduction of part of the Ni^{2+} to zero valent nickel by prereduction of the alumina-pillared NiSMM by heating with hydrogen at 450°C for 16 hr in 1000 psi H_2 did not change the conversion (77%) in the reaction with bibenzyl. The product distribution was changed very little. These data indicate that the catalyst is effectively reduced during the hydrocracking reaction, and prereduction is not needed to achieve high hydrocracking activity.

Sulfidation of the alumina-pillared NiSMM resulted in a substantial increase in catalytic activity for hydrocracking. The reaction of bibenzyl with presulfided alumina-pillared NiSMM gave a significantly higher conversion (87%) than NiSMM. The product distribution was similar to that of the NiSMM reaction. Somewhat larger amounts of benzene and alkylbenzenes were converted into cyclohexane and cycloalkanes, compared with the nonsulfided reactions. This may result from a more efficient transfer of hydrogen via nickel sulfide-catalyzed hydrogenation of the single aromatic rings. The increase in hydrocracking activity on sulfidation might be attributed to a more effective hydrogen addition to some intermediate involved in hydrogen (hydride) transfer.

TABLE 1

Hydrocracking Activity of Pillared Synthetic Clays with Bibenzyl

Reaction Temp. = 350°C, Reaction Time = 3 hr, H_2 = 1000 psi, Catalyst wt.:Substrate wt. = 0.5

Catalyst	Bibenzyl (mmol)	Conv. (%)	Major Products (mmol)
NiSMM	2.72	67	Benzene (1.46) Toluene ((0.11) Ethylbenzene (0.41)
APNiSMM[a]	2.80	77	Benzene (1.63) Toluene (0.11) Ethylbenzene (0.55) Methylcyclohexane (0.03) Tetralin (0.07)
ZPNiSMM[b]	2.80	74	Benzene (1.54) Toluene (0.10) Ethylbenzene (0.47) Cyclohexane (0.14) Methylcyclohexane (0.02) Ethylcyclohexane (0.05) Tetralin (0.1)
APNiSMM (reduced)	2.76	77	Benzene (1.75) Toluene (0.14) Ethylbenzene (0.61) Methylcyclohexane (0.03) Tetralin (0.10)
APNiSMM (sulfided)	2.83	87	Benzene (2.21) Toluene (0.19) Ethylbenzene (1.0) Methylcyclohexane (0.04) Tetralin (0.11)
MoAPNiSMM[c] (sulfided)	2.76	85	Benzene (0.34) Toluene (0.05) Ethylbenzene (0.11) Cyclohexane (0.13) Methylcyclohexane (0.16) Ethylcyclohexane (0.06) Tetralin (0.14)
MoZPNiSMM[d] (sulfided)	2.78	97	Benzene (0.50) Toluene (0.07) Ethylbenzene (0.18) Cyclohexane (0.20) Methylcyclohexane (0.16) Ethylcyclohexane (0.07)
Co–Mo–Al$_2$O$_3$	2.78	99	Methylcylcohexane (0.03) Ethylcyclohexane (0.01)

[a] Alumina-pillared NiSMM.
[b] Zirconia-pillared NiSMM.
[c] Molybdenum–alumina-pillared NiSMM.
[d] Molybdenum–zirconia-pillared NiSMM.

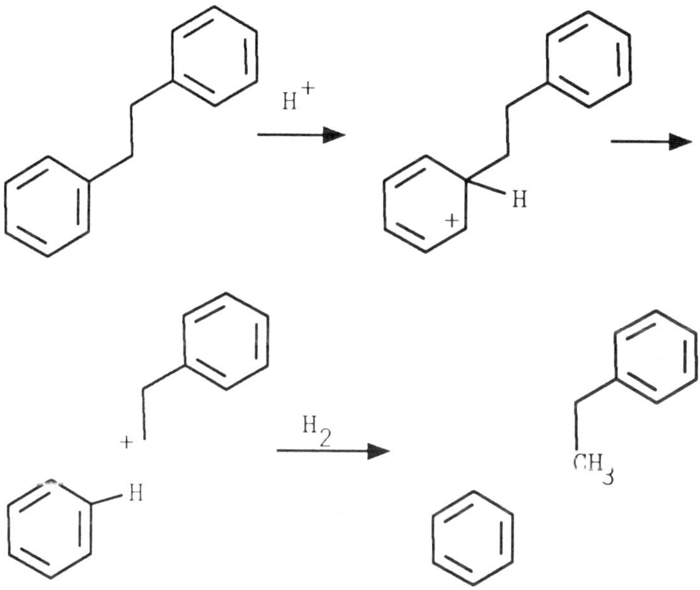

Catalytic Hydrotreating with Synthetic Clays

Figure 1a. Mechanism for Brønsted acid-catalyzed reaction of bibenzyl.

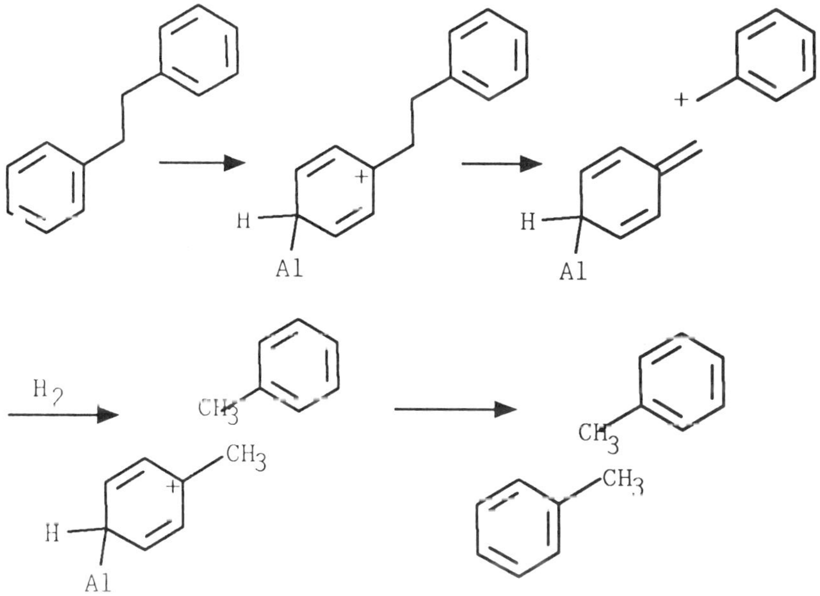

Figure 1b. Mechanism for Lewis acid-catalyzed reaction of bibenzyl.

To increase the hydrogen-activating function of the catalyst, molybdenum was impregnated into the pillared NiSMM, and the catalyst was sulfided. In hydrocracking tests with this catalyst, very high conversions of bibenzyl and somewhat different product distributions were obtained. Benzene and ethylbenzene were primary products from Brønsted acid-catalyzed cleavage, and a small amount of phenylethylcyclohexane was produced from hydrogenation of bibenzyl. The ratios of benzene:cyclohexane and ethylbenzene:ethylcyclohexane were similar, which indicates that cyclohexane and ethylcyclohexane result mainly from hydrogenation of the primary products, benzene and ethylbenzene. Both cyclohexane and ethylcyclohexane are further hydrocracked to alkylcyclopentanes and isoparaffins, but these secondary reactions occur more slowly and do not significantly change the ratio of aromatic to cycloalkane.

More interesting is the low toluene:methylcyclohexane ratio, resulting from the formation of a relatively larger amount of methylcyclohexane, compared with the reactions with catalysts that did not contain molybdenum. Since much more methylcyclohexane is produced relative to cyclohexane and ethylcyclohexane, there must be another route to methylcyclohexane other than toluene hydrogenation. A likely source of the methylcyclohexane is the hydrocracking reaction of the intermediate phenylethylcyclohexane (see Figure 2). The reaction produces equimolar amounts of toluene and cyclohexane, and further hydrogenation of the toluene would give the observed low toluene:methylcyclohexane ratio. The proposed cracking mechanism involves formation of the tertiary carbonium ion on the cylcohexyl ring, via Lewis acid-catalyzed hydride abstraction or partial hydrogenation of a protonated aromatic ring. Cleavage of the tertiary carboniun ion forms the benzyl carbonium ion and *exo*-methylenecyclohexane, which are hydrogenated to toluene and methylcyclohexane, respectively.

For comparison purposes, bibenzyl was hydrotreated under the same conditions with a sulfided commercial cobalt–molybdenum–alumina (cobalt–moly) catalyst. The conversion of bibenzyl was very high; however, only a small amount of cracking to smaller products (methylcyclohexane, ethylcyclohexane) was observed. This is consistent with the much lower acidity of the catalyst support in this system. The major products were the ring-hydrogenated products, phenylethylcyclohexane (0.03 mmoles) and dicyclohexylethane (0.22 mmoles), as well as the hydrogenated isomerized products, such as perhydrophenanthrenes (0.12 mmoles), and octahydrophenanthrenes and other m/e 186 components (0.02 mmoles). Thus, the molybdenum sulfide is very efficient at hydrogenation, and these isomerized

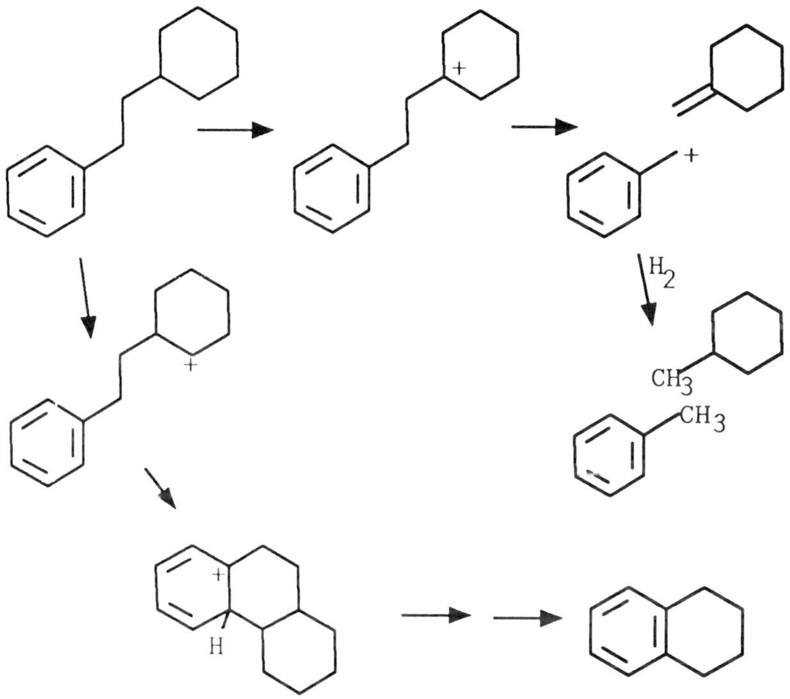

Figure 2. Mechanism for reactions of phenylethylcyclohexane.

products result from rearrangements of hydrogenated precursors catalyzed by the Lewis acidity of the alumina support (Figure 2).

B. Hydrogenation

Pyrene was used as a test compound to investigate hydrogenation activities of nonpillared and pillared synthetic clay catalysts with and without molybdenum sulfide. The reactions were carried out by heating the substrate and the desired catalyst at 350° – 400°C for 3 hr in the presence of 1000 psi of molecular hydrogen. The reaction conditions, conversion data, and product distribution are given in Table 2.

TABLE 2

Hydrogenation Reactions of Pyrene
Reaction Temp. = 350°C, Reaction Time = 3 hr, H_2 = 1000 psi
Catalyst wt.:Substrate wt. = 0.5

Catalyst (g)	Substrate (mmol)	Conversion (%)	Major Products (mmol)
NiSMM (sulfided)	2.43	90	Hexadecahydropyrene (0.07) (3 isomers) Decahydropyrene (0.36) (3 isomers) Hexahydropyrene (0.64) (2 isomers) 4,5,9,10-Tetrahydropyrene (0.06) 4,5-Dihydropyrene (0.21)
APNiSMM	2.47	81	Hexadecahydropyrene (0.02) (3 isomers) Decahydropyrene (0.22) (3 isomers) Hexahydropyrene (0.69) (2 isomers) 4,5,9,10-Tetrahydropyrene (0.08) 4,5-Dihydropyrene (0.34)
APNiSMM (sulfided)	2.44	89	Hexadecahydropyrene (0.07) (3 isomers) Decahydropyrene (0.45) (3 isomers) Hexahydropyrene (0.77) (2 isomers) 4,5,9,10-Tetrahydropyrene (0.09) 4,5-Dihydropyrene (0.25)
MoAPNiSMM (sulfided)	2.41	92	Hexadecahydropyrene (0.44) (3 isomers) Decahydropyrene (0.53) (3 isomers) Hexahydropyrene (0.60) (2 isomers) 4,5,9,10-Tetrahydropyrene (0.05) 4,5-Dihydropyrene (0.14)
MoZPNiSMM (sulfided)	2.61	90	Hexadecahydropyrene (0.19) (3 isomers) Decahydropyrene (0.39) (3 isomers) Hexahydropyrene (0.72) (2 isomers) 4,5,9,10-Tetrahydropyrene (0.07) 4,5-Dihydropyrene (0.23)
HDN-30[a] Ni–Moly	2.47	69	Hexadecahydropyrene (0.03) (3 isomers) Decahydropyrene (0.12) (3 isomers) Hexahydropyrene (0.66) (2 isomers) 4,5,9,10-Tetrahydropyrene (0.24) 4,5-Dihydropyrene (0.64)

[a] Commercial catalyst.

Catalytic Hydrotreating with Synthetic Clays

The reaction of pyrene with sulfided NiSMM at 350°C gave a high conversion (87%) of pyrene into products (15). Most of the products were di-, hexa-, and decahydropyrenes, but small amounts of hexadecahydro- and tetrahydropyrenes were also formed. Other components were formed by hydrocracking and rearrangement reactions, but hydrocracking is not prevalent at 350°C (15).

Pillaring of NiSMM with polyoxyaluminum cations resulted in a very similar hydrogenation activity for the sulfided catalyst (Table 2). The slightly higher activity for formation of hexa- and decahydropyrene can be attributed either to a shape selectivity factor or to increased kinetic reactivity resulting from the higher acidity of the pillared clay.

The alumina-pillared NiSMM without sulfiding was somewhat less effective in catalyzing pyrene hydrogenation (81% conversion). Only a very small amount of the hexadecahydropyrene is formed with the nonsulfided catalyst. A similar effect of nickel sulfidation on hydrogenation was demonstrated for the bibenzyl case discussed above.

Molybdenum sulfide supported on zirconia-pillared NiSMM gave the same conversion (90%) of pyrene into products as the sulfided NiSMM. However, the product distribution included significantly larger amounts of hexadeca- and decahydropyrenes. The amounts of dihydropyrene and tetrahydropyrene were correspondingly lower. It is clear that the hydrogen-activating component (molybdenum sulfide) is highly effective in achieving deep hydrogenation when supported on the synthetic clay. Substantially larger amounts of the hexadecahydro- and decahydropyrene were produced in the reaction of pyrene with the molybdenum-loaded alumina-pillared NiSMM.

The reaction of pyrene with commercial Ni–Moly (HDN-30) catalyst at the same temperature gave considerably less conversion (69%) of pyrene. Major products were hexa-, tetra-, and dihydropyrenes. In comparison with synthetic clay catalysts, Ni–Moly gave considerably large amounts of tetra- and dihydropyrenes, but the amounts of hexadeca- and decahydro-pyrenes were much smaller. The smaller amounts of more deeply hydrogenated pyrenes implies that there is some kinetic factor for the formation of the hexadeca- and decahydropyrenes that is more favorable for the molybdenum-impregnated synthetic clays. This factor may indicate some difference in the mechanism of hydrogenation on this catalyst. At higher temperatures (400°C), more extensive hydrocracking of the hydropyrenes occurs. Further studies at the higher temperatures are in progress.

V. CONCLUSIONS

Pillaring synthetic clay catalysts generates additional acidic sites that result in higher hydrocracking activities. Sulfidation decreases the acidity substantially, yet increases hydrocracking activity by creating better hydrogen-activating sites. Molybdenum incorporation has a similar but larger effect. High hydrogenation activities were observed for pillared NiSMM, which improved further with sulfidation and molybdenum incorporation. The molybdenum sulfide-loaded zirconia-pillared NiSMM was a significantly better catalyst than the molybdenum sulfide-loaded alumina-pillared NiSMM for bibenzyl hydrocracking (97% conversion). However for the pyrene hydrogenation, the molybdenum sulfide-loaded alumina-pillared catalyst was more active. The combined effects of high acidity and hydrogen-activating metal sites resulted in powerful catalysts for the hydrotreatment of aromatic materials.

VI. LITERATURE CITED

1. B.C. Gates, J.R. Katzer, and G.C.A. Schuit. Chemistry of Catalytic Processes, McGraw-Hill, New York, 1979, p. 1

2. T.J. Pinnavaia. Science 220: 365 (1983).

3. J. Shabtai. U.S. Patents 4,216,188 (1980) and 4,238,364 (1980).

4. C.E. Vaughn. U.S. Patents 4,176,090 (1979), 4271,043 (1981) and 4,248,739 (1981).

5. M.L. Occelli and R.J. Renard. Catal. Today 2:339 (1988).

6. M.L. Occelli, in Keynotes in Energy-Related Catalysis (S. Kaliaguine, ed.) Elsevier, Amsterdam, 1988, pp. 101–137.

7. M.L. Occelli, P.A. Peaden, G.P. Ritz, P.S. Iyer, and M. Yokoyma. Macroporous Materials 1:99 (1993).

8. R.K. Sharma and E.S. Olson. Preprints, ACS Div. Fuel Chem. 39, 702: (1994).

9. J. Jaffe and J.R. Kittrell. U.S. Patent 3,535,229 (1970).

10. B.F. Mulaskey. U.S. Patent 3,682,811 (1972).

11. S.M. Csicsery and B.F. Mulaskey. U.S. Patent 3,655,798 (1972).

12. H.E. Swift and E.R. Black. Ind. Eng. Chem. Prod. Res. Dev. 13:106 (1974).

13. J. Gaaf, R. van Santen, A. Knoester, and B. van Wingerden. J. Chem. Soc. Chem. Commun. 655 (1983).

14. J.J.L. Heinerman, I.L.C. Freriks, J. Gaaf, G.T. Pott, and J.G.F. Coqlegem. Journal of Catalysis 80, 145 (1983).

15. E.S. Olson and R.K. Sharma. Preprints, ACS Div. Fuel Chem. 39:706 (1994).

16. R.K. Sharma, E.S. Olson, and J.W. Diehl. Preprints, ACS Div. Fuel Chem. 36:578 (1991).

Residual Oil Hydrotreating Catalyst Rejuvenation by Leaching of Foulkant Metals: Effect of Metal Leaching on Catalyst Characteristics and Performance

A. Stanislaus, M. Marafi, and M. Absi-Halabi

Petroleum Technology Department, Petroleum, Petrochemicals & Materials Division, Kuwait Institute for Scientific Research, P.O.Box 24885, 13109-Safat-Kuwait

I. ABSTRACT

Due to environmental concerns increasing emphasis has been placed in recent years on the development of processes for the rejuvenation of spent residual oil hydrotreating catalysts, which are deactivated by deposition of metals(e.g. vanadium) and coke. In the present work, the influence of foulant metal removal by chemical leaching on catalyst characteristics and performance was investigated as part of a research program on the rejuvenation of spent residual oil hydrotreating catalysts. The results revealed that the surface area and pore volume increased substantially with increasing vanadium extraction and the HDS activity showed a parallel increase. More than 85% of the catalyst initial HDS activity was recovered by removal of about 35% vanadium from the catalysts, although all carbon deposits remained on the catalyst. Extraction of more vanadium up to 90% by repeated extraction did not lead to further appreciable increase in surface area and activity. These results have been explained in terms of the location of vanadium as determined by electron microprobe analysis in the spent and treated catalysts and its role in catalyst deactivation.

II. INTRODUCTION

Catalysts used in the hydroprocessing of petroleum residues are deactivated by coke and metal deposits originating from the heavy feedstock (1 3). The foulant metals are usually concentrated near the outer surface of the pellet, blocking pore mouths and reducing markedly the active surface area available within the inner pores of the catalyst (4, 5). Regeneration of the deactivated residue hydrotreating catalysts is not an easy task (6, 7).

Regeneration by conventional procedures using nitrogen-air or steam-air under controlled conditions does not result in complete reactivation of the catalyst. While the carbon deposit is removed completely, the metallic impurities remain on the catalyst and act as an diffusion barrier for the reactants. If the contaminant metals can be removed selectively by chemical treatment without significantly affecting the chemical and physical characteristics of the original catalyst, then the catalyst may be rejuvenated and reactivated (6, 8).

Rejuvenation of spent heavy oil hydrotreating catalysts by leaching of foulant metals has been the subject of some investigation in this laboratory (9-12). In a previous study, we compared the effectiveness of leaching foulant metals from coked (sulfide) and decoked (oxide) forms of spent residue hydrotreating catalyst using oxalic acid and oxalic acid mixed with an oxidizing agent such as H_2O_2 (9). Oxalic acid alone showed very poor activity for leaching of metals from coked spent catalyst, but in presence of an oxidizing agent it was very effective. Metal leaching kinetics and the influence of inorganic salt additives on the leaching efficiency of different organic acids were also studied (10-12).

In this paper the effect of metal leaching on catalyst characteristics are reported and discussed. Since it is important that no major damage is done to the useful properties of the catalyst during the leaching process, information on the characteristics of the treated catalyst would be useful to assess the effectiveness of the leaching procedure. Samples of spent catalyst and chemically treated catalysts were characterized for surface area, porosity and chemical composition. The influence of chemical treatment on the distribution of the foulant metals (V and Ni) and the active metals (Co and Mo) within the catalyst pellet was examined by electron microprobe analysis. The samples were also tested for HDS activity recovery as a result of the rejuvenation process and the relationship between the amount of vanadium removed and the recovery of surface area, pore volume and HDS activity was examined.

III. EXPERIMENTAL

Spent catalyst was obtained from the atmospheric residue hydrodesulfurization unit of Kuwait National Petroleum Company (KNPC). For consistency, the spent catalyst for all the experimentation was collected from one batch. The catalyst was in the form of cylindrical extrudates with average diameter of 1 mm and average length of 4.3 mm. The catalyst contained residual oil, sulfur, carbon, vanadium and nickel deposits in addition to the catalyst metals (Co and Mo) originally present (Table 1). The contaminating residual oil was removed by thoroughly washing with naphtha in a mechanical shaker. The cleaned catalyst was then dried in an oven at 120°C for 24 hours.

Leaching experiments were conducted in a specially constructed system. The system consisted of a tubular stainless steel reactor (ϕ = 2 cm, L = 50 cm), variable flow rate diaphragm pump with polypropylene head, and a reservoir tank for the reagent. In a typical experiment, the reactor was packed with 100g of the catalyst in extrudate form in between layers of glass beads and glass wool. The reagent was pumped continuously through the catalyst bed from the bottom of the reactor (up-flow), at a flow rate of 3 lit./h. The liquid product was collected in the reagent vessel and circulated continuously through the catalyst bed for the requisite periods of time by the pump. All experiments were conducted at ambient temperatures (25°C ± 1°C). Samples from the reagent tank were taken for analysis at different time intervals. The concentration of various metals (V, Ni, Mo and Co) present in the leach liquid and catalyst samples were determined by inductively coupled Argon Plasma (ICAP) spectroscopy. The surface areas of the catalysts were determined by nitrogen adsorption (BET method) using a Quantasorb adsorption unit manufactured by Quantachrome Corporation, U.S.A. Pore volume was determined using water by the pore filling method. The distribution profile of the metals within the catalyst pellet was measured using a scanning electron x-ray microprobe analyzer (Camebax) equipped with an energy dispersive x-ray analyzer. X-ray diffraction patterns of the catalysts were obtained using a Philips PW 1410 x-ray spectrometer operated at 30 KV and 20 MA with Cu K$\propto$ radiation.

TABLE 1. Characteristics of Fresh and Spent Residue Hydrotreating Catalysts.

Catalyst Property	Fresh Catalyst	Spent Catalyst
Chemical Analysis		
Mo (wt %)	8.80	5.40
Ni (wt %)	0.00	3.07
V (wt %)	0.00	14.90
C (wt %)	0.00	15.60
S (wt %)	0.00	5.30
Co (wt %)	3.20	1.90
Fe (wt %)	0.00	0.80
Physical Properties		
Surface area (m^2/g)	240	52
Pore volume (ml/g)	0.48	0.12
Bulk density (g/ml)	0.73	1.18

Activity tests were conducted in a fixed bed microreactor using 5 ml catalyst charge. Atmospheric gas oil containing 2 wt.% sulfur was used as feed. The operating conditions were: pressure 40 bars; temperature 350°C; H_2/oil ratio 400 ml/ml; LHSV 6h^{-1}. The catalyst samples were presulfided in the reactor before use. The presulfiding procedure involved passing 3% wt CS_2 in atmospheric gas oil over the catalyst which was maintained at 300°C and 40 bars pressure. The LHSV was adjusted to 6h^{-1} and presulfiding was continued for 3h. A Princeton Gamma Tech. Model 100 sulfur analyzer was used to measure the sulfur content in the feed and product. The reproducibility of the leaching experiments was determined to be on the order of ±2% by repeating selected experiments under identical conditions.

IV. RESULTS AND DISCUSSIONS

The spent catalyst used in this study contained 14.9 wt.% vanadium and 15.6 wt.% carbon. The surface area and pore volume of the spent catalyst were, respectively, 78% and 75% lower than that of the fresh catalyst.

Leaching studies were conducted on coked spent catalyst using oxalic acid and oxalic acid+hydrogen peroxide mixed reagent as previously reported (9). Catalysts with varying degrees of vanadium removal were obtained by varying the leaching duration and reagent concentration. In some experiments repeated leaching was performed to extract up to 90% vanadium from the spent catalyst. Samples of spent and treated catalysts were characterized for surface area, porosity and HDS activity, and the relationship between the amount of vanadium removed and the recovery of surface area, pore volume and HDS activity was examined.

The surface area and pore volume of treated spent catalysts are plotted as a function of (%) vanadium extracted in Figure 1. It is noticed that both the surface area and pore volume increase substantially with increasing vanadium extraction. The HDS activity shows a parallel increase (Fig. 2).

An interesting point to note in Figure 1. is that even an initial extraction of 5-10% V from the spent catalyst leads to remarkable increase in surface area. The increase in surface area is particularly high up to about 30% vanadium removal. Extraction of more vanadium above this level (up to 90%) does not result in appreciable increase in surface area.

Activity data presented in Figure 2 shows that about 85% of the catalyst's initial HDS activity is regained by removal of about 30% vanadium from the spent catalyst, although all carbon deposits remained on the catalyst. No appreciable increase in HDS activity is noticed with further extraction of vanadium from the spent catalyst (Fig. 2). This trend is similar to that observed for the surface area improvement with vanadium removal (Fig. 1).

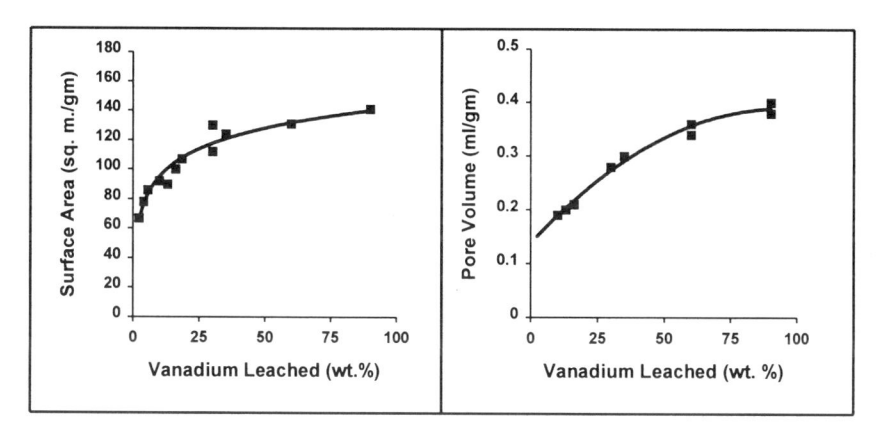

Residual Oil Hydrotreating Catalyst Rejuvenation

Figure 1. Relation between vanadium removal and surface area and pore volume.

Some explanation of these observations can be advanced in terms of the location of the vanadium in the spent catalyst. Electron microprobe analysis of the spent catalyst showed that high concentrations of vanadium were present near the edge of the catalyst pellet (Fig. 3). Such large accumulations may block the pore mouths and make the surface within the pores inaccessible to the reactants. In other words, the catalyst is deactivated by pore mouth plugging (4, 5).

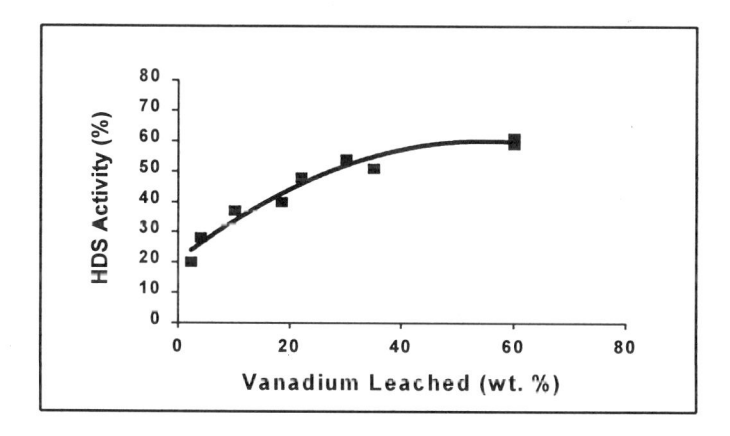

Figure 2. Effect of vanadium leaching on desulfurization activity

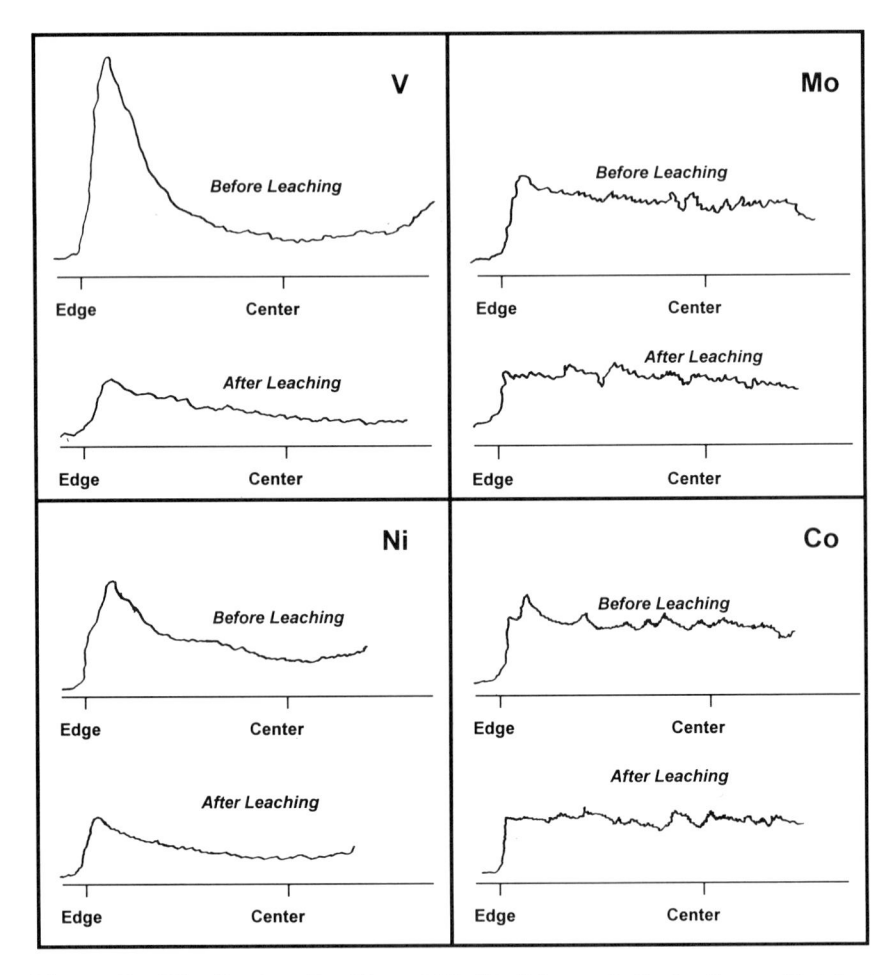

Figure 3. Distribution Profiles of V, Ni, Mo, and Co in Spent Catalyst before and after Leaching.

A simple model representing pore plugging by foulant metal deposition is shown in Figure 4. It is possible that during the catalyst life on-stream, the small pores (C) become plugged fast simply because of the small pore mouth diameter, and as the catalyst use continues progressively larger pores become plugged until the catalyst performance reaches the lowest permissible limit. Obviously at this stage some large pores (as represented by A) that lead to the pellet's interior are still accessible to the feed. This is evidenced by the 52 m^2/g surface area of the spent catalyst. If the catalyst pellet had been totally plugged the surface area would have been less than 1 m^2/g.

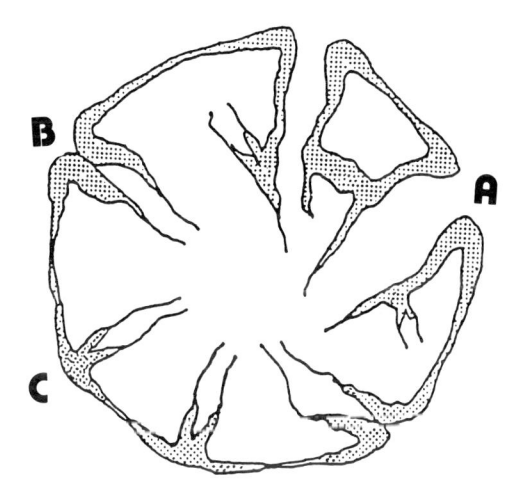

Figure 4. Representation of Pore Plugging by Foulant Metal Deposits. (A: large pores; B: medium-sized pores; C: micropores)

Leaching the foulant metal deposits from the pore mouths should open the pores and consequently increase the surface area of the catalyst. A real gain in surface area and activity will be achieved only when the amount of vanadium removed is sufficient to open the small and medium sized pores providing access to the active sites within the internal pores. It appears that less than 30% of the total vanadium deposited on the catalyst was responsible for pore mouth plugging and the obstruction of much of the catalyst surface area within the pores. It should be noted that the improvement in surface area and HDS activity of the catalyst are substantially high up to about 30% vanadium removal from the spent catalyst.

Figure 3 shows the effect of leaching on the distribution profiles of vanadium and other metals within a pellet of coked spent catalyst: the untreated spent catalyst contains large concentrations of vanadium near the edge, whereas the treated catalyst, contains a considerably decreased amount of vanadium near the edge. The results indicate that during the chemical treatment, the vanadium accumulated near the edge at the pore mouths is leached out gradually, thus opening up the pores. In agreement with this, the pore volume and catalyst surface area were found to increase with increasing vanadium removal. The unleached vanadium remaining in the treated catalyst is not concentrated at the pore mouths, but is distributed uniformly within the pellet.

Nickel has penetrated further than the vanadium into the interior of the pellet. Nickel extraction by chemical reagents changed the distribution pattern; the concentration of nickel both at the edge and on the interior surfaces are decreased by leaching. The unleached nickel appears to be well distributed within the catalyst pellet. The distribution profiles of catalyst metals (Co - Mo) in coked spent catalyst pellets before and after leaching are also included in Figure 3. In the coked samples, both cobalt and molybdenum concentrations are slightly higher near the edge than at the center, whereas after leaching, the metals appear to be redistributed.

XRD patterns of fresh, spent and treated catalysts were measured to determine whether any structural or phase changes had occurred in the catalyst and in the γ-alumina support as a result of the leaching and decoking (rejuvenation) treatments. The results showed identical XRD patterns for the fresh and rejuvenated catalysts. The γ- alumina phase structure of the support was retained in the spent and leached catalysts. In the unleached spent catalyst, a diffuse peak (d value = 2.59) showed that vanadium was present as V_2S_3 either as small crystallites or amorphous material. No peaks corresponding to nickel, molybdenum or cobalt sulfides or other phases were detected. It is likely that these metals were well dispersed in the form of small crystallites.

V. SUMMARY AND CONCLUSIONS

As part of an extensive research program on the rejuvenation of spent residual oil hydrotreating catalyst by leaching of foulant metals, we have investigated the effect of metal leaching on catalyst characteristics. Samples of spent and chemically treated catalysts were characterized for surface area, porosity and chemical composition. The samples were also tested for HDS activity recovery as a result of the rejuvenation process, and the relationship between the amount of vanadium removed and the recovery of surface area, pore volume and HDS activity was examined. The results of the studies revealed that the surface area and pore volume increased substantially with increasing vanadium extraction from the spent catalyst and the HDS activity showed a parallel increase. Improvements in surface area and HDS activity were substantial up to 30% vanadium removal but did not significantly improve for vanadium extraction above this level. More than 85% of the catalyst's initial HDS activity was recovered by removal of about 35% vanadium from the catalyst, although all carbon deposits remained on the catalyst.

These results have been explained in terms of the location of vanadium in the spent and treated catalysts and its role in catalyst deactivation. Electron microprobe analysis was used to determine the distribution profiles of vanadium and other metals (Mo, Co and Ni) within pellets of treated and untreated spent catalysts. The untreated spent catalyst contained large concentrations of vanadium near the outer edge, whereas the treated catalyst pellets contained substantially decreased amount of vanadium near the edge. Redistribution V, Co, Mo and Ni within the pellet as a result of leaching was observed. X-ray diffraction pattern of the rejuvenated catalyst was identical to that of fresh catalyst. No peaks corresponding to alumina phases other than γ-alumina was noticed indicating the absence of phase changes in the γ-alumina was noticed indicating the absence of phase changes in the γ-alumina support as a result of the leaching treatment.

ACKNOWLEDGMENT

Financial support provided by the Kuwait Foundation for Advancement of Sciences (KFAS) and the Kuwait Institute for Scientific Research is gratefully acknowledged.

REFERENCES

1. M. Absi Halabi, A. Stanislaus, and D. L. Trimm. Appl. Catal. 72, 193 (1991).

2. D. S. Thakur and M. G. Thomas. Appl. Catal. 15, 197 (1985).

3. P. W. Tamm, H. F. Harnsberger, and A. G. Bridge. Ind. Eng. Chem. Process Des. Dev. 20, 262 (1981).

4. R. J. Quan, R. A. Ware, C. W. Hung, and J. Wei. Advances in Chemical Engineering 14, 95 (1988).

5. S. T. Sie, in Catalyst Deactivation (B. Delmon and G. F. Froment, Editors), Elsevier Science Publishing, Amsterdam, The Netherlands, 1980, p. 545.

6. D. L. Trimm, in Catalysts in Petroleum Refining 1989, (D. L. Trimm, S. Akasha, M. Absi-Halabi and A. Bishara, Editors), Elsevier Science Publishing, Amsterdam, The Netherlands, 1990, pp. 41 - 60.

7. E. Furimsky and F. Massoth. Catalysis Today 17, 537 (1993).

8. M. Marafi. The Regeneration of Deactivated Hydrotreating Catalyst. M.Sc. Thesis. The University of Aston, Birmingham, 1988.

9. A. Stanislaus, M. Marafi, and M. Absi-Halabi. Appl. Catal. 105, 195 (1993).

10. M. Marafi, A. Stanislaus, C. J. Mumford, and M. A. Fahim, in <u>Catalysts in Petroleum Refining 1989</u>, (D. L. Trimm, S. Akasha, M. Absi-Halabi and A. Bishara, Editors), Elsevier Science Publishing, Amsterdam, The Netherlands, 1990, p. 213.

11. M. Marafi, A. Stanislaus, and M. Absi-Halabi. Appl. Catal. 47, 85 (1989).

12. Marafi, A. Stanislaus, and C. J. Mumford. Catalysis Letters 18, 141 (1993).

Characterization of Coke on Spent Resid Catalyst from Ebullating Bed Service and its Effect on Activity

Per Zeuthen[1], Barry H. Cooper[1], Fred Clark[2] and David Arters[2]

[1] Haldor Topsøe Research Laboratories, DK-2800 Lyngby, Denmark
[2] Amoco R&D, Naperville, Illinois

ABSTRACT

Two series of bimodal $CoMo/Al_2O_3$ catalyst samples have been characterized using temperature-programmed oxidation, NMR spectroscopy and nuclear microprobe analysis. The samples are withdrawn from the first- and third-stage reactors of an ebullating-bed resid hydrotreating pilot unit after varying times on stream of up to 120 days. The results show that coke and metals are deposited rather rapidly. The coke deposited on the third-stage catalyst is shown to be more aromatic than the coke on first-stage samples. Temperature-programmed oxidation studies show that vanadium catalyzes the oxidation of coke and that the coke contains significant amounts of sulfur and nitrogen. With increasing time on stream, the coke in both the first- and third-stage samples becomes more hydrogen and sulfur deficient as evidenced by increasing concentration of aromatic carbon (from NMR), falling H/C ratio (from nuclear microprobe) and a lowering in concentration of S associated with the coke (from TPO). There is evidence to the effect that the sulfur in the coke is associated with the upper layers of coke and that nitrogen may adsorb preferentially during the initial coke laydown. Model compound activity measurements carried out on the samples showed that the HDS activity is less sensitive to the effects of coke deposition than the HDN and hydrogenation activities. It was also shown that the initial metal deposits have a stronger effect on the loss of activity for HDS than for HDN and hydrogenation, indicating that different sizes are involved in these hydrotreating reactions.

INTRODUCTION

Catalysts for resid upgrading operate under severe process conditions which require the catalyst to be active for heteroatom removal while undergoing deposition of relatively large amounts of carbon and feed metals. Deactivation of a resid catalyst is a complex process which has been studied by Fleisch et al. and others (1-7).

This paper describes the nature of carbon and feed metals deposition on aged resid catalyst. To this end, a series of aged resid hydrotreating catalysts from a three stage expanded bed pilot plant has been studied using nuclear microprobe temperature programmed oxidation, solid state NMR spectroscopy and elemental analyses. These techniques combined with model compound activity studies have been used to obtain more insight into the properties of the deposits on the working catalyst as well as to characterize the relationship between activity functions and the carbonaceous and metal deposits.

EXPERIMENTAL

The series of aged Co/Mo-promoted alumina resid hydrotreating catalysts was obtained from a 100 kg/day three stage expanded bed pilot plant. The catalysts consisted of 4.5% MoO_3 and 0.7% CoO supported on a high surface area alumina. On-stream catalyst withdrawals from each reactor were performed during the first 21 days of the run and at the end of the run on day 120. Initially, the three stage pilot plant was charged with fresh catalyst in each reactor.

The details of the used techniques have been described elsewhere (8-11).

RESULTS AND DISCUSSION

Chemical analyses of the aged catalyst samples have been carried out and are shown in a previous paper (11). Apart from the day-120 sample, the catalysts from the third reactor only contained small amounts of deposited Ni and V. The coke levels and coke aromaticities (by ^{13}C NMR) were generally higher for the third reactor samples.

The carbon level increased logarithmically with catalyst age for third stage samples but increased only slowly after day 1 for the first stage samples.

Characterization of Coke on Spent Resid Catalyst

Presumably, this is associated with the higher level of metal deposits in the first reactor samples.

Another result of interest concerns changes in nitrogen and hydrogen content of the coke as a function of reactor stage and catalyst age. Initially, the N/C ratio of both first stage and third stage catalysts is high compared with that of the feed, indicating that nitrogen is strongly adsorbed on acid sites or on the active phases on the catalyst surface. As the catalyst ages, differences are noted, however. In the first stage, the N/C ratio stays reasonably constant, whereas in the third stage it is reduced to a level typically found for product asphaltenes (12).

On the other hand, the changes in hydrogen content of the coke as a function of time indicate the presence of a coke "hardening" mechanism in both first and third stage samples. A high initial H/C ratio in the coke drops to a level that is lower than that of typical H/C ratios found for product asphaltenes (12).

Nuclear microprobe analyses (NMA) have been used to measure light elements and deposited metal distributions in the used catalysts. Hydrogen and carbon are determined by nuclear reaction analysis, whereas vanadium is measured by particle induced X-ray emission (PIXE).

Figures 1a and b show the carbon distribution across pellets of different ages for reactor 1 and reactor 3, respectively.

The distributions for samples from reactor 3 indicate that the carbon is present in two concentrations, a high concentration peripheral zone and a lower concentration central zone. For reactor 1 samples, the transition is much smoother.

The hydrogen profiles have the same shape as the carbon profiles. Figure 2a and b show the H/C ratio, which clearly shows that the latter is reduced with increased time on stream, indicating graphitization.

^{13}C NMR spectra show that spent catalysts from the third stage have consistently higher concentrations of aromatic carbon as compared with spent catalysts from the first stage after an equivalent time on stream (11). It is also shown that the aromaticity of the coke on both first and third stage samples increases with time on steam. This is in good agreement with the NMA data.

Reactor 1

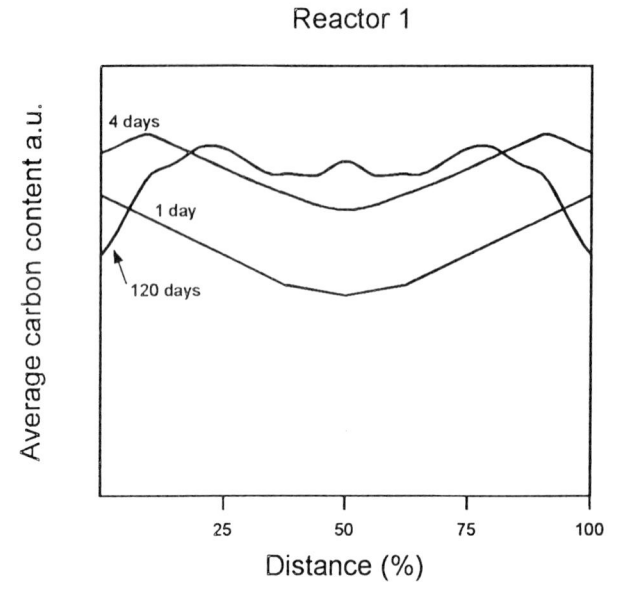

Figure 1a   Average normalized carbon distribution across the catalyst pellet sampled after increasing time on stream. Reactor 1.

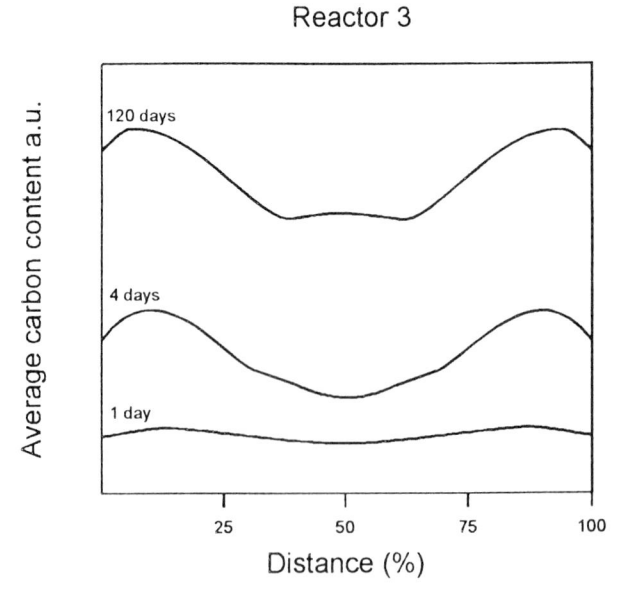

Reactor 3

Figure 1b   Average normalized carbon distribution across the catalyst pellet sampled after increasing time on stream. Reactor 3.

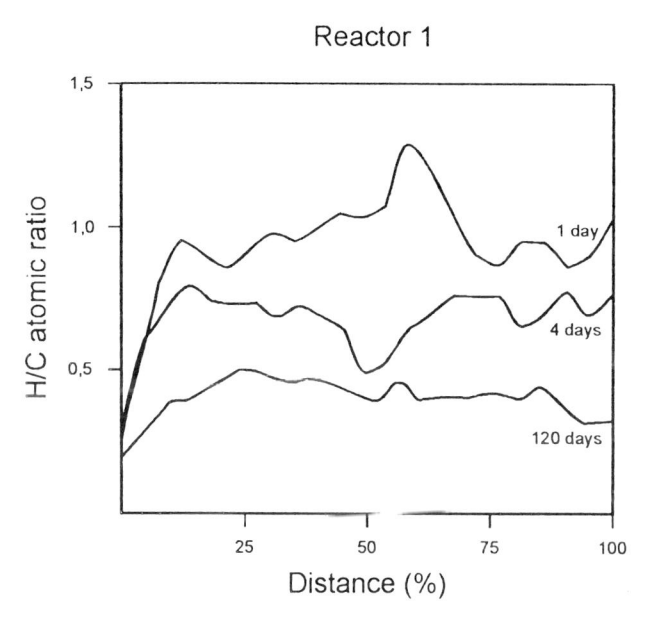

Characterization of Coke on Spent Resid Catalyst 341

Reactor 1

Figure 2a   Variations in H/C atomic ratio across the catalyst pellet exposed to the feed for different periods of time. Reactor 1.

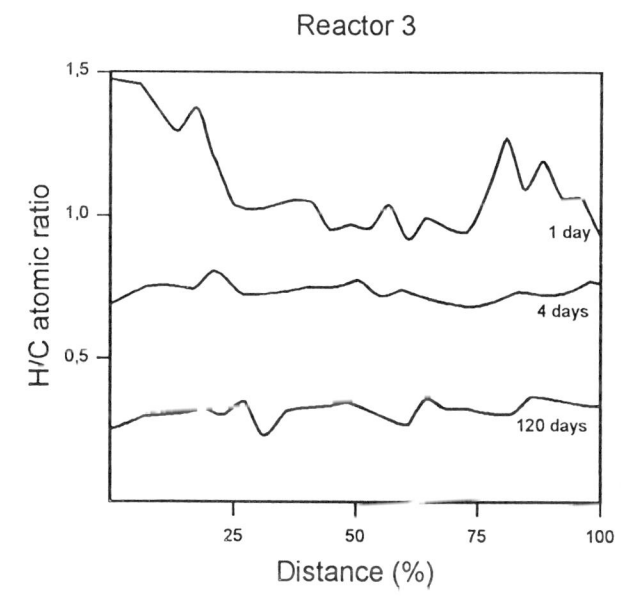

Reactor 3

Figure 2b   Variations in H/C atomic ratio across the catalyst pellet exposed to the feed for different periods of time. Reactor 3.

Temperature-programmed oxidation studies were carried out in order to characterize the deposits on the spent catalysts. Oxidation of carbon gave CO_2 and CO, oxidation of nitrogen and sulfur gave NO and SO_2, respectively. NO_2 was not formed.

The oxidation of carbon on the samples typically gave a single broad peak for both CO and CO_2. The position of the peak maximum for the first reactor samples varied from 380°C to 460°C, with lower temperatures being obtained, the older the sample. For samples from the third reactor, the position of the peak maximum occurred at about 510-520°C except for the day-120 sample, where the peak maximum was 410°C. See also (11).

The explanation of this seems to be that vanadium catalyzes the oxidation of the coke, and in fact a good correlation exists between the temperature of coke burn-off and the amount of vanadium on the catalyst. These results are consistent with those of Massoth (13).

TPO of the sulfided fresh catalyst shows that the oxidation of Co/Mo sulfides to SO_2 occurs between 200 and 250°C. The spent catalysts, however, also exhibit a high-temperature SO_2 peak occurring between 400 and 500°C, which is not present in the fresh catalyst. The position of the peak coincides with the front edge of the corresponding CO_2 peak. The peak heights are larger for reactor 1 samples than reactor 3 samples, but the differences are less pronounced than for the low-temperature peak.

Both these facts indicate that the high-temperature SO_2 peak in samples containing less than 1 wt% Ni + V is due to oxidation of organic sulfur associated with the coke rather than inorganic sulfur bound to deposited metals. The organic sulfur is thought to be only associated with the upper layer(s) of coke, since the evolution of SO_2 occurs at the same temperature as the initial oxidation of carbon.

Figure 3a and b show the activities of the samples compared with those of the fresh catalyst as a function of age. In these three-dimensional plots, the relative activities (HDS and HDN) are shown as a function of the carbon content (normalized to day-21 samples) and the vanadium content (logarithmic scale). The figures also show the activities of decoked (regenerated) first stage samples. Because of the somewhat higher activities of these samples, the activity tests were carried out at 350°C instead of at 390°C.

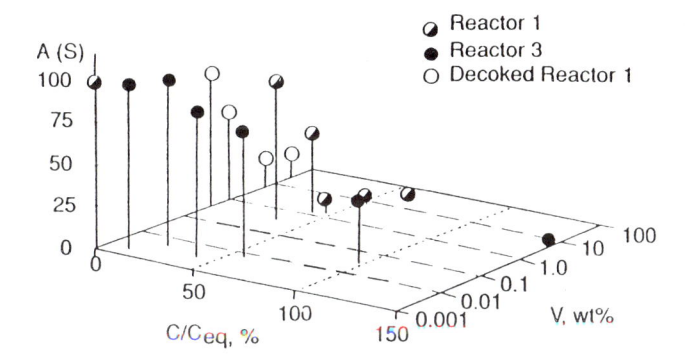

Characterization of Coke on Spent Resid Catalyst

Figure 3a   Relative fresh catalyst activities with carbon content as normalized to 21 days of exposure and vanadium deposits (logarithmically). Sulfur removal activity.

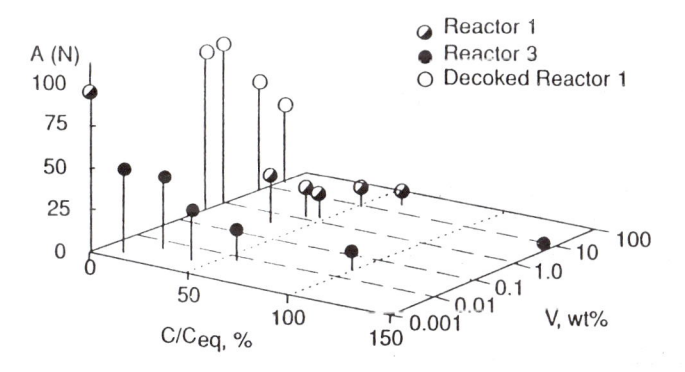

Figure 3b   Relative fresh catalyst activities with carbon content as normalized to 21 days of exposure and vanadium deposits (logarithmically). Nitrogen removal activity.

The deactivating effect of coke is not the same for both reactions. This is seen most clearly for the third reactor samples, days 0-8, which contain very small amounts of deposited metals. The HDN function decreases rapidly as the coke level increases. The curves can be fitted by an expression of the type: activity = $f(\exp(-\%C))$, which suggests a fouling-type deactivation mechanism by the coke. The HDS function deactivates quite differently with only slight (approx. 20%) deactivation up to a coke level of 55 wt% followed by an extremely sharp decline in activity as the coke level increases. The HDS activity fits an exp(-age) type function reasonably well.

The difference in activity decline for HDS and for HDN indicates that there are two different sites for these reactions as also discussed by previous authors (see e.g. Muralidhar et al. (14)).
For the reactor 1 samples, the situation is more complex. The carbon level increases only slightly after day 3, and the spent catalyst samples contain much larger amounts of deposited metals. The deactivation is more rapid for all functionalities.

REFERENCES
1. T.H. Fleisch, B.L. Meyser, J.B. Hall, and G.L. Ott, J. Catal. 86 (1): 147 (1984).
2. S.J. Khang, and J.F. Mosby, Ind. Eng. Chem. Proc. Des. Dev. 25: 437 (1986).
3. E.E. Wolf, and F. Alfani, Catal. Rev.-Sci. Eng. 24 (3): 329 (1982).
4. P.N. Hannerup, and A.C. Jacobsen, ACS Preprints, "Div. Petrol. Chem.", 28: 576 (1983).
5. D.S. Thakur, and M.G. Thomas, Appl. Catal. 15: 197 (1985).
6. F. Diaz, B.C. Gates, J.T. Miller, D.J. Sajkowski, and S.G. Kukes, Ind. Eng. Chem. Res. 29: 1999 (1990).
7. J. Bartholdy, and B.H. Cooper, ACS Preprints, Div. Petrol. Chem., 38: 386 (1993).
8. T.E. Myers, F.S. Lee, B.L. Meyers, T. Fleisch, and G.W. Zajac, in Fundamentals of Resid Upgrading, (R.H. Heck and T.F. Degnan, Eds), AIChE Symposium Series No. 273, 1989, p. 21-31.
9. P. Zeuthen, P. Stoltze, and U.B. Pedersen, Bull. Soc. Chim. Belg. 96: 985 (1987).
10. P.R. Fozard, J.W. McMillan, and P. Zeuthen, accepted by J. Catal.
11. P. Zeuthen, B.H. Cooper, F. Clark, and D. Arters, accepted for publication in Industrial & Engineering Chemistry Research.
12. Amoco unpublished data.
13. F.E. Massoth, Fuel Processing Technology 4: 63 (1981).
14. G. Muralidhar, H.T. Chung, F.E. Massoth, Ind. Eng. Chem. Res. 30: 29 (1991).

Microbial Removal of Organic Sulfur from Fuels: A Review of Past and Present Approaches

Matthew J. Grossman
Exxon Research and Engineering Co.
Clinton, New Jersey 08801

I. INTRODUCTION
A. Sulfur in Petroleum
Crude oils contain significant quantities of sulfur, largely in the form of organic sulfur compounds. Sulfur in fuels causes corrosion of transportation and processing equipment, fouling of processing catalysts, and upon combustion, air pollution, in the form of particulate and acidic gas emissions (1). The concentration of sulfur in crude oil is typically between 0.05 and 5.0 wt. %, however, values as high as 13.95 wt. % have been reported (2). These values are impressive, considering that a crude oil with specific gravity of 0.9, and a sulfur content of 1.25 wt. %, contains 2 tons of elemental sulfur per thousand barrels of oil (2). In general, the distribution of sulfur in crude oil is such that the proportion of sulfur increases along with the boiling point of the distillate fraction. Organic sulfur compounds in the lower-boiling fractions of petroleum, e.g. the gasoline range, are mainly thiols, sulfides, and thiophenes. Higher-boiling fractions, e.g. the diesel and fuel oil range, contain thiols and sulfides as well, and in addition, significant amounts of benzothiophenes and dibenzothiophenes (DBTs).

Hydrodesulfurization (HDS) is the current method used by the refining industry to remove sulfur. HDS involves the catalytic reaction of hydrogen and feed, at pressures between 150 and 3000 psi, and temperatures between 290 and 455°C, depending upon the feed and level of desulfurization required (1). Thiols, sulfides, and thiophenes, are readily removed by HDS. However, benzothiophenes and DBTs are considerably more difficult to remove by HDS. Further, DBTs bearing alkyl substitutions adjacent to the sulfur atom (referred to as sterically hindered compounds), are the most resistant to HDS, and represent a significant barrier to reaching very low sulfur levels in fuels (3). Due to the high cost and inherent chemical limitations associated with HDS, alternatives to this technology are of interest to the petroleum industry. Current trends toward stricter regulations on the content of sulfur in fuels, is an impetus for a continued search for improved desulfurization processes.

B. Sulfur in Biology

Sulfur is essential for all living things, playing a key role in a variety of cellular activities. For example, as a component of the amino acids methionine and cysteine, and a variety of coenzymes, sulfur is involved in determining the structure and function of proteins. In addition, for some organisms, sulfur compounds act as electron donors and acceptors in energy generating reactions. Microbiologists have long sought to exploit the enzymatic processes involved in sulfur metabolism in search of economically attractive systems that can be applied to petroleum desulfurization. This is evidenced by Maliyantz, who in the 1930's described a reductive/anaerobic system, employing *Spirilum aestuarii*, whereby organic sulfur was converted to H_2S, and an oxidative/aerobic system, employing "thioacid bacteria", whereby organic sulfur was converted to sulfuric acid (4,5).

Clearly, oxidative and reductive systems yield quite different end products. The distinction between oxidative and reductive systems is a useful way to divide the world of microbial desulfurization, as it reflects a basic difference in the underlying chemistry. In this paper, I will describe the key developments in reductive and oxidative microbial desulfurization research. Dibenzothiophene (DBT) has frequently been used as model compound in desulfurization research, because it represents those compounds that are abundant in the higher boiling fractions of crude oils, and the most difficult to remove by HDS. I have used DBT as a model compound throughout this paper when presenting examples of different microbial desulfurization systems.

II. REDUCTIVE DESULFURIZATION

A. Overview of Reductive Systems

Reductive microbial desulfurization is perhaps the most attractive conceptually, because the sulfur end product, H_2S, is easily removed from the reaction mixture and readily recovered by existing refinery operations, which handle H_2S produced via HDS. In addition, the absence of oxygen prevents the nonspecific oxidation of hydrocarbons to colored, acidic and/or gum forming products. One approach, which has generated a considerable amount of interest, uses sulfate reducing bacteria (SRB) for the reductive desulfurization of organic sulfur compounds. These organisms have the unique ability to obtain energy by reducing sulfate to H_2S, which is then excreted (dissimilatory sulfur reduction). A general pathway for dissimilatory reductive microbial desulfurization is

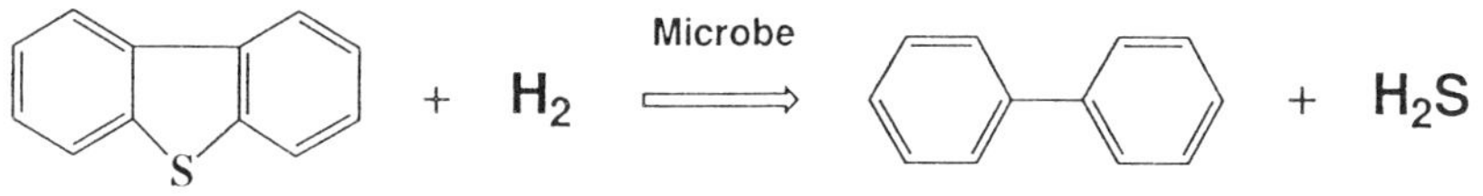

$$\text{dibenzothiophene} + H_2 \xrightarrow{\text{Microbe}} \text{biphenyl} + H_2S$$

Figure 1. Dissimilatory reductive microbial desulfurization. Reductants other than H_2 can be used, e.g. lactate, or possibly electrons directly.

shown in figure 1. A biologically more common activity is the reduction of sulfate to H_2S for use in the biosynthesis of organic sulfur compounds (assimilative sulfur reduction). Assimilative sulfur reduction has not been identified as a candidate for the purpose of fuel desulfurization, most likely because the organic sulfur compounds in fuels are not substrates for the enzyme systems involved. In fact, substrate range is a central issue for all microbial desulfurization systems. This is likely the result of requiring the enzymes involved to recognize and act on compounds that are not their physiological substrates. Taking this into consideration, it is not surprising that all of the known microbial desulfurization systems have some limitation on the range of organic sulfur compounds they will act on.

B. Dissimilatory Sulfur Reduction

ZoBell was among the first to describe a process for removing organic sulfur from petroleum by dissimilatory sulfur reduction (6). In the described process, petroleum was reacted with hydrogen, in the presence of one or more microorganisms or isolated enzymes capable of activating hydrogen, such that sulfur reduction and removal occurred. The principle enzyme required for hydrogen activation was claimed to be hydrogenase, which allows sulfur reducing bacteria to use hydrogen as a source of electrons (Fig. 2). The SRB, *Desulfovibrio desulfuricans*, was suggested as a preferred organism, because of its noted hydrogenase activity. Unfortunately, supporting evidence for the efficacy of this process was lacking. However, the correlation of hydrogenase activity with desulfurization activity was later claimed by others.

The ability of SRB to remove sulfur from organic sulfur compounds was demonstrated by Kohler et al. (7). Working with *Desulfovibrio*, and hydrogen

$$H_2 \xrightleftharpoons{\text{HYDROGENASE}} 2H^+ + 2e^-$$

Figure 2. Hydrogenase.

as the reductant, they demonstrated H$_2$S production from the model compounds dibenzylsulfide (DBS) and DBT. The products of DBS degradation were toluene, benzylmercaptan and H$_2$S. In contrast, dibenzyldisulfide (DBDS), butylsulfide, octylsulfide and thianaphthene were not appreciably attacked. Corroborating ZoBell's suggestion, Kohler et al. also correlated desulfurization activity with hydrogenase activity. In later work, Miller obtained different results, employing SRB mixed cultures and isolated SRB, obtained from pond sediment, swine waste holding pits, petroleum waste sites and methanogenic sewage digesters (8). Miller used lactate instead of hydrogen as an electron donor, and found that DBDS was attacked with the formation of toluene and H$_2$S. In addition, she observed the production of H$_2$S from coal upon treatment with SRB. Interestingly, unlike the observations of Kohler, DBT, DBT sulfone and phenyl sulfone were not attacked. Miller speculated that the lack of activity toward DBT was a result of the very low energy yield obtained from its reductive desulfurization. Lizima et al., demonstrated the ability of three different SRB to grow on DBT as their sole sulfur source and sole electron acceptor (9). This group used hydrogen, lactate and butyric acid as electron donors and observed the release of H$_2$S as the product of dissimilatory sulfur reduction. They suggested that the reduction of DBT to H$_2$S may be a trait common to all SRB.

C. Novel Approaches to Reductive Desulfurization

Recently two novel approaches to reductive sulfur removal have been reported. In place of hydrogen or reduced organic compounds, Kim et al. have developed a bioelectrochemical process to deliver electrons, via an electrochemical cell, to a highly active SRB strain, designated *Desulfovibrio desulfuricans* M6 (10,11). The reaction was carried out with the cathode in contact with a mixture of the sulfur containing substrate, the microorganism, and methyl viologen as an electron mediator. Studies with DBT demonstrated the production of biphenyl and H$_2$S. Further, experiments run with crude oil and diesel oil resulted in 20% reduction in total sulfur. As mentioned above, virtually all the work done on reductive desulfurization has used SRB. Finnerty has broken this trend, reporting on an entirely novel approach using a Gram positive aerobic bacteria, designated FE-9 (12). Finnerty used a non-aqueous system, contacting FE-9 with DBT in 100% dimethylformamide under a hydrogen atmosphere, and reported the conversion of DBT to biphenyl and H$_2$S. It remains to be seen whether commercial applications of these systems can be achieved.

III. OXIDATIVE DESULFURIZATION
A. Overview of Oxidative Systems

Analogous to reductive microbial desulfurization, oxidative microbial desulfurization is conveniently divided into dissimilatory and assimilatory systems. The sulfur end products from these systems are oxidized compounds, e.g. sulfate, which are typically water soluble. Dissimilatory sulfur oxidation

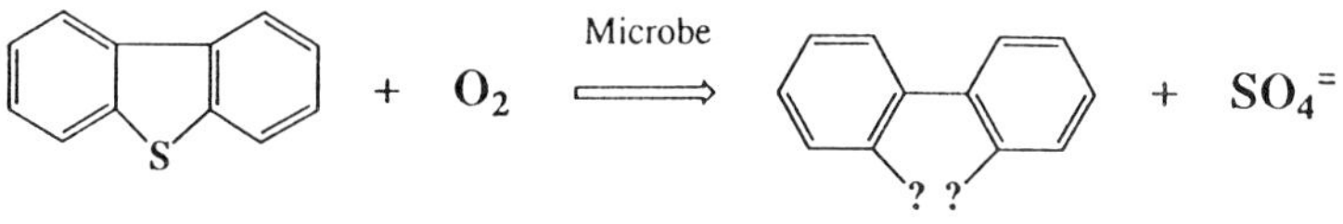

Microbial Removal of Organic Sulfur from Fuels 349

$$\text{DBT} + O_2 \xrightarrow{\text{Microbe}} \text{biphenyl}(??) + SO_4^=$$

Figure 3. Dissimilatory oxidative microbial desulfurization. Organic sulfur compounds used as electron donors in energy yielding reactions.

occurs when reduced sulfur compounds are used as electron donors in energy yielding reactions. Assimilatory sulfur oxidation is used to obtain sulfur in a form available for the biosynthesis of organic sulfur compounds. These systems involve the direct (or specific) oxidation of the sulfur moiety in the compound being oxidized. In contrast to sulfur specific oxidation systems, are systems which use organic sulfur compounds as the source of carbon and energy (and in some cases sulfur), via the oxidative mineralization of the hydrocarbon component of the molecule. Because the sulfur is not oxidized, or only oxidized along with hydrocarbon oxidation, these systems will be referred to as non-specific oxidative microbial desulfurization systems. The result of this metabolism is the production of biomass and partially oxidized organic products, and/or the complete mineralization of the compound to CO_2 and $SO_4^=$. In either case the products are typically water soluble.

B. Dissimilatory Oxidative Desulfurization

Sulfur oxidizing bacteria are frequently associated with acid mine drainage sites (e.g. *Thiobacillus*), or hot springs (e.g. *Sulfolobus*). A possible pathway for the desulfurization of DBT by sulfur oxidizing bacteria is shown in figure 3. Kirshenbam was one of the first to describe a process using this approach (13). In this process *Thiobacillus* was the recommended organism and rapid sulfur removal from petroleum was claimed; no analytical data or supporting evidence to substantiate these claims was presented. Kargi and Robinson subsequently reported on the ability of the thermophilic organism *Sulfolobus* to use DBT and coal as the sole source of sulfur and carbon (14,15). Desulfurization of both substrates was claimed on the basis of an increase in $SO_4=$ concentration; no additional analysis was performed on the fate of the organic sulfur compounds. Although intriguing, further work is required to determine if these systems can in fact remove organic sulfur from compounds found in petroleum.

C. Non-specific Oxidative Desulfurization

Non-specific oxidative microbial desulfurization systems were extensively studied prior to the discovery of organisms which use assimilatory sulfur oxidation to obtain sulfur from DBT. Figure 4 illustrates the general pathway for non-specific oxidative microbial desulfurization. Contemporary with

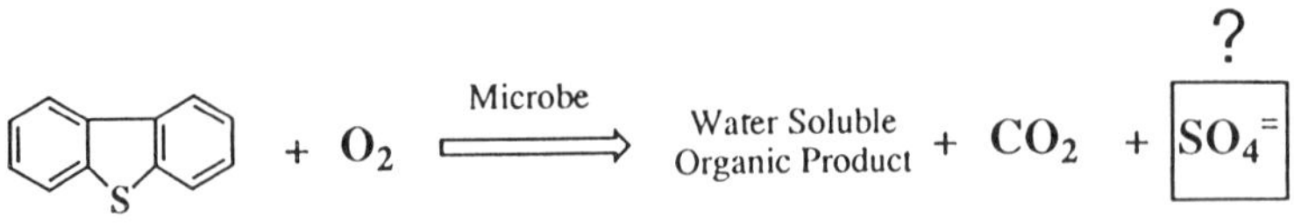

$$\text{DBT} + O_2 \xrightarrow{\text{Microbe}} \text{Water Soluble Organic Product} + CO_2 + \boxed{SO_4^{=}} \; ?$$

Figure 4. Non-specific oxidative microbial desulfurization. Organic sulfur compounds used as a carbon, energy and possibly sulfur source.

ZoBell's work on reductive microbial desulfurization, Raymond Strawinski described processes based on oxidative microbial desulfurization (16,17). Strawinski recognized the propensity of microorganisms which degraded organic sulfur compounds to also indiscriminately degrade hydrocarbons. He sought to overcome this problem by providing petroleum as the sole source of sulfur in the presence of "diverters", such as sugars, that would serve as a more attractive food source than the petroleum. Using a *Pseudomonas* strain designated 31, obtained from enrichment cultures using crude oil as the sole carbon and sulfur source, and 2 wt. % glucose as "diverter", Strawinski demonstrated a 12.5% reduction in the sulfur content of an Arabian crude. The effect on the hydrocarbon content of the crude was not reported. Although the use of "diverters" may not have achieved the desired goal, the concomitant degradation of hydrocarbons with sulfur compounds was clearly recognized as a significant problem. The reason for why this occurred would become evident, when the pathways involved in the mineralization of organic sulfur compound degradation were elucidated.

Kodama et al. performed the seminal work elucidating the pathway of DBT degradation (18,19). Their research focused on the metabolism of DBT by *Pseudomonas*, a bacterial genus whose members are well known for their ability to degrade a wide variety of hydrocarbons and related compounds. The ring cleavage products, trans-4[2-(3-hydroxy)-thianaphthenyl]]-2-oxo-3-butenoic acid (a highly colored red substance) and 3-hydroxy-2-formyl-benzothiophene, both with unmodified sulfur atoms, were identified as intermediates in the degradation of DBT. Kodama et al. recognized that these compounds were analogous to intermediates in the pathway of naphthalene degradation by other *Pseudomonas* strains (20). Based on these findings, a similar pathway was proposed for DBT degradation, which terminated with 3-hydroxy-2-formyl-benzothiophene. An additional compound, DBT sulfoxide, was also identified as a dead end product.

Following the work of Kodama et al., the pathway for DBT degradation was pieced together by the combined effort of a number of research groups working with variety of different organisms (21,22,23). In most cases studied, the final product of DBT degradation was 3-hydroxy-2-formyl-benzothiophene. In contrast, Malik and Clause, isolated *Rhizobium* sp. and *Acinetobacter* sp.,

Microbial Removal of Organic Sulfur from Fuels

able to utilized DBT as a sole carbon, energy and sulfur source for growth, indicating that the molecule was degraded sufficiently to release sulfur from the thiophene ring (**21**,24). Monticello et al., and Yen and Gunsalus studied the genetics of DBT degradation by *Pseudomonas* (25,26,27). They demonstrated that disruption of naphthalene degradation genes, by Tn5-insertions, eliminated both the ability to grow on naphthalene and the ability to oxidize DBT. This provided genetic evidence that DBT and naphthalene are metabolized by enzymes encoded by the same genes, and hence by a common pathway (Fig. 5).

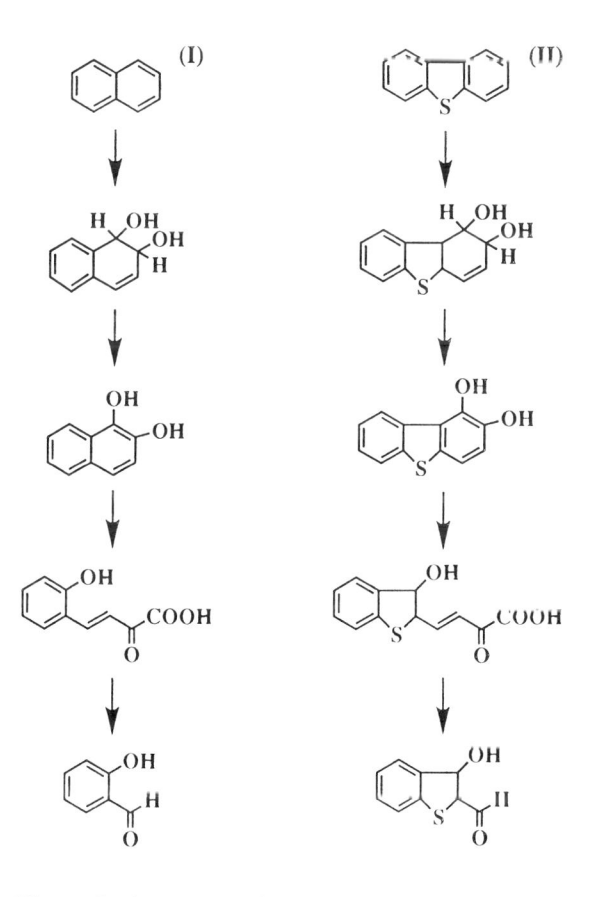

Figure 5. Common pathway of naphthalene (I) and DBT (II) oxidative degradation. (Adapted from Refs. 19, 20 and 24.)

Overall, the analysis of the DBT degradation pathway, demonstrated that DBT was oxidized via attack on the hydrocarbon component of the molecule. Therefore, desulfurization using this approach, would result in the production of water soluble sulfur containing compounds which could be separated form the oil. Unfortunately, due to the lack of specificity for sulfur containing compounds, hydrocarbons of similar structure would also be degraded, resulting in a significant loss of fuel. Finnerty, attempted to overcome this draw back by isolating Gram negative bacteria (designated DBT-2,3 and 4) able to oxidize DBT, but unable to oxidize paraffinic or aromatic hydrocarbons (28). The oxidation products were water soluble and highly colored, indicating that the compounds were degraded via a sulfur non-specific pathway, similar to that described for naphthalene. Even though only organic sulfur compounds were attacked, the removal of the hydrocarbon associated with these compounds still represented a significant fuel loss. Using DBT as an example, for every gram of sulfur removed almost 5 grams of hydrocarbon goes with it.

Although significant progress was made in the understanding of oxidative DBT degradation, and selectivity toward organic sulfur compounds appeared attainable, no commercial process has been developed based on this technology. Hartdegen et al., evaluated the economics of a desulfurization process based on the use of the DBT degrading organisms reported by Finnerty (29). They assumed the use of whole cells, in a packed bed, bounded by a permeable membrane and sandwiched between an oil stream and a aqueous stream. This design was assumed to lower processing costs, due to the elimination of downstream steps otherwise required to separate the oil from the sulfur containing water. In addition, the authors assumed that the oxidized products could be economically converted to marketable benzene and phenol, a process, as pointed out by Foght et al. (30), that has not been demonstrated. Nonetheless, it was concluded, that the process was uneconomical in comparison to the cost of HDS or the direct purchase of low sulfur crude. The cost of cell (biocatalyst) production, reactor residence time and biocatalyst lifetime, were identified as key process parameters limiting the economics of this system. Bharda et al., came to similar conclusions, pointing out the need for improved microbial strains, having increased efficiency, substrate specificity, and lower nutritional requirements, coupled with the use of energy efficient bio-reactors utilizing water recycle (31). Assuming the above improvements were met, it was noted that the price differential between high-sulfur and low sulfur crude oil would be the most important economic variable.

D. Sulfur Specific Oxidative Desulfurization

The poor economics associated with the sulfur non-specific oxidative systems effectively eliminated them from further consideration. This encouraged the search for novel systems, which could remove sulfur form organic sulfur containing compounds without concomitant hydrocarbon degradation.

In 1961, Knecht reported on an *Arthrobacter* isolate which oxidatively metabolized DBT, as a sole carbon, energy and sulfur source (32). Based upon the substrate range of the organism, which seemed to correlate with how readily the sulfur moiety was oxidized, he suspected the sulfur atom was the initial site of attack. If true, this was likely the first DBT oxidation pathway identified which involved sulfur specific oxidation. Aromatic hydrocarbons, including naphthalene, where not substrates for this isolate, and growth on DBT occurred only in the presence of a co-isolated *Pseudomonas*. This suggested the possibility that a symbiotic arrangement existed in which the *Arthrobacter* provided sulfur from DBT and the *Pseudomonas* provided carbon from DBT.

Isbister and Kobylinski, were the first to provide analytical evidence of an oxidative sulfur specific microbial desulfurization pathway (33). While searching for microorganisms which could remove sulfur from coal, they isolated an organism, which, after mutagen treatment, possessed a high activity toward the oxidation of DBT when grown on benzoate as the carbon source (34). The organism was identified as a member of the Gram negative genus *Pseudomonas* and was designated CB1. CB1 demonstrated significant desulfurization activity toward organic sulfur in coal, reducing up to 47% of the organic sulfur in a < 100 mesh size Homer City coal sample. A unique characteristic of CB1, was its use of a sulfur specific desulfurization mechanism resulting in the production of 2,2'-dihydroxybiphenyl and $SO_4^=$ from DBT. Thus, CB1 possessed the highly desirable trait of organic sulfur removal without concomitant hydrocarbon degradation. CB1 was apparently lost, but, the net result was the birth of a new concept, the "4S" pathway (35). The "4S" pathway drives its name from the inferred oxidation sequence for DBT, involving the sulfoxide, sulfone, sulfonate, and the final sulfur product, $SO_4^=$, hence the name "4S".

After the loss of CB1, a second search for organisms possessing the "4S" pathway was launched, via another round of funding by the US DOE. Working under this funding, Kilbane et. al., reported the isolation of a *Rhodococcus rhodochrous* strain, designated IGTS8, which possessed a stable sulfur specific oxidative desulfurization pathway (36). IGTS8 produced 2-hydroxybiphenyl and $SO_4^=$ from DBT, in contrast to the 2,2'-dihydroxybiphenyl and SO_4^- reported for CB1, suggesting a modified 4S pathway was in operation. DBT sulfoxide and sulfone were not detected as products of DBT oxidation by IGTS8, although they did serve as sulfur sources. The desulfurization activity was found to be shut down by sulfate, and rich bacterial growth medium, indicating sulfur starvation was necessary for its expression. A variety of organic sulfur compounds, in addition to DBT, were shown to be substrates for desulfurization by IGTS8, including thianthrene, phenyl sulfoxide, trithiane, and benzyldisulfide. In contrast, IGTS8 was unable to metabolize thianaphthene, phenyl disulfide and 2-thiophene-carboxylic acid (37), demonstrating a similar substrate profile to the *Arthrobacter* isolate previously described by Knecht. In 1991, IGT granted exclusive license to the patents covering IGTS8 to Energy Biosystems Corp. (EBC). EBC is currently mounting an aggressive program to develop a commercial microbial desulfurization process based on this technology.

Following the isolation of IGTS8, Gallagher et al. performed additional studies on the sulfur specific nature of the its DBT desulfurization pathway (38). Two pathways were proposed based on the metabolites observed and the physiological state of the organism. During growth the pathway followed the sequence: DBT sulfoxide, sulfone, 2'-hydroxybiphenyl-2-sulfonate, and 2,2'-dihydroxybiphenyl. In contrast, when DBT was metabolized during stationary phase, 2-hydroxybiphenyl was identified as the end product, and 2'-hydroxybiphenyl-2-sulfinate was observed as an intermediate, instead of the sulfonate. Denome et al. and Piddington et al., recently cloned the genes encoding DBT desulfurization in IGTS8, referred to as *sox* and *dsz* genes by the respective groups (39,40). Three genes were identified which were necessary and sufficient to conferred the desulfurization activity to recipient *Rhodococcus* and *Escherichia coli* strains. When individually expressed in *Escherichia coli*, the gene product of *soxC* (*dszC*) converted DBT to the sulfone and the combined activity of the gene products of *soxA* and *B* converted DBT sulfone into 2-hydroxybiphenyl. The pathway interpreted from these results is shown in figure 6. These data indicate that DBT sulfoxide may not be a true intermediate of the IGTS8 desulfurization pathway.

E. Sulfur Specific Desulfurization of Sterically Hindered DBTs

As mentioned earlier, sterically hindered aromatic organic sulfur compounds (e.g. dibenzothiophenes bearing alkyl substitutions adjacent to the sulfur atom) are the most resistant to HDS. The removal of these compounds becomes important when very low sulfur levels are desired. The 1990 Clean Air Act mandated that, by year end 1993, the maximum allowable sulfur content of on-road diesel be restricted to 500 ppm, down from the previous 2000 ppm (41). Similar restrictions are likely to be applied to other fuels, and are also gradually being adopted in Canada, Europe and Japan. Stricter sulfur regulations place an increased demand on the currently available HDS capacity, and challenge the ability of HDS to meet the required low sulfur levels. As a result, alternatives to conventional HDS, which can remove sterically hindered DBTs from fuels, may become necessary to meet the low sulfur levels required by future regulations.

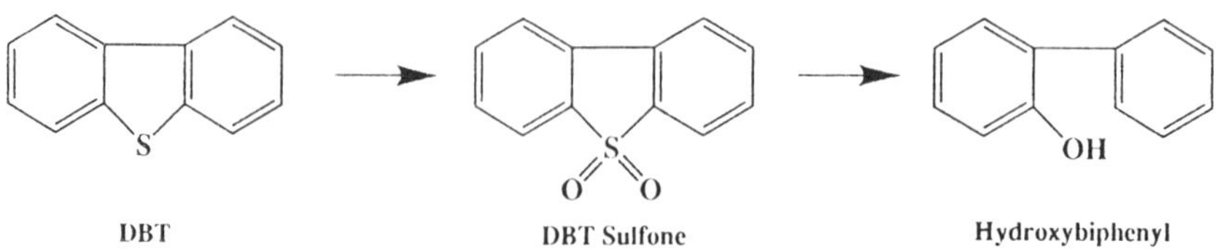

Figure 6. Sulfur specific microbial desulfurization pathway of Rhodococcus sp. IGTS8. (Adapted from Ref. 40.)

A number of Gram positive organisms with pathways similar to IGTS8 have been isolated, including, *Rhodococcus erythropolis* strains N1-36 and D-1 (42,43) and *Corynebacterium* strain SY1 (44). However, the ability to desulfurize sterically hindered analogous of DBT has not been demonstrated by any of the reported sulfur specific organisms. Microbial desulfurization research at Exxon Research and Engineering has focused on obtaining isolates which can remove sulfur from sterically hindered analogs of DBT. From enrichment cultures using the sterically hindered compound 4,6-diethyl DBT as a sole sulfur source, an *Arthrobacter* species was isolated . The isolate, designated ECRD-1, was shown to be capable of desulfurizing this compounds as well as DBT, 4,6-dimethyl DBT, 4-methyl DBT, benzylphenyl sulfide, and other related compounds. Desulfurization proceeded through a sulfur specific pathway, similar to that of IGTS8, producing the substituted analogs of 2-hydroxybiphenyl in the case of the sterically hindered DBT compounds. To test the ability of ECRD-1 to treat real feeds, it was grown with a middle distillate cut of Oregon Basin crude oil as the sole source of sulfur. Treatment of the middle distillate fraction with ECRD 1, resulted in a 30% reduction in total sulfur content (Fig. 7). These results demonstrated that the oxidative sulfur specific desulfurization pathway could be extended to those compounds known to be the most resistant to conventional HDS, and suggests that petroleum is a suitable substrate for desulfurization applications.

IV. CONCLUSION

The field of microbial desulfurization has made significant progress, and recent activities have provided us with reason to be somewhat optimistic that a commercial process may yet be developed. Certainly, the emergence of a company dedicated to this reality is evidence of this possibility. However, significant improvements still must be made. Among these improvements is the development of biocatalysts with significantly increased activity and stability, and a decrease in the associated cost of biocatalyst production and maintenance (e.g. cofactor regeneration). Additionally, improved bioreactor configurations, which efficiently handle the contacting and separation of biocatalyst, gases, and the non-aqueous feed, will likely be required. Nonetheless, the future of biodesulfurization has never been brighter.

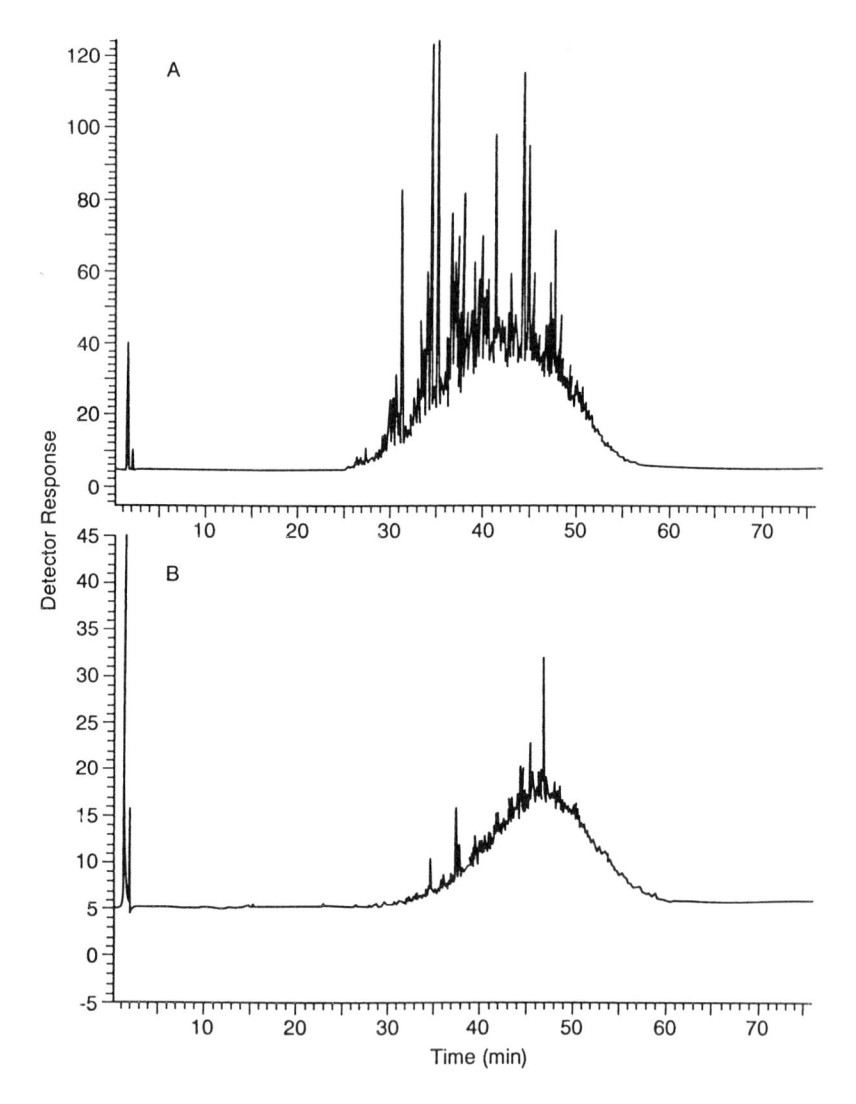

Figure 7. Desulfurization of Oregon Basin crude oil, 232 - 343°C middle distillate cut, by ECRD-1. Gas Chromatography, with Sulfur Chemiluminescence Detector (sulfur selective), of extracted oils from the sterile control (A) and ECRD-1 (B) cultures. The scale of the chromatograms were normalized to reflect an equivalent carbon content between samples.

REFERENCES

1. J.G. Speight, in The Desulfurization of Heavy Oils and Residua (Heinz Hienemann, ed.), Marcel Dekker, New York (1981).

2. H.T. Rall, C.J. Thompson, H. J. Coleman, and R.L. Hopkins. U. S. Bureau of Mines, Bulletin 659 (1972).

3. T. Kabe, A. Ishihara, and H. Tajima. Ind. Eng. Chem. Res. 31:1577 (1992).

4. A.A. Maliyantz. Azerbaidzhanskoe Netyanoe Khoz 15:89 (1935).

5. A.A. Maliyantz. Azerbaidzhanskoe Nefyanoe Khoz 15:36 (1936).

6. C.E. ZoBell, U.S. Patent 2,641,564 to Texaco Development Corporation (1953).

7. M. Kohler, I.L. Genz, B. Schicht and V. Eckart. Zentralbl. Mikrobiol. 139:239 (1984).

8. K.W. Miller. Appl. Environ. Microbiol. 58:2176 (1992).

9. H.M. Lizima, L.A. Wilkens, and T.C. Scott. Biotech. Lett. (in press).

10. T.S. Kim, H.Y. Kim, and B.H. Kim. Biotechnol. Letts. 12:757 (1990).

11. B.H. Kim, T.S. Kim, and H.Y. Kim, U.S. Patent 4,954,229 to Korea Advanced Institute of Science and Technology (1990).

12. W.R. Finnerty. Fuel 72:1631 (1993).

13. I. Kirshenbaum, U.S. Patent 2,975,103 to Esso Research and Engineering Company (1961).

14. F. Kargi, and J. M. Robinson. Biotech. and Bioeng. 26:687 (1984).

15. F. Kargi, and J. M. Robinson. Fuel 65:397 (1986).

16. R.J. Strawinski, U.S. Patent 2,521,761 to Texaco Development Corporation (1950).

17. R.J. Strawinski, U.S. Patent 2,574,070 to Texaco Development Corporation (1951).

18. K. Kodama, S. Nakatani, K. Umehara, K. Shimizu, Y. Minoda, and K. Yamada. Agr. Biol. Chem. 34:1320 (1970).

19. K. Kodama, K. Umehara, K. Shimizu, S. Nakatani, Y. Minoda, and K. Yamada. Agr. Biol. Chem. 37:45 (1973).

20 C.E. Cerniglia, in Petroleum Microbiology (R. Atlas, ed.), Macmillan, New York, 1984, pp. 99-128.

21. C. T. Hou, and A. I. Laskin. Dev. Ind. Microbiol. 17:351 (1976).

22. K. A. Malik, and D. Claus, in Abstracts of the Fifth Int. Ferment. Symp., Berlin, 1976, pp.421.

23. A. Laborde, and D.T. Gibson. Appl. Environ. Microbiol. 34:783 (1977).

24. K. A. Malik. Process Biochem. 13(9):10 (1978).

25. D.J. Monticello, D. Bakker, and W.R. Finnerty. Appl. Environ. Microbiol. 49:756 (1985).

26. D.J. Monticello, and W.R. Finnerty. Ann. Rev. Microbiol. 39:371 (1985).

27. K.M. Yen, and I.C. Gunsalus. Proc. Natl. Acad. Sci. 79:874 (1982).

28. W.R. Finnerty, in Energy technology IX: "Energy efficiency in the eighties" (R.F. Hill, ed.). Government Institutes, Inc., Maryland, 1982, pp. 883-890.

29. F.J. Hartdegen, J.M. Coburn, and R.L. Roberts. Chem. Eng. Progress 80:63 (1984).

30. J.M. Foght, P.M. Fedorak, M.R. Gray and D.W.S. Westlake, in Microbial Mineral Recovery H.L. (Ehrlich and C.L. Brierley, ed.), McGraw-Hill, New York, 1990, pp. 379-407.

31. A. Bhadra, J.M. Scharer, and M. Moo-Young. Biotechnol. Adv. 5:1 (1987).

32. A.T. Knecht Jr., Ph.D. Thesis, Louisiana State University (1961).

33. J. D. Isbister, and E. A. Kobylinski, in Coal science and technology series, No. 9 (Y.A. Attia (ed.), Elsevier, Amsterdam, 1985, p. 627-641.

Microbial Removal of Organic Sulfur from Fuels

34. Arctech Inc. Archtech Inc. Final Report, submitted to U.S. DOE, Pittsburgh Energy Technology Center. Report No. DE89 001572 (1988).

35. I.M. Campbell. American Chemical Society, Division of Petroleum Chemistry, Preprints 38(2):275 (1993).

36. J.J. Kilbane, and B.A. Bielaga. CHEMTECH 20:747 (1990).

37. K.J. Kayser, B.A. Bielaga-Jones, K. Jackowski, O. Odusan, and J.J. Kilbane II. J. Gen. Micro. 139:3123 (1993).

38. J.R. Gallagher, E.S. Olson, and D.C. Stanley. FEMS Micro. Lett. 107:31 (1993).

39. S.A. Denome, C. Oldfield, L.J. Nash, and K.D. Young. J. Bacteriol. 176:6707 (1994)

40. C.S. Piddington, B.R. Kovacevich, and J. Rambosek. Appl. Environ. Microbiol. 61: 468 (1995).

41. Federal Register, August 22, 1989.

42. P. Wang, and S. Krawiec. Arch. Microbiol. 161:266 (1994).

43. Y. Izumi, T. Ohshiro, H. Ogino, Y. Hine, and M. Shimao. Appl. Environ. Microbiol. 60: 223 (1994).

43. T. Omori, L. Monna, Y. Saiki, and T. Kodama. App. Environ. Micro. 58:911 (1992).

Index

A

Additives, impregnation stabilizers, hydrodesulfurization activity, 211-234, 215t-216t, *217*
- additives, 216t, 218-219
- effect of additives and stabilizers on NiMo/alumina catalysts, 216t
- impregnation stabilizers, *217,* 219-220, 222t, 223-224, *224, 225*
- overview, 212-214
- phosphorus
  - on hydrodesulfurization activities, CoMo catalysts on alumina supports, 221t
  - stabilizer, 218
    - on CoMo catalysts, 220-223, 221t
- sulfided catalysts, characterization of, 221t, 225-233, *226-229,* 230t, *231*
- XPS analysis of sulfided form of NiMo and CoMo catalysts, 230t

AFM. *See* Atomic force microscopy

Aniline, nickel single crystal surfaces, hydrogen in hydrogenation and hydrogenolysis of, 253-262, *255-260*
- overview, 253-254, 261

Aromatics hydrogenation, alumina supported Mo and NiMo hydrotreating catalysts, 277-290, *281, 288*
- catalysts tested, composition of, 278t
- experiment, 278t, 278-279
- hydrogen disulfide, partial pressure, influence of, 280-282, *280-282,* 282t
- kinetic studies, 283t, 283-285, *285*
- overview, 277-278, 289

Atomic force microscopy, scanning tunneling microscopy, transition metal chalcogenides, 71-83
- atomic force microscopy, *74,* 74-75
- ReS$_2$ (2DFFT filtered), *76*
- scanning tunneling microscopy data, 76-77, *76-77*
- atomic force microscopy schematic, *74*
- defects, 80-82, *80-82*

361

edges, steps, 80-82, *80-82*
instrumentation, 72-75
model calculations, 77-79, *78-79*
molybdenum disulfide particle, on graphic substrate, scanning tunneling microscopy image of, *82*
molybdenum disulfide step edge, scanning tunneling microscopy image of, *81*
overview, 71-72
ReS_2
   basal plane, *75,* 75-79
   images, calculated, *78*
scanning tunneling microscopy, 72-74, *73*
   schematic, *73*
SPM studies, 75-82
step edge on graphite, scanning tunneling microscopy, *80*
tunnel junction schematic, *73*

C

C_6H_5N, chemisorption complex, hydrotreating constituent portion of, *22*
specific phases as, active, portion of, *22*
Carbazole on alumina-supported molybdenum nitride-catalyst, hydrodenitrogenation of, 263-275
   activation energies, adsorption constants for Model A and B, 272t
   contact time, 265-267, *266-268*
   experiment, 264
   kinetics, 268-274, *269-270,* 272t
   mass transfer limitation, 268
   overview, 263-264, 274

Catalytic sites, heterolytic reaction mechanisms in, hydrotreating catalysis
addition, reaction mechanism, nd substitution reaction mechanisms, 34-36, 35t, *36-37*
addition and substitution reactions, kinetics of, 35t, 37-42, 39t, *41*
C-N bond cleavage, 35t
C-S and C-N bond cleavage, 35t
chemical formula, of hexagonal molybdenum disulfide slab containing 127 Mo ions, 32t
elimination
   kinetics of, 35t, 37-42, 39t, *41*
   reaction mechanism, 34-36, 35t, *36-37*
first hydrogenation step, aromatic ring by sequence H + SH, *41*
hydrogenation, C-S bond cleavage, 35t
ions in ideal hexagonal molybdenum disulfide slab *vs.* slab size, number of, 31t
Mo ions in ideal hexagonal molybdenum disulfide slab *vs.* slab size, number of, 31t
molybdenum disulfide edge planes, surface species on, 31-32t, 31-34
nucleophilic substitution, catalytic cycle for hydrodenitrogenation by, *36*
overview, 29-30, 42-43
sequence H + SH, catalytic cycle for hydrogenation by, *37*
substitution reaction
   kinetics of, 35t, 37-42, 39t, *41*
   mechanisms, 34-36, 35t, *36-37*

Index

Co-sulfide species, in sulfided
    supported Co and CoMo
    catalysts, 95-96, 95-113, *106*
  implications, 108-111
  influence of support, 106-107
  overview, 96-97, 105-106
  relation between Q.S. value and
    particle size of formed
    'Co-sulfide' species, 97t,
    100t, 103-105, *106*
  on similarity of Co/C and CoMo/C,
    according to MES, 97t,
    97-106, *98-99*
  structural model, 108
  sulfidability of Co/C and CoMo/C
    catalysts, 97t, 100t, 101-103
  temperature dependence of MES
    parameters of 'Co-Mo-S'
    doublet in Co/C, 101
Coke, resid catalyst, ebullating bed
    service and   effect on
    activity, characterization of,
    337-338, 337-344, 338-344,
    340-341, 343
  experiment, 338
  overview, 338
CoMo/AL$_2$O$_3$, sulfided, NiNo/AL$_2$O$_3$
    catalysts in
    hydrodesulfurization of,
    comparison, gas oil
    fractions, 183, 186-191, *188,*
    191-194
  alkyl DBTs, hydrodesulfurization
    networks of, *188*
  catalysts, chemical composition,
    physical properties of, 185t
  experiment, 184-185t, 184-186
  gas oil, composition, distribution
    fractions, 184t
  hydrodesulfurization activities,
    CoMo, NiMo catalysts,
    comparison of, *186*
  model compounds
    hydrodesulfurization product
    distribution of, 188t
  sulfur, hydrodesulfurization of,
    188, 188t, 188-191, 190t
  overview, 183-184
  rate constants of DBT, 4,6 DMDBT
    for hydrogenolysis and
    hydrogenation, 190t
  sulfur compound
    existing in each fraction, *187,*
    187-188
    in gas oil and
        distribution, *187*
CoMoS/NiMoS, hydrotreating
    constituent
  comparison with, *20,* 25-26
  specific phases as, active,
    comparison with, *20,* 25-26
CoMoS$_4$, hydrotreating constituent,
    18-20, 19-20
  Mo environment in, *20*
  model HDN cycles, 21-25, *22-24*
  projection, along b axis, *18*
  specific phases as, active, *18-20,*
    19-20

D

Delaminated, pillared interlayered
    clays as supports for
    hydrotreating catalysts,
    291-292, 291-311, 295-309
  basal spacings, crystallite sizes,
    pillared and delaminated
    clays, 297t
  catalyst
    characterization, 294
    deactivation studies, 305-308,
    306t, *307*
  coal-derived liquid, catalyst
    deactivation studies,
    294-295

coprocessed heavy gas oil, catalyst activity studies using, 295
delaminated clays, preparation, 293
experiment, 293-295
feedstocks, properties of, 306t
heteroatom conversion, catalyst activity studies for, *308,* 309
Ni-Mo catalyst
  chemical, physical properties of, 300t
  extrudates, preparation, 293-294
overview, 292-293, 309-310
physical properties, catalyst materials, 295-299, *296, 297t, 298,* 300t
pillared clays, preparation, 293
pyridine/FTIR adsorption, determination of surface acidity by, 299-302, *301-302*
transmission electron microscopy studies, *298,* 302-305, *303, 304*

E

Electron acceptor additives, molybdenum disulfide-based hydrotreating catalysts modified by, 115-116, 115-127, 118-122, 122t, 122-125, 125t
  catalytic tests, 117
  chemical analyses, XPS of sulfided catalysts, 118t, 118-120, *119*
  experiment, 117
  morphology of supported molybdenum disulfide phase, 120, 120t
  overview, 116, 125
  performances, catalytic, 120t, 120-122, *121,* 122t
  physical characterization, 117
  preparation, catalyst, 117
  S/Mo atomic ratio, 118t

Electron donor or electron acceptor additives, molybdenum disulfide-based hydrotreating catalysts modified by, 115-116, 115-127, 118-122, 122t, 122-125, 125t
  catalytic tests, 117
  chemical analyses, XPS of sulfided catalysts, 118t, 118-120, *119*
  experiment, 117
  morphology of supported molybdenum disulfide phase, 120, 120t
  overview, 116, 125
  performances, catalytic, 120t, 120-122, *121,* 122t
  physical characterization, 117
  preparation, catalyst, 117
  S/Mo atomic ratio, 118t

F

Foulant metals, leaching of, residual oil hydrotreating catalyst rejuvenation, 327-336, 330-334, *331-333*
  experiment, 328-330, 329t
  fresh, spent residue hydrotreating catalysts, characteristics of, 329t
  overview, 327-328, 334-335
Fuels, organic sulfur, microbial removal from, 345-346, 345-359
  dissimilatory sulfur reduction, *347,* 347-348
  non-specific oxidative desulfurization, 349-352, *350-351*
  overview, 355
  oxidative desulfurization, 348-355
    dissimilatory, 349, *349*

oxidative systems, overview of, 348-349
reductive desulfurization, 346-348
  novel approaches to, 348
reductive systems, overview of, 346-347, *347*
sulfur
  in biology, 346
  in petroleum, 345-346
sulfur specific desulfurization
  oxidative, 352-354, *354*
  sterically hindered DBTs and, 354-355, *356*

G

Gas oil fractions, sulfided CoMo/AL$_2$O$_3$, NiNo/AL$_2$O$_3$ catalysts in hydrodesulfurization of, comparison, 183, 186-191, *188,* 191-194
alkyl DBTs, hydrodesulfurization networks of, *188*
catalysts, chemical composition, physical properties of, 185t
experiment, 184-185t, 184-186
gas oil, composition, distribution fractions, 184t
hydrodesulfurization activities, CoMo, NiMo catalysts, comparison of, *186*
model compounds
  hydrodesulfurization product distribution of, 188t
  sulfur, hydrodesulfurization of, *188,* 188t, 188-191, 190t
overview, 183-184
rate constants of DBT, 4,6 DMDBT for hydrogenolysis and hydrogenation, 190t
sulfur compounds
  existing in each fraction, *187,* 187-188
  in gas oil, distribution, *187*

H

Heterolytic reaction mechanisms, hydrotreating catalysis, catalytic sites and, 31
addition, reaction mechanism, 34-36, 35t, *36-37*
chemical formula, of hexagonal molybdenum disulfide slab containing 127 Mo ions, 32t
elimination
  kinetics of, 35t, 37-42, 39t, *41*
  reaction mechanism, 34-36, 35t, *36-37*
first hydrogenation step, aromatic ring by sequence H + SH, *41*
hydrogenation, C-S bond cleavage, 35t
ions in ideal hexagonal molybdenum disulfide slab vs. slab size, number of, 31t
Mo ions in ideal hexagonal molybdenum disulfide slab vs. slab size, number of, 31t
molybdenum disulfide edge planes, surface species on, 31-32t, 31-34
nucleophilic substitution, catalytic cycle for hydrodenitrogenation by, *36*
overview, 29-30, 42-43
sequence H + SH, catalytic cycle for hydrogenation by, *37*
Hydrodesulfurization catalysis with sulfido bimetallic clusters, modeling, 129-145
cluster complexes, nature of, 134
Mo/Co/S models for hydrodesulfurization catalysts, 134-141
organometallic chemistry, 131-134
overview, 129-134

reactivity of Mo/Co/S clusters, homogeneous, 134-140, *137, 140*
relevance of homogeneous cluster reactivity to heterogeneous hydrodesulfurization catalysts, *137,* 140-141
Hydrogen disulfide, as active species in mechanism, thiophene hydrodesulfurization, 147
  base catalyst, 148
  confidence interval, optimized parameters with, 154t
  experiment, 148
  kinetic expression, derivation of, 151-154, 154t
  mechanism, 155-156
  other models, comparison with, 155-156
  overview, 148
  partial pressure
    H2S, effect of, 149-151, *151*
    hydrogen, effect of thiophene and, 149
    stimulation, of experimental data, *150-151,* 154-155, *155*
  sulfidation and thiophene reaction, 148-149
  temperature, effect of, 149, *150*
  thiophene reaction, 148-149
Hydrogen in hydrogenation and hydrogenolysis of aniline on nickel single crystal surfaces, 253-262, 254-261, *255-260*
  overview, 253-254, 261
Hydrogenation, carbon-heteroatom bond cleavage, by EAS reactions, rate laws for, 39t
Hydrotreating catalysis, catalytic sites and heterolytic reaction mechanisms in, 29-45, 31
  addition, reaction mechanism, 34-36, 35t, *36-37*
  chemical formula, of hexagonal molybdenum disulfide slab containing 127 Mo ions, 32t
  elimination
    kinetics of, 35t, 37-42, 39t, *41*
    reaction mechanism, 34-36, 35t, *36-37*
  first hydrogenation step, aromatic ring by sequence H + SH, *41*
  hydrogenation, C-S bond cleavage, 35t
  ions in ideal hexagonal molybdenum disulfide slab *vs.* slab size, number of, 31t
  Mo ions, in ideal hexagonal molybdenum disulfide slab, *vs.* slab size, number of, 31t
  molybdenum disulfide edge planes, surface species on, 31-32t, 31-34
  nucleophilic substitution, catalytic cycle for hydrodenitrogenation by, *36*
  overview, 29-30, 42-43
  sequence H + SH, catalytic cycle for hydrogenation by, *37*
Hydrotreating constituent
  active, specific phases as, 17-18, 17-27
    chemisorption complex, *22, 23, 24*
    CoMoS/NiMoS, comparison with, *20,* 25-26
    CoMoS$_4$, *18-20,* 19-20
      Mo environment in, *20*
      projection, along b axis, *18*
    model HDN cycles, with CoMoS$_4$, 21-25, *22-24*
    molybdenum disulfide, projection, along axis, *19*

Index

NiMo$_3$S$_4$, 20, *21*
  projection, along c axis, 21
  overview, 18-19
  prognosis, 26-27
molybdenum disulfide, projection, along axis, *19*
NiMo$_3$S$_4$, 20, *21*
  projection, along c axis, 21
  overview, 18-19
  prognosis, 26-27
specific phases as, active, 17-18, 17-27
  chemisorption complex, *22, 23, 24*
  CoMoS/NiMoS, comparison with, *20*, 25-26
  CoMoS$_4$, *18-20*, 19-20
    Mo environment in, *20*
    projection, along b axis, *18*
  model HDN cycles, with CoMoS$_4$, 21-25, *22-24*
  molybdenum disulfide, *18-20*, 19-20
    projection, along axis, *19*
  NiMo$_3$S$_4$, 20, *21*
    projection, along c axis, 21
    overview, 18-19
    prognosis, 26-27

I

Impregnation stabilizers, hydrodesulfurization activity, 211-234,
  additives, 216t, 218-219
  effect of additives and stabilizers on NiMo/alumina catalysts, 216t
  overview, 212-214, 233
  phosphorus on hydrodesulfurization activities, CoMo catalysts on alumina supports, 221t
  phosphorus stabilizer, 218
    on CoMo catalysts, 220-223, 221t

sulfided catalysts, characterization of, 221t, 225-233, *226-229*, 230t, *231*
XPS analysis of sulfided form of NiMo and CoMo catalysts, of, 230t

M

Metal sulfide hydrodesulfurization catalysts, periodic and promotion effects in transition
  activity parameter, 13-14
    factors influencing value of, 12-13
  hydrodesulfurization mechanism provided by, 10-12
  alternative activity parameter, definition of, *4*, 7-9, *8*
  experimental activities, *I*, correlation between, *3, 8, 9, 10*
  metal oxidation state, 6t
  metal-sulfur bond length, 6t
  MS$_6^{-6}$ clusters, molecular orbital diagram, spin-restricted, *4*
  overview, 14
  pi electrons, number of, 6t
  promotion effects, 13-14
  sigma electrons, number of, 6t
  theoretical activity parameter *I*, vs. position of metal atom, periodic table, *10*
  unified theory, transition, 1-15
    activity parameter, Harris, Chianelli, 5-7
    MS$_6^{-n}$ clusters, electronic structure of, 3-5, *4*, 6t
    nonmetalic TMS, dibenzothiophene, 3
    overview, 1-2, *3*

Modeling hydrodesulfurization catalysis with sulfido bimetallic clusters, 129-145
  cluster complexes, nature of, 134
  Mo/Co/S models for hydrodesulfurization catalysts, 134-141
  organometallic chemistry, 131-134
  overview, 129-134
  reactivity of Mo/Co/S clusters, homogeneous, 134-140, *137, 140*
  relevance of homogeneous cluster reactivity to heterogeneous hydrodesulfurization catalysts, *137,* 140-141
Molten salt preparation, mixed transition metals on zirconia application of hydrotreating reactions, 235-251, 238-249, 244t
  atomic ratios, metals in sulfided samples, determined by XPS and chemical analysis, 247t
  BET surface areas, 241-242, 242t
  binding energies, 245-248, 246t
  catalytic activity, 248, 249t
  catalytic tests, 238, 238t
  characterization, 237-238
  chemical composition of samples, 241t
  DRS UV-vis spectroscopy, 242-245, *243,* 244t
  experiment, 237-238
    conditions of catalytic tests, 238t
  intensities, 247t, 248
  overview, 236-237, 249
  preparation, 237-238
  stoichiometries, 247t, 248
  XPS characterizations, 245-248
  XRD patterns, 238-241, *240,* 241t
Molybdenum disulfide, 63t
  atomic resolution, *90*
  experiment, 86
  with Fe, 91, *92, 92-93, 93*
    molybdenum disulfide:Co, HREM characterization of, 87, *87-88*
    atomic resolution, *90*
    experiment, 86
    hexagonal arrangements, molybdenum disulfide crystalline structures, *91*
    molybdenum disulfide, *87, 88*
    with Fe, 91, *92-93*
    molybdenum disulfide:Co, crystal edges, *89*
    molybdenum disulfide:Co system, 87-91, *89-91*
    overview, 85-86, 91
    tilting sequence of, molybdenum disulfide stacks disappearring, with tilting angle, *92, 93*
  hexagonal arrangements, molybdenum disulfide crystalline structures, *91*
  hydrotreating constituent
    projection, along axis, *19*
    specific phases as, active, *18-20,* 19-20
Molybdenum disulfide:Co, 87-91, *89-91*
  crystal edges, *89*

N

Naphtha, transition metal sulfide catalysts, nitrogen and sulfur compounds in, competitive conversion of, 197, 200, 201-207
  analytical data for commercial catalysts, 205t
  catalyst, 199t, 200-201
  preparation, 198, 199t

composition, catalyst prepared, 199t
elemental composition, naphthas, 199t
experiment, 198-201
extras of hydrodesulfurization and HDN, comparison of, *204*
hydrodenitrogenation, *202-204,* 202-205
hydrodesulfurization, *201,* 201-202
hydrotreatment, 200
overview, 197-198, 208
reactor system, 200
ruthenium sulfide, commercial catalysts, comparison of, 205t, 205-207, *206-208*
Nickel catalysts, role of sulfiding procedure on activity and dispersion of, 159-167, 161-167, *162, 164,* 165t, *166*
characteristics of investigated catalysts, 160t
experimental, 160t, 160-161
oxidic catalyst precursor, sulfided precursor, measured XPS intensities of, *166*
stoichiometry of nickelsulfide from TPS, *165*
TPR profiles of oxidic nickel catalyst precursor, *162*
TPS profiles of oxidic nickel catalyst precursor, *164*
Nickel single crystal surfaces, aniline, hydrogen in hydrogenation and hydrogenolysis of, 253-262, 254-261, *255-260*
overview, 253-254, 261
NiMo/AL$_2$O$_3$ catalysts, gas oil fractions, sulfided CoMo/AL$_2$O$_3,$ hydrodesulfurization of, comparison, 183, 186-191, *188,* 191-194

alkyl DBTs, hydrodesulfurization networks of, *188*
catalysts, chemical composition, physical properties of, 185t
experiment, 184-185t, 184-186
gas oil, composition, distribution fractions, 184t
hydrodesulfurization activities, CoMo, NiMo catalysts, comparison of, *186*
model compounds
hydrodesulfurization of, *188,* 188t, 188-191, 190t
hydrodesulfurization product distribution of, 188t
overview, 183-184
rate constants of DBT, 4,6 DMDBT for hydrogenolysis and hydrogenation, 190t
sulfur compound
existing in each fraction, *187,* 187-188
in gas oil and distribution, *187*

O

Oil hydrotreating catalyst rejuvenation, by leaching of foulant metals, residual, 327-336, 330-334, *331-333*
experiment, 328-330, 329t
fresh, spent residue hydrotreating catalysts, characteristics of, 329t
overview, 327-328, 334-335
Organic sulfur, microbial removal from fuels, 345-346, 345-359
dissimilatory sulfur reduction, *347,* 347-348
non-specific oxidative desulfurization, 349-352, *350-351*

overview, 355
oxidative desulfurization, 348-355
  dissimilatory, 349, *349*
  oxidative systems, overview of, 348-349
reductive desulfurization, 346-348
  novel approaches to, 348
  reductive systems, overview of, 346-347, *347*
sulfur
  in biology, 346
  in petroleum, 345-346
  specific desulfurization, 352-354, *354*
    sterically hindered DBTs and, 354-355, *356*

P

Periodic, promotion effects, transition metal sulfide hydrodesulfurization catalysts
  activity parameter, hydrodesulfurization mechanism provided by, 10-12
  alternative activity parameter, definition of, *4*, 7-9, *8*
  experimental activities, *I*, correlation between, *3, 8*, 9, *10*
  metal oxidation state, 6t
  metal-sulfur bond length, 6t
  MS_6^{-6} clusters, molecular orbital diagram, spin-restricted, *4*
  overview, 14
  pi electrons, number of, 6t
  sigma electrons, number of, 6t
  theoretical activity parameter *I, vs.* position of metal atom, periodic table, *10*
  unified theory
    activity parameter, Harris, Chianelli, 5-7

    MS_6^{-n} clusters, electronic structure of, 3-5, *4*, 6t
  nonmetalic TMS, dibenzothiophene, 3
  overview, 1-2, *3*
Pi orbital for RhS_6^{-9}, antibonding between metal and sulfur atoms, but bonding between neighboring sulfur atoms, topology, *8*
Pillared, delaminated interlayered clays as supports for hydrotreating catalysts, 291-292, 291-311, 295-309
  basal spacings, crystallite sizes, pillared and delaminated clays, 297t
  catalyst
    characterization, 294
    deactivation studies, 305-308, 306t, *307*
  coal-derived liquid, catalyst deactivation studies, 294-295
  coprocessed heavy gas oil, catalyst activity studies using, 295
  delaminated clays, preparation, 293
  experiment, 293-295
  feedstocks, properties of, 306t
  heteroatom conversion, catalyst activity studies for, *308*, 309
  Ni-Mo catalyst
    chemical, physical properties of, 300t
    extrudates, preparation, 293-294
  overview, 292-293, 309-310
  physical properties, catalyst materials, 295-299, *296*, 297t, *298*, 300t
  pillared clays, preparation, 293

pyridine/FTIR adsorption, determination of surface acidity by, 299-302, *301-302*
transmission electron microscopy studies, *298,* 302-305, *303, 304*
Promotion effects, periodic and, transition metal sulfide hydrodesulfurization catalysts
  alternative activity parameter, definition of, *4,* 7-9, *8*
  experimental activities, *I,* correlation between, *3, 8,* 9, *10*
  metal oxidation state, 6t
  metal-sulfur bond length, 6t
  MS_6^{-6} clusters, molecular orbital diagram, spin-restricted, *4*
  overview, 14
  pi electrons, number of, 6t
  sigma electrons, number of, 6t
  theoretical activity parameter *I,* vs. position of metal atom, periodic table, *10*
  unified theory
    activity parameter, Harris, Chianelli, 5-7
    MS_6^{-n} clusters, electronic structure of, 3-5, *4,* 6t
    nonmetalic TMS, dibenzothiophene, 3
  overview, 1-2, *3*
Pyrene, 315-316, 316-323
  catalyst characterization, 315
  hydrocracking, 316-321, 318t, *319, 321*
  hydrogenation, 321-323, 322t
  prereduction, 315

R

Residual oil hydrotreating catalyst rejuvenation, by leaching of foulant metals, 327-336, 330-334, *331-333*
  experiment, 328-330, 329t
  fresh, spent residue hydrotreating catalysts, characteristics of, 329t
  overview, 327-328, 334-335
Ruthenium sulfide base catalysts, 169-182, 170-178
  active sites
    characterization, interaction with hydrogen, 170-174
    nature of, 171-173, *172*
  alumina supported catalysts, 175t, 175-176
  catalytic activities, 177t
  catalytic properties of, 175
  hydrogen, interaction with, 173, *174*
  lattice parameters, *179*
  $Ni_xRu_{1-x}S_{22},$ alumina supported, *180,* 180-181
  overview, 169-170, 181
  particle sizes, catalytic activities, ruthenium sulfide supported on alumina catalyst, 175t
  ternary sulfide phases, 178-181
  thiophene, hydrodesulfurization of, *179*
  unsupported system, 178-179, *179*
  zeolites, ruthenium sulfide dispersed in, 176-178, 177t

S

Scanning tunneling microscopy, atomic force microscopy, transition metal chalcogenides, 71-83, 72-74, *73*
  atomic force microscopy, *74,* 74-75, 76-77, *76-77*
    ReS_2 (2DFFT filtered), *76*
  atomic force microscopy schematic, *74*

defects, 80-82, *80-82*
edges, steps, 80-82, *80-82*
instrumentation, 72-75
model calculations, 77-79, *78-79*
molybdenum disulfide
    particle, on graphic substrate, scanning tunneling microscopy image of, *82*
    step edge, scanning tunneling microscopy image of, *81*
overview, 71-72
ReS_2
    basal plane, *75,* 75-79
    images, calculated, *78*
SPM studies, 75-82
step edge on graphite, scanning tunneling microscopy, *80*
tunnel junction schematic, *73*
STM. *See* Scanning tunneling microscopy
Sulfided supported Co and CoMo catalysts, co-sulfide species in, 95-96, 95-113, *106, 108-111*
    influence of support, 106-107
    overview, 96-97, 105-106
    relation between Q.S. value and particle size of formed 'Co-sulfide' species, 97t, 100t, 103-105, *106*
    on similarity of Co/C and CoMo/C, according to MES, 97t, 97-106, *98-99*
    structural model, 108
    sulfidability of Co/C and CoMo/C catalysts, 97t, 100t, 101-103
    temperature dependence of MES parameters of 'Co-Mo-S' doublet in Co/C, 101
Sulfido bimetallic clusters, modeling hydrodesulfurization catalysis with, 129-145

cluster complexes, nature of, 134
Mo/Co/S models for hydrodesulfurization catalysts, 134-141
organometallic chemistry, 131-134
overview, 129-134
reactivity of Mo/Co/S clusters, homogeneous, 134-140, *137, 140*
relevance of homogeneous cluster reactivity to heterogeneous hydrodesulfurization catalysts, *137,* 140-141
Sulfur, organic, microbial removal from fuels, 345-346, 345-359
    dissimilatory sulfur reduction, *347,* 347-348
    non-specific oxidative desulfurization, 349-352, *350-351*
    overview, 355
    oxidative desulfurization, 348-355
        dissimilatory, 349, *349*
    oxidative systems, overview of, 348-349
    reductive desulfurization, 346-348
        novel approaches to, 348
    reductive systems, overview of, 346-347, *347*
    sulfur
        in biology, 346
        in petroleum, 345-346
    sulfur specific desulfurization
        oxidative, 352-354, *354*
        sterically hindered DBTs and, 354-355, *356*

T

Thiophene hydrodesulfurization, hydrogen disulfide as active species in mechanism of, 147-157

Index

base catalyst, 148
confidence interval, optimized parameters with, 154t
experiment, 148
kinetic expression, derivation of, 151-154, 154t
mechanism, 155-156
other models, comparison with, 155-156
overview, 148
partial pressure
  H2S, effect of, 149-151, *151*
  hydrogen, effect of thiophene and, 149
simulation, of experimental data, *150-151*, 154-155, *155*
sulfidation and thiophene reaction, 148-149
temperature, effect of, 149, *150*
thiophene reaction, 148-149
Transition metal sulfide
catalysts, nitrogen and sulfur compounds in, competitive conversion of, naphtha, 197, 200, 201-207
  analytical data for commercial catalysts, 205t
  catalyst preparation, 198, 199t
  catalysts, 199t, 200-201
  composition, catalyst prepared, 199t
  elemental composition, naphthas, 199t
  experiment, 198-201
  extras of hydrodesulfurization and HDN, comparison of, *204*
  hydrodenitrogenation, *202-204*, 202-205
  hydrodesulfurization, *201*, 201-202
  hydrotreatment, 200
  naphtha, 199
  overview, 197-198, 208
  reactor system, 200
  ruthenium sulfide, commercial catalysts, comparison of, 205t, 205-207, *206-208*
hydrodesulfurization catalysts, periodic and promotion effects in
  activity parameter, 13-14
    factors influencing value of, 12-13
  hydrodesulfurization mechanism provided by, 10-12
  alternative activity parameter, definition of, *4*, 7-9, *8*
  experimental activities, *I*, correlation between, *3, 8, 9, 10*
  metal oxidation state, 6t
  metal-sulfur bond length, 6t
  MS_6^{-6} clusters, molecular orbital diagram, spin-restricted, *4*
  overview, 14
  pi electrons, number of, 6t
  promotion effects, 13-14
  sigma electrons, number of, 6t
  theoretical activity parameter *I*, vs. position of metal atom, periodic table, *10*
  unified theory, 1-15
    activity parameter, Harris, Chianelli, 5-7
    MS_6^{-n} clusters, electronic structure of, 3-5, *4*, 6t
    nonmetalic TMS, dibenzothiophene, 3
    overview, 1-2, *3*
Tungsten disulfide, 63t, 64t
  catalysts, microstructural characterization of, 47-48,

　　　　47-69, 49-66, *52-62,* 63-64t,
　　　　　65
　　experiment, 49
　　high resolution micrograph of
　　　　WS$_2$ prepared at 673K,
　　　　before computer processing,
　　　　57
　　molybdenum disulfide catalyst
　　　　prepared at 673 K, high
　　　　resolution micrograph of, *53*
　　overview, 48-49
　　thiophene,selectivity of
　　　　molybdenum disulfide
　　　　catalyst in
　　　　hydrodesulfurization
　　　　of,function of time on
　　　　stream, 64t
　　WS$_2$
　　　catalyst
　　　high resolution micrographs of,
　　　　55
　　　HREM micrograph of, seen
　　　　along [131] zone axis, *61*

　　　Fourier transform method,
　　　　computer-processed HREM
　　　　micrograph, *58, 59*
　　　HREM micrographs of,
　　　　periodicity of gray intensities
　　　　in, *56*
　　molybdenum disulfide, catalysts,
　　　　microstructural
　　　　characterization of, 47-48,
　　　　47-69, 49-66, *52-62,* 63-64t,
　　　　65
　　experiment, 49
　　high resolution micrograph of
　　　　WS$_2$ prepared at 673K,
　　　　before computer processing,
　　　　57
　　molybdenum disulfide catalyst
　　　　prepared at 673 K, high
　　　　resolution micrograph of, *53*
　　overview, 48-49
　　reaction rates for molybdenum
　　　　disulfide catalyst, before,
　　　　after hydrodesulfurization
　　　　reaction, 63t